METHODS IN MOLECULAR BIOLOGY

For further volumes:
http://www.springer.com/series/7651

For over 35 years, biological scientists have come to rely on the research protocols and methodologies in the critically acclaimed *Methods in Molecular Biology* series. The series was the first to introduce the step-by-step protocols approach that has become the standard in all biomedical protocol publishing. Each protocol is provided in readily-reproducible step-by-step fashion, opening with an introductory overview, a list of the materials and reagents needed to complete the experiment, and followed by a detailed procedure that is supported with a helpful notes section offering tips and tricks of the trade as well as troubleshooting advice. These hallmark features were introduced by series editor Dr. John Walker and constitute the key ingredient in each and every volume of the *Methods in Molecular Biology* series. Tested and trusted, comprehensive and reliable, all protocols from the series are indexed in PubMed.

Protein Self-Assembly

Methods and Protocols

Edited by

Jennifer J. McManus

Department of Chemistry, Maynooth University, Maynooth, Co. Kildare, Ireland

Humana Press

Editor
Jennifer J. McManus
Department of Chemistry
Maynooth University
Maynooth, Co. Kildare, Ireland

ISSN 1064-3745 ISSN 1940-6029 (electronic)
Methods in Molecular Biology
ISBN 978-1-4939-9677-3 ISBN 978-1-4939-9678-0 (eBook)
https://doi.org/10.1007/978-1-4939-9678-0

This Humana imprint is published by the registered company Springer Science+Business Media, LLC, part of Springer Nature.
The registered company address is: 233 Spring Street, New York, NY 10013, U.S.A.

Preface

Protein self-assembly describes many different pathways leading to a range of condensed states of proteins that include concentrated protein droplets, reversible and irreversible amorphous aggregates, fibrils, viral capsids, protein nanocages, and crystals. These condensed states are important in understanding fundamental features of biology and several industrial processes. Protein condensation is associated with many diseases, including sickle-cell anemia, cataract disease, and several amyloid-associated diseases including Alzheimer's disease and Parkinson's disease. The observations of protein de-mixing in mammalian cells leading to the formation of transient non-membrane-bound organelles are revealing significant new insights in cell biology, RNA processing, and possibly even the origins of life. Drug development relies on the availability of high-resolution protein structures, and the vast majority of protein structures are determined from X-ray diffraction of protein crystals. Many industrial processes also rely on a fundamental understanding of protein self-assembly. The production of biopharmaceuticals, food, cosmetics, and even some electronics all involve protein self-assembly.

Understanding protein self-assembly is incredibly difficult. If or how protein self-assembly occurs depends on a wide range of factors, many relate to the characteristics of the protein itself and others relate to the external environment. Proteins can self-assemble in folded or unfolded states by several different mechanisms and on different time scales, and these assemblies often exist in non-equilibrium states. The assembled forms of the protein can also be challenging to characterize, due to the wide range of sizes over which they form—from several nanometers to tens of microns—and this creates an analytical burden. For these reasons, it is often necessary to employ several different techniques and approaches to measure the assembly process. In this volume, experimental and computational approaches to measure the most widely studied protein assemblies, including condensed liquid phases, aggregates, and crystals, are described.

Understanding the protein-protein interactions that direct self-assembly is far from straightforward. The most basic approach is to view proteins as small particles that have an interaction energy, which has both attractive (van der Waals forces, hydrophobic effect, dipole-dipole) and repulsive (electrostatic repulsion, hydration) contributions that arise from the amino acids on the protein surface. The net interaction potential is the sum of these contributions. If it is averaged over the surface of the protein, then the effective potential resembles that of a small colloidal particle, and we can use what we know about colloidal science to understand protein assembly. This view and approach works reasonably well for some proteins, particularly for small globular proteins at low concentrations, and can predict some general features of protein self-assembly. However, for many proteins, this simplified model fails to describe the protein-protein interactions, particularly when protein concentration is increased or there are small modifications to the protein surface. It has become clear that protein-protein interactions are directional in nature, and this anisotropy in the interaction potential is key to explain protein assembly. In Part I of this volume, the techniques to measure protein-protein interactions and equilibrium protein phases are described for both dilute and concentrated proteins.

Protein aggregation is perhaps the most widely studied self-assembly pathway. However, aggregation is a very generic term that describes a range of pathways, including self-association, cluster formation, fibrillation, amorphous aggregate formation, gelation (sometimes), and precipitation. There are a wide range of techniques used to measure both the kinetics of aggregation and the assembled state once formed, and in Part II, several are described. In general, a combination of techniques that rely on different analysis methods are used to measure protein aggregation. Protocols describing analytical ultracentrifugation, electrophoresis, chromatography, calorimetry, light scattering, imaging, fluorescence spectroscopy, and NMR are included here. This comprehensive set of protocols allow the analysis of assembled states ranging in size from dimer and small oligomers to large amorphous aggregates across a range of protein concentrations.

A major goal in the field is to develop models that will allow protein-protein interactions and protein assembly pathways to be predicted ab initio. While predictability is currently difficult, several computational approaches to understand protein self-assembly are providing valuable insights and are described in Part III. For peptides and very small proteins, all-atom simulations with explicit solvent are possible if some structural information is already available. For larger proteins, or for simulations that require several protein molecules, computational resources are still not sufficiently powerful to perform all-atom simulations, and some coarse-graining is required. Some of the most successful models to date are those based on colloids that include anisotropic interaction potentials or "patchiness." The details of how these patches are modelled vary, with some tuning interactions to match experimental data, while others incorporate molecular-level details from crystallographic data to precisely describe the protein-protein interactions. Using these coarse-grained models in combination with simulations, experimental data can be described and explained. As these models become more sophisticated, and computational resources increase, even greater insights will be possible.

Much progress has been made in understanding protein self-assembly, but obstacles remain. Detailed knowledge about all of the phases and states of proteins exists for only a relatively small number of proteins, i.e., those that are available in sufficient purity and scale to allow experiments to be performed. As more experiments on a greater number of proteins are performed, and computational tools become faster and more sophisticated, further insights and possibly even control over protein self-assembly will emerge.

Maynooth, Co. Kildare, Ireland *Jennifer J. McManus*

Contents

Contributors

IREM ALTAN • *Department of Chemistry, Duke University, Durham, NC, USA*
NEER ASHERIE • *Department of Physics, Yeshiva University, New York, NY, USA; Department of Biology, Yeshiva University, New York, NY, USA*
ALICE BLUMLEIN • *Department of Chemistry, Maynooth University, Maynooth, Kildare, Ireland*
ANDRÉ BRODKORB • *Teagasc Food Research Centre, Fermoy, Cork, Ireland*
KARA M. BUZZA • *Department of Pharmacy and Optometry, Faculty of Biology Medicine and Health, School of Health Sciences, University of Manchester, Manchester, UK*
JORDAN W. BYE • *School of Chemical Engineering and Analytical Science, The University of Manchester, Manchester, UK*
CESAR CALERO-RUBIO • *Department of Chemical and Biomolecular Engineering, University of Delaware, Newark, DE, USA*
PATRICK CHARBONNEAU • *Department of Physics, Duke University, Durham, NC, USA*
SABRI CHERRAK • *Department of Chemical Sciences, Bernal Institute, University of Limerick, Limerick, Ireland; Department of Biology, University Abou-Bekr Belkaid, Tlemcen, Algeria*
DANIEL CORBETT • *School of Chemical Engineering and Analytical Science, The University of Manchester, Manchester, UK*
NICOLE R. CUNNINGHAM • *Department of Biology, Haverford College, Haverford, PA, USA*
ROBIN A. CURTIS • *School of Chemical Engineering and Analytical Science, The University of Manchester, Manchester, UK*
ELISA FADDA • *Department of Chemistry, Hamilton Institute, Maynooth University, Maynooth, Kildare, Ireland*
ROBERT FAIRMAN • *Department of Biology, Haverford College, Haverford, PA, USA*
DANIEL L. FORTUNATI • *Trinity Biomedical Sciences Institute (TBSI), School of Biochemistry and Immunology, Trinity College Dublin, Dublin, Ireland*
ERIC M. FURST • *Department of Chemical and Biomolecular Engineering, University of Delaware, Newark, DE, USA*
SOPHIE JEANNE GASPARD • *Teagasc Food Research Centre, Moorepark, Fermoy, Cork, Ireland; School of Food and Nutritional Sciences, University College Cork, Cork, Ireland*
NICOLETTA GNAN • *CNR-ISC, UOS Sapienza, Roma, Italy*
MARK GRACE • *Department of Chemistry, Maynooth University, Maynooth, Kildare, Ireland*
ANNE MARIE HEALY • *Synthesis and Solid State Pharmaceutical Centre, School of Pharmacy and Pharmaceutical Sciences, Trinity College Dublin, Dublin, Ireland*
MATTHEW R. JACOBS • *Department of Chemistry, Maynooth University, Maynooth, Kildare, Ireland*
PANCHAM S. KANDIYAL • *Trinity Biomedical Sciences Institute (TBSI), School of Biochemistry and Immunology, Trinity College Dublin, Dublin, Ireland*
JI YOON KIM • *Trinity Biomedical Sciences Institute (TBSI), School of Biochemistry and Immunology, Trinity College Dublin, Dublin, Ireland*
BASHKIM KOKONA • *Department of Biology, Haverford College, Haverford, PA, USA*
RAMIL F. LATYPOV • *Sanofi, Framingham, MA, USA*

Kate McComiskey • *Synthesis and Solid State Pharmaceutical Centre, School of Pharmacy and Pharmaceutical Sciences, Trinity College Dublin, Dublin, Ireland*
Jennifer J. McManus • *Department of Chemistry, Maynooth University, Maynooth, Co. Kildare, Ireland*
Deniz Meneksedag-Erol • *Department of Chemical and Physical Sciences, University of Toronto Mississauga, Mississauga, ON, Canada; Department of Physics, University of Toronto, Toronto, ON, Canada*
Judith J. Mittag • *Department of Chemistry, Maynooth University, Maynooth, Kildare, Ireland*
K. H. Mok • *Trinity Biomedical Sciences Institute (TBSI), School of Biochemistry and Immunology, Trinity College Dublin, Dublin, Ireland*
Matthew G. Nixon • *Department of Chemistry, Hamilton Institute, Maynooth University, Maynooth, Kildare, Ireland*
Alain Pluen • *Department of Pharmacy and Optometry, Faculty of Biology Medicine and Health, School of Health Sciences, University of Manchester, Manchester, UK*
Jeanne M. Quinn • *Department of Biology, Haverford College, Haverford, PA, USA*
Zahra Rattray • *Strathclyde Institute of Pharmacy and Biomedical Sciences, University of Strathclyde, Glasgow, UK*
Sarah Rauscher • *Department of Chemical and Physical Sciences, University of Toronto Mississauga, Mississauga, ON, Canada; Department of Physics, University of Toronto, Toronto, ON, Canada; Department of Chemistry, University of Toronto, Toronto, ON, Canada*
Christopher J. Roberts • *Department of Chemical and Biomolecular Engineering, University of Delaware, Newark, DE, USA*
Francesco Sciortino • *Dipartimento di Fisica, "Sapienza" Università di Roma, Roma, Italy*
Svenja Sladek • *Synthesis and Solid State Pharmaceutical Centre, School of Pharmacy and Pharmaceutical Sciences, Trinity College Dublin, Dublin, Ireland*
Orla Slattery • *Department of Chemical Sciences, Bernal Institute, University of Limerick, Limerick, Ireland; Department of Biopharmaceutical and Medical Science, Galway-Mayo Institute of Technology, Galway, Ireland*
Tewfik Soulimane • *Department of Chemical Sciences, Bernal Institute, University of Limerick, Limerick, Ireland*
Lidia Tajber • *Synthesis and Solid State Pharmaceutical Centre, School of Pharmacy and Pharmaceutical Sciences, Trinity College Dublin, Dublin, Ireland*
Ying Wang • *Department of Chemistry and Biochemistry, University of North Carolina at Wilmington, Wilmington, NC, USA*
Mahlet A. Woldeyes • *Department of Chemical and Biomolecular Engineering, University of Delaware, Newark, DE, USA*
Emanuela Zaccarelli • *CNR-ISC, UOS Sapienza, Roma, Italy*
Egor Zindy • *Faculty of Biology Medicine and Health, School of Biological Sciences, University of Manchester, Manchester, UK*

Part I

Measuring Protein-Protein Interactions and Protein Phase Diagrams

Chapter 1

Measuring Nonspecific Protein–Protein Interactions by Dynamic Light Scattering

Daniel Corbett, Jordan W. Bye, and Robin A. Curtis

Abstract

Dynamic light scattering has become a method of choice for measuring and quantifying weak, nonspecific protein–protein interactions due to its ease of use, minimal sample consumption, and amenability to high-throughput screening via plate readers. A procedure is given on how to prepare protein samples, carry out measurements by commonly used experimental setups including flow through systems, plate readers, and cuvettes, and analyze the correlation functions to obtain diffusion coefficient data. The chapter concludes by a theoretical section that derives and rationalizes the correlation between diffusion coefficient measurements and protein–protein interactions.

Key words Biopharmaceuticals, Second virial coefficients, Protein crystallization, Protein aggregation, Osmometry

1 Introduction

Nonspecific protein–protein interactions directly relate to protein solution properties such as crystallization propensity, phase behavior and solution opalescence, or to a lesser extent, protein aggregation kinetics and rheological characteristics of concentrated protein solutions. For these reasons, there has been much interest in measuring protein–protein interactions across disciplines ranging from structural biology, biomaterials, cell biology, biopharmaceuticals, and medicine. The most direct approach is to characterize them from dilute solution thermodynamic properties such as the osmotic second virial coefficient termed B_{22} measurable through static light scattering [1–3], osmometry [4, 5], or sedimentation equilibrium by analytical ultracentrifugation [2, 6]. Alternatively, protein–protein interactions can be quantified in terms of an interaction parameter termed k_D obtainable from diffusion coefficient measurements by dynamic light scattering [7–9]. This approach, which has grown in popularity, generally requires less protein material, is more user

Jennifer J. McManus (ed.), *Protein Self-Assembly: Methods and Protocols*, Methods in Molecular Biology, vol. 2039, https://doi.org/10.1007/978-1-4939-9678-0_1,

friendly, and is amenable to medium and high throughput measurements using multiwell plates.

Because k_D is determined from protein diffusion behavior, it contains contributions from both thermodynamic and hydrodynamic interactions. The thermodynamic term is directly related to B_{22}, but the link to hydrodynamics, although established analytically, is less well known. As such, many studies rely on the established correlation between k_D and B_{22} as sufficient evidence that protein–protein interactions are directly probed by DLS [10–12].

The first half of this report presents the theory for obtaining diffusion coefficients from the measured intensity autocorrelation function. In addition, the theory underpinning the relationship between protein–protein interactions and the hydrodynamic contribution to diffusion coefficients is given. The derived equations provide a clear way to interpret measurements of k_D in terms of excluded volume forces, short-ranged attractions, and longer-ranged electrostatic repulsions between proteins. This delineation is often required for establishing the relationships to solution properties of interest. Short-ranged protein–protein attractions, under moderate ionic strength conditions, often dictate the phase behavior [13–16], and crystallization propensity [17–21], while long-ranged electrostatic repulsions have been correlated with protein aggregate growth kinetics [22–24].

In the second half of the report is provided a protocol with a step-by-step procedure for sample preparation and measurement by commonly used experimental set-ups (plate readers, cuvettes, and syringe injection systems). Detailed notes are provided for analyzing the processed data and quantifying data quality.

1.1 Theory

1.1.1 Dynamic Light Scattering

The primary quantity measured in a dynamic light scattering experiment is the light scattering intensity $I(q, t)$ as a function of time t at a given scattering angle θ which defines the magnitude of the scattering vector $q = (4\pi/\lambda)\sin(\theta/2)$ where λ is the wavelength of incident and scattered light [25]. The measurement is processed in real time using a correlator to determine the intensity autocorrelation function $G_2(q, t)$ from averaging over an acquisition time T_a

$$G_2(q,\tau) = \langle I(q,0)I(q,\tau)\rangle = \frac{1}{T_a}\int_0^{T_a} I(q,t)I(q,t+\tau)\mathrm{d}t, \quad (1)$$

where τ is the delay time and the angular brackets indicate a time-average. The scattered light intensity is related to the amplitude of the scattered electric field $E(q, t)$ by $I(q, t) = |E(q, t)|^2$. The property of interest is the electric field correlation function $G_1(q, \tau)$ which is related to the cumulative protein diffusion coefficient D_c according to

$$G_1(q,\tau) \sim \exp\left(-D_c q^2 \tau\right) \quad (2)$$

The two correlation functions are related to each other by

$$g_2(q,\tau) = B + \beta[g_1(q,\tau)]^2, \tag{3}$$

where $g_2(q, \tau) = G_2(q, \tau)/\langle I(q)\rangle^2$, $g_1(q, \tau) = G_1(\tau)/\langle I(q)\rangle$, and $\langle I(q)\rangle$ is the average light scattering intensity. Here B corresponds to the baseline reading of $\langle I(q, 0)I(q, \tau)\rangle/\langle I(q)\rangle^2)$, which is equal to 1 because the light has become decorrelated at long delay times, while β depends on the experimental configuration.

Analysis of light scattering data is made more complicated because there will always exist in the illuminated light scattering volume a distribution of different sized species that contribute to the overall decay in the correlation function, in which case, $g_1(q, \tau)$ is an averaged quantity given by

$$g_1(q,\tau) = \int_0^\infty G(\Gamma)\exp(-\Gamma\tau)\mathrm{d}\Gamma, \tag{4}$$

where $G(\Gamma)$ is the normalized distribution of particles with decay times by $(\Gamma = D_c q^2)^{-1}$, or equivalently the probability distribution function for finding a particle with a diffusion coefficient D_c. In determination of protein–protein interactions, any particles much larger than the protein molecule of interest need to be removed through careful filtration, which is essential as the light scattering signal is weighted by larger particles. After filtration, there will still exist in solution a distribution of particles with similar sizes to the protein, which could correspond to small aggregates of the protein or other impurities. In this case, the most appropriate analysis is based on the method of cumulants, which assumes that the sample contains a mononodal size distribution of protein particles.

The method of cumulants is used to extract the moments of $G(\Gamma)$ from the cumulant generating function $K(-\tau, \Gamma) \equiv \ln[g_1(q, \tau)]$. The mth cumulant κ_m is given by the derivative

$$\kappa_m = \left(\frac{\mathrm{d}^m K(-\tau,\Gamma)}{\mathrm{d}(-\tau)^m}\right)_{-\tau\to 0}. \tag{5}$$

The first cumulant is related to the mean of the distribution $\overline{\Gamma}$

$$\kappa_1 = \overline{\Gamma} = \int_0^\infty \Gamma G(\Gamma)\mathrm{d}\Gamma, \tag{6}$$

where the overline indicates a z-weighted average. The higher order cumulants correspond to the moments about the mean

$$\kappa_m = \int_0^\infty (\Gamma - \overline{\Gamma})^m G(\Gamma)\mathrm{d}\Gamma. \tag{7}$$

The cumulants also correspond to the coefficients in a Taylor series expansion for $\ln[g_1(q, \tau)]$

$$\ln[g_1(q,\tau)] = -\bar{\Gamma}\tau + \frac{\kappa_2}{2!}\tau^2 - \frac{\kappa_3}{3!}\tau^3 + \ldots \tag{8}$$

Due to experimental noise, only the first two terms in the cumulant expansion can be determined with reasonable accuracy from fitting to the measured correlation function $g_2(q, \tau)$. The fitting equation is obtained by substituting Eq. 8 into Eq. 3 to give

$$\ln[g_2(q,\tau) - B] = A - 2\bar{\Gamma}\tau + \kappa_2\tau^2, \tag{9}$$

where the fit parameters are Γ, κ_2, A, which is an adjustable constant, and the parameter B, which should be equal to 1. The fit value for the second cumulant κ_2 is often reported in terms of a polydispersity index defined as

$$P = \frac{\kappa_2}{\bar{\Gamma}^2} = \frac{\overline{D^2} - \bar{D}^2}{\bar{D}^2}. \tag{10}$$

The polydispersity corresponds to the width of the diffusion coefficient distribution divided by the mean and provides a measure of the monodispersity in the sample, which is often used as an indicator of the sample quality.

1.2 Background to Interpreting Protein–Protein Interaction Measurements

1.2.1 Determination of k_D

The protein–protein interaction parameter is determined from a plot of the diffusion coefficient versus protein mass concentration

$$D_c = D_0\left(1 + k_D^{(c)} c_p\right), \tag{11}$$

where D_0 is the infinite-dilute value for the cumulative diffusion coefficient, which is equivalent to the infinite dilution value of the self-diffusion coefficient. A superscript c is used to denote the interaction parameter $k_D^{(c)}$ is determined from a plot against protein mass concentration c_p, in which case the parameter has units of inverse mass concentration. The dimensionless form of the parameter, denoted here as k_D, is defined in a similar way to Eq. 11,

$$D_c = D_0\left(1 + k_D\phi_p\right), \tag{12}$$

but using volume fraction ϕ_p as the protein concentration variable. The two parameters are related to each other by $k_D^{(c)} = k_D v_p$, where v_p is the protein partial specific volume. The fit value of D_0 is used to determine the hydrodynamic radius of the protein $R_{H,0}$ according to the Stokes Einstein relation

$$R_{H,0} = \frac{k_B T}{6\pi\eta D_0}, \tag{13}$$

where η is the viscosity of the solvent, k_B is Boltzmann's constant, and T is temperature. Because protein molecules are not spheres, the measured hydrodynamic size corresponds to the radius of a sphere that has the same diffusion coefficient as the protein.

1.2.2 Protein–Protein Interactions

A direct measure of nonspecific protein–protein interactions is the osmotic second virial coefficient, B_{22}^{v}, which is obtainable via osmotic pressure measurements by an osmometer, or more commonly, from the osmotic compressibility measured through static light scattering. The osmotic second virial coefficient is rigorously defined within the McMillan-Mayer framework [26] as

$$B_{22}^{\mathrm{v}} = -\frac{1}{2}\int_0^{\infty}[\exp(-w(r)/k_{\mathrm{B}}\mathrm{T}) - 1]4\pi r^2 \mathrm{d}r, \qquad (14)$$

where r is the protein center-to-center separation. $w(r)$ is the protein–protein interaction free energy (commonly referred to as the two-body potential of mean force), which has been averaged over the relative orientations between a pair of proteins. The subscript v is used to denote the osmotic virial coefficient has units of volume, while often the parameter measured by an experiment has units of volume-moles per mass squared, which is referred to here as B_{22}. The parameters are related to each other by $B_{22} = B_{22}^{\mathrm{V}} N_{\mathrm{A}}/M_{\mathrm{p}}^2$, where N_{A} is Avogadro's number and M_{p} is protein molecular weight.

The meaning of B_{22}^{v} can be understood by considering the relationship between $w(r)$ and the pair distribution function at infinite dilution $g(r)$

$$g(r) = \exp[-w(r)/k_{\mathrm{B}}T]. \qquad (15)$$

$g(r)$ is related to the microscopic structure of the protein solution according to the relation $g(r) = \rho(r)/\rho_{\mathrm{p}}$, where $\rho(r)$ corresponds to the averaged "local" protein density at a separation r about a tagged protein molecule and ρ_{p} is the bulk protein density. All protein molecules exhibit excluded volume forces, which prevent the overlap of surfaces due to the Pauli exclusion principle. If protein molecules are modelled as spheres, with a diameter equal to σ, these forces can be represented by a step function, where $g(r) = 0$ or $w(r) = \infty$ when $r < \sigma$ and $g(r) = 1$ or $w(r) = 0$ when $r > \sigma$. As such, surrounding a tagged protein, there will be a spherical zone of exclusion with a diameter equal to 2σ and volume equal to $8V_{\mathrm{p}}$ where $V_{\mathrm{p}} = \pi\sigma^3/6$ is the protein sphere volume. According to Eq. 14, the excluded volume contribution to the virial coefficient is given by $B_{22}^{\mathrm{v,ex}} = 4V_{\mathrm{p}}$, which is the volume of exclusion divided by two, as this region needs to be shared between two spheres when determining thermodynamic properties. Positive values of B_{22}^{v} thus reflect the volume excluded to centres of protein molecules about a tagged protein. Conversely, negative values of B_{22}^{v} correspond to the case where the average local concentration of proteins about a tagged molecule is greater than the bulk concentration.

Predicting protein solution behavior requires delineating between the excluded volume forces and all other contributions, which are collectively referred to as *soft* protein–protein interactions. These can include repulsive terms such as electric double layer forces or hydration interactions, or short-ranged attractions, which are still poorly understood, but likely related to a combination of hydrophobic interactions, dispersion forces, electrostatic attractions, and hydrogen bonding effects [27]. The net contribution of *soft* interactions is often characterized in terms of a reduced virial coefficient defined as

$$B_{22}^{*} = \frac{B_{22}^{\mathrm{V}} - B_{22}^{\mathrm{V,ex}}}{B_{22}^{\mathrm{V,ex}}}. \tag{16}$$

A negative value of B_{22}^{*} indicates a net short-ranged protein–protein attraction, while a positive value indicates the presence of a protein–protein repulsion that extends beyond contact between protein surfaces. While it is not possible to measure directly $B_{22}^{\mathrm{V,ex}}$, calculations using all-atomistic representations to describe protein shapes have shown $B_{22}^{\mathrm{V,ex}}$ can be approximated by its value for a sphere with the same hydrodynamic radius of the protein [28]. The net contribution of soft protein–protein interactions can thus be estimated from dynamic light scattering using the Stokes Einstein relation (Eq. 13).

1.2.3 The Relationship Between k_D and B_{22}

The diffusion coefficient obtained from dynamic light scattering by applying the cumulant analysis is equivalent to the gradient diffusion coefficient, which controls the decay of macroscopic gradients in protein concentration according to Fick's law [29, 30]. The gradient diffusion coefficient is controlled by two competing factors, which is best illustrated by the relation

$$D_{\mathrm{c}} = \frac{1}{\xi}\left(\frac{d\Pi}{\mathrm{d}\rho_{\mathrm{p}}}\right). \tag{17}$$

There is a thermodynamic term related to the osmotic compressibility ($\mathrm{d}\Pi/\mathrm{d}\rho_{\mathrm{p}}$) (where Π is the osmotic pressure of the protein solution), which arises because diffusion is driven by chemical potential gradients, and a hydrodynamic term due to the drag force exerted on the protein by the solvent. The drag force is proportional to the frictional coefficient ξ, which, at infinite dilution, is given by the Stokes law $\xi_0 = 6\pi\eta a$, where a is the protein radius ($a = \sigma/2$). A hydrodynamic function H is defined as $H \equiv \xi_0/\xi$ to reflect changes to the drag force from alterations in the solvent flow field brought about by surrounding proteins. This definition is usually used in Eq. 17 to give

$$D_{\mathrm{c}} = D_0\frac{H}{k_{\mathrm{B}}T}\left(\frac{\mathrm{d}\Pi}{\mathrm{d}\rho_{\mathrm{p}}}\right). \tag{18}$$

The hydrodynamic function can be determined directly from the protein sedimentation velocity U_{sed} measurable by analytical ultracentrifugation experiments according to $H = U_{sed}/U_{sed,\,0}$, where $U_{sed,\,0}$ is the infinite dilution value.

Analysis of protein–protein interactions is usually carried out at low protein concentrations where deviations from infinite-dilution behavior occur only due to two-body interactions. In this limit, the hydrodynamic function and osmotic compressibility can be expanded in a power series of protein concentration to first order

$$H = 1 - k_H \phi, \tag{19}$$

where k_H is two-body hydrodynamic function and

$$\frac{1}{k_B T}\left(\frac{d\Pi}{d\rho_p}\right) = 1 + 2\left(B_{22}^{v}/V_p\right)\phi = 1 + k_V \phi, \tag{20}$$

where $k_V \equiv 2\left(B_{22}^{v}/V_p\right)$. Combining these relations into Eq. 18 gives

$$\frac{D_c}{D_0} = 1 + (k_V - k_H)\phi = 1 + k_D \phi, \tag{21}$$

where the protein–protein interaction parameter k_D has been decomposed into the sum of a thermodynamic and a hydrodynamic term according to $k_D = k_V - k_H$.

In order to express k_D in terms of protein–protein interactions requires linking k_H to the potential of mean force $w(r)$ (or equivalently $g(r)$). Expressions for k_H have been derived by solving the fluid flow problem (known as the Navier Stokes equations) between two spheres under an applied external field to determine the average sedimentation velocity, U_{sed}. As is the case for B_{22}, the contributions of excluded volume interactions can be separated from the effects of *soft* protein–protein interactions using

$$k_H = k_H^{ex} + k_H^{soft}. \tag{22}$$

The excluded volume contribution is given by [31].

$$k_H^{ex} = 5.5 - 0.5 + 1.55 = 6.55. \tag{23}$$

The first term (5.5) is due to the backflow of the solvent, which arises predominantly because the volumetric flux of a sedimenting particle creates a flux in the opposite direction from the solvent displaced by the particle. In addition, the sedimenting sphere drags along some of the solvent, which must be balanced by an equivalent upward flux of the solvent in the regions further away from the sphere. The second term (−0.5) is hydrostatic in origin; the addition of particles to the fluid creates an additional gradient in the hydrostatic pressure, which increases the sedimentation velocity. The last term is due to direct hydrodynamic interactions between the proteins, that is, the influence of the drag force on sphere 1 due to the presence of sphere 2 and vice versa.

The effects of soft protein–protein interactions are only manifested in the direct hydrodynamic interactions. The result is [32, 33]

$$k_{\mathrm{H}}^{\mathrm{soft}} = \int_{2a}^{\infty} [A_{11} + A_{12} + 2(B_{11} + B_{12})][g(r) - 1]\frac{r^2}{a^3}\,\mathrm{d}r, \tag{24}$$

where the terms A_{ij} and B_{ij} are the coefficients in the mobility tensor, which relate the instantaneous protein velocities to the applied forces. The coefficients depend on the sphere size and their center-to-center separation r. Analytical expressions for A_{ij} and B_{ij} exist in the far field limit, $r \gg 2a$, and are given by

$$\begin{aligned} A_{11} &= 1 - \frac{15}{4}\left(\frac{a}{r}\right)^4 + O(r^{-6}) \\ A_{11} &= 1 - \frac{15}{4}\left(\frac{a}{r}\right)^4 + O(r^{-6}) \\ B_{11} &= 1 + O(r^{-6}) \\ B_{12} &= \frac{3}{4}\left(\frac{a}{r}\right) + \frac{1}{2}\left(\frac{a}{r}\right)^3 + O(r^{-7}) \end{aligned} . \tag{25}$$

The near field forms corresponding to when r is just greater than contact separation between surfaces ($2a$) are more problematic. A_{11} and A_{12} vary smoothly from their far field forms toward their values $A_{11} = A_{12} = 0.7750$ at contact ($r = 2a$), but there is a logarithmic divergence at contact for the terms B_{11} and B_{12}. Approximate expressions are given by [34]

$$\begin{aligned} A_{11} &= 1 - \frac{15}{4}\left(\frac{a}{r}\right)^4 + O(r^{-6}) \\ B_{11} &= 0.891 - \frac{0.388}{\log(r/a - 2)} \\ B_{12} &= 0.490 + \frac{0.144}{\log(r/a - 2)} \end{aligned} . \tag{26}$$

The effects of protein–protein interactions are manifested in the expression for $k_{\mathrm{H}}^{\mathrm{soft}}$ in terms of the radial distribution function $g(r)$ (*see* Eq. 24). In order to quantify the effect in a simple way, an approximate relationship to B_{22}^* can be derived by noting that the mobility term appearing in the integrand of Eq. 24 ($A_{11} + A_{12} + 2(B_{11} + B_{12})$) is a slowly varying function of r and only changes by about 6% as r increases from $2a$ to $2.25a$. Under moderate ionic strength conditions, where long-ranged electrostatic forces are screened between proteins, the contributions of all soft protein–protein interactions have a range on the order of one to two solvent layers between protein surfaces [13, 20]. In this limiting case, the mobility term can be factored out of the integrand to give

$$k_{\mathrm{H}}^{\mathrm{soft}} = [A_{11} + A_{12} + 2(B_{11} + B_{12})] \int_{2a}^{\infty} [g(r) - 1] \frac{r^2}{a^3} \mathrm{d}r \approx 3.52 B_{22}^{*}, \tag{27}$$

where the second equality follows from using the values at contact for the mobility coefficients ($A_{11} = A_{12} = 0.7750$ and $B_{11} + B_{12} = 1.381$). Equation 27 provides a clear link between the magnitude of the net protein–protein interactions, as often characterized in terms of B_{22}, and the sedimentation behavior. Accordingly short-ranged attractions will enhance sedimentation, for instance, an attractive interaction that balances the hard sphere repulsion will reduce k_{H} by less than one half of its hard sphere value. The reason why attractive interactions enhance sedimentation is that there is an increased likelihood of finding proteins at contact, in which case, there is less surface exposed to the drag than the pair in isolation, while the sedimenting force is essentially doubled. Conversely, repulsive protein–protein interactions have the opposite effect of decreasing the sedimentation velocity.

The equation for the protein–protein interaction parameter k_{D} can be written as

$$k_{\mathrm{D}} = k_{\mathrm{V}}^{\mathrm{ex}} + k_{\mathrm{V}}^{\mathrm{soft}} - \left(k_{\mathrm{H}}^{\mathrm{ex}} + k_{\mathrm{H}}^{\mathrm{soft}}\right), \tag{28}$$

where the hard sphere and *soft* contributions to the thermodynamic term are given by $k_{\mathrm{V}}^{\mathrm{ex}} = 8$ and $k_{\mathrm{V}}^{\mathrm{soft}} = 8B_{22}^{*}$. Substitution of these terms into Eq. 28 combined with the expressions for k_{H} (Eqs. 23 and 24) leads to a linear relationship between k_{D} and B_{22}

$$k_{\mathrm{D}} = (8 + 8B_{22}^{*}) + (-6.55 - 3.52B_{22}^{*}) = 1.45 + 4.48B_{22}^{*}. \tag{29}$$

The linearity has been observed experimentally for monoclonal antibodies in particular. The predicted correlation given by Eq. 29 is plotted along with experimental data for lysozyme [8] and for a monoclonal antibody [35] in Fig. 1. Good agreement is obtained between the model and the experimental data without using any adjustable parameters. The only parameter required by the calculation is the equivalent spherical diameter of the protein, which has been set equal to the measured hydrodynamic diameter. The first bracketed term on the right side of Eq. 29 corresponds to the contribution from thermodynamic effects, while the second bracketed term is due to hydrodynamics. These contributions always occur in the opposite direction to each other. Excluded volume interactions increase the chemical potential driving force for diffusion ($k_{\mathrm{V}}^{\mathrm{ex}} = 8$), which is balanced by an increased drag force that slows down diffusion predominantly due to the backflow effect ($k_{\mathrm{H}}^{\mathrm{ex}} = -6.55$). Conversely, the thermodynamic consequence of attractive protein–protein interactions is a slowing down of diffusion, while hydrodynamic effects enhance the diffusion rate. An important reference point is the hard sphere contribution $k_{\mathrm{D}}^{\mathrm{ex}}$ equal

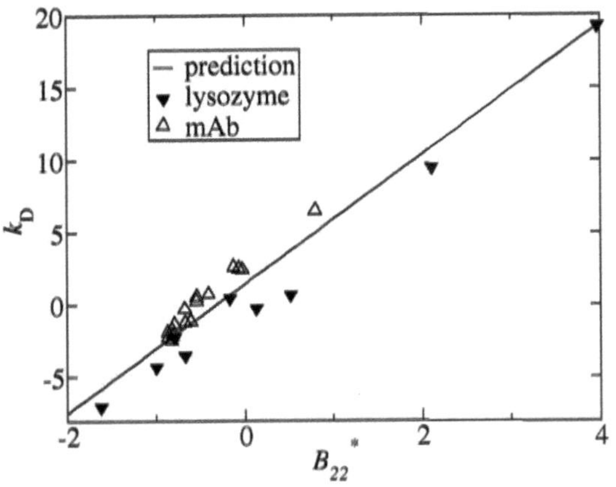

Fig. 1 The correlation of k_D with B_{22}^*. The line is the prediction according to Eq. 29 of the text. The closed symbols are experimental data for lysozyme in solutions at pH 4.5, while the open symbols correspond to a monoclonal antibody over a range of pH and ionic strength in solutions containing sodium chloride

to 1.45. Measured $k_D$values greater than k_D^{ex} indicate the presence of a longer-ranged repulsion, while short-ranged protein–protein attractions are reflected by values of k_D less than 1.45.

Further delineating between the different contributions to protein–protein interactions requires fitting simplified interaction models, usually that provide an approximation for the pair potential of mean force $w(r)$, to the interaction measurements. While the fitting procedure is generally done for interpreting B_{22} measurements through applying Eq. 1, a similar approach can be applied by fitting $w(r)$ to k_D measurements using Eq. 28 combined with expressions for k_H given by Eqs. 23 and 24. A number of studies have shown measured values of k_D provide similar fittings to model interaction potentials as happens when fitting using B_{22} values [35, 36].

Nevertheless, using simplified models to describe proteins needs to be done with care as there is considerable uncertainty toward the molecular basis for protein–protein interactions. Some success has been achieved in predicting the pH and ionic strength patterns of protein–protein interactions in terms of electric double layer forces based on modelling proteins as uniformly charged spheres under low ionic strength conditions [8, 35–37]. Because electrostatic forces are repulsive and longer-ranged (greater than 1 nm for ionic strengths below 100 mM), there is no orientational biasing of the protein–protein interactions so that the averaging process involved in determining B_{22} or equivalently k_D reflect only the averaged protein surface properties. This contrasts with shorter-ranged attractive forces that are orientation-dependent and specific to the protein surfaces buried by the interacting configurations. Capturing these types of anisotropic interactions requires all-atomistic representations of proteins and the surrounding solvent medium, which remains beyond current capabilities [38, 39]. As such, measurements of B_{22} or k_D, at the very least, provide an averaged attractive protein–protein interaction after subtracting out the contributions from excluded volume forces and electrostatics.

2 Materials

Dynamic light scattering measurements can be carried out using cuvette-based systems such as the DynaPro NanoStar from Wyatt Technology or any of the Zetasizer Nano series from Malvern Panalytical Instruments or by Wyatt flow-through systems such as the miniDAWN Treos or DAWN HELEOS detectors equipped with a QELS (DLS) module. Higher throughput can be achieved in a multiwell format using the Wyatt DynaPro Plate Reader or the Malvern Panalytical Zetasizer APS.

The method provides an example procedure for solutions of lysozyme containing 10 mM sodium phosphate buffer at pH 7.0. The salts used in the experiment, sodium phosphate monobasic and sodium phosphate dibasic, should be of high purity (≥98%) and dissolved in at least Milli-Q grade water that has a resistivity of 18.2 MΩ cm. Use only high quality formulations of lyophilized or crystalline lysozyme with greater than 95% purity that contains a minimal amount of high molecular weight aggregates.

3 Methods

Determining the interaction parameter k_D requires measuring a series of eight to ten samples with incremental increases in protein concentration at fixed solvent composition (here the solvent refers to all solution components other than the protein). The series of samples should be prepared by dilution of a concentrated protein sample with its dialysate. The required volumes of buffers and protein solutions depend on whether the measurement is carried out using a cuvette, a multi-well plate or a flow through configuration. The method below is modifiable by scaling all volumes by the same factor to achieve the final sample requirement. All sample preparations need to be done carefully with particular attention to minimizing dust, irreversibly aggregated protein, and bubbles, all of which lower data quality.

3.1 Sample Preparation

The experiment described below requires ten 1 mL samples ranging in protein concentration between 1 and 10 mg/mL. Two stock solutions are prepared, a 20 mL sample of 10 mg/mL lysozyme in

10 mM sodium phosphate buffer at pH 7.0, and 4 L of 10 mM sodium phosphate at pH 7.0 to be used for dialysis.

1. Volumetrically prepare three 4 L batches of 10 mM sodium phosphate buffer at pH 7.0. For each preparation, dissolve 2.03 g (or 0.0169 mol) of sodium phosphate monobasic and 3.28 g (or 0.0231 mol) of sodium phosphate dibasic in approximately 3.8 L of Milli-Q water. Check the pH after the salts have dissolved and adjust the total volume to 4 L by adding Milli-Q water.
2. Filter the dialysis buffer through a 0.22 μm filter to remove any larger particulates. Prerinse glassware with a small amount of filtered dialysis buffer to remove any particulates in the glassware.
3. Dissolve 300 mg of lysozyme in 20 mL of dialysis buffer (*see* **Notes 1** and **2**).
4. Place the lysozyme solution in a dialysis bag or cassette with an appropriate molecular weight cutoff (MWCO), which should be the highest cutoff value that is less than the protein molecular weight.
5. Dialyze the lysozyme solution against 4 L of fresh dialysis buffer twice for 4 h each time and then redialyze in fresh buffer overnight (*see* **Note 3**). Carry out each dialysis step at 4 °C with gentle stirring to increase mass transport and solvent equilibrium across the membrane.
6. After dialysis is complete, remove the protein sample from a dialysis cassette using a needle and syringe or by decanting out the dialysis bag into a falcon tube.
7. Check the pH of the lysozyme solution and readjust to 7.0 if necessary by adding diluted acid or base. Check for any protein loss during dialysis by measuring the lysozyme concentration using a UV-spectrophotometer.
8. Degas the protein solution using an appropriate method (*see* **Note 4**).
9. Filter the lysozyme solution using a 0.1 μm pore size syringe top filter that has been prerinsed with 30 mL of dialysate to remove any particulates (*see* **Note 5**). Apply the same procedure to filter 20 mL of the degassed dialysate.
10. Measure the protein concentration after filtration to check for any protein loss due to adsorption or dilution with the dialysate contained in the dead volume of the filtration unit. If necessary, dilute the lysozyme solution to achieve the target protein concentration of 10 g/L.

11. Prepare ten samples with lysozyme concentrations between 1 and 10 mg/mL by serial dilutions of the 10 mg/mL lysozyme using the filtered dialysate as the diluent (*see* **Note 6**).

3.2 Sample Measurement

1. Carry out measurements on the solvent first. Analyze the light scattering data to confirm that the light scattering cell and the sample are clean (*see* **Note 7**).
2. Measurements on protein samples are carried out in order of either ascending or descending protein concentrations (*see* **Note 8**).
3. Set the run parameters for the dynamic light scattering experiment. The number of acquisitions and acquisition time depend on the scattering properties of the sample. Default settings for protein solutions are ten acquisitions each for 30 s. These might need to be adjusted to improve data quality (*see* **Note 9**).

Details that are specific for the experimental set-up are given in **Notes 10–12** for cuvette readers, **Notes 13–16** for plate readers, and **Notes 17–19** for flow cells.

3.3 Data Analysis

1. Analyze the quality of the light scattering data and include only the data that meets specified guidelines (*see* **Notes 20–24**).
2. Plot the measured diffusion coefficient against protein mass concentration c_p and fit the data to a linear function given by Eq. 11. The *y*-intercept of the line is equal to D_0 and the slope is equal to $k_D^{(c)} D_0$.
3. Calculate the hydrodynamic radius of the protein using the Stokes-Einstein relation given by Eq. 13 (*see* **Note 25**).

4 Notes

1. If the protein is obtained as a solid formulation, it is dissolved directly into the dialysis buffer at a concentration of 20–50% greater than the maximum concentration required for the experiment. This allowance is required to overcome small decreases in protein concentration that occur due to protein dilution during dialysis. Here, a 15 g/L sample is prepared by dissolving 300 mg of lysozyme in 20 mL of dialysis buffer.
2. Proteins obtained as liquid formulations require a buffer exchange step, which can be achieved using either a desalting column or by diafiltration using centrifugation or pressure-driven filtration units. These last steps can also be used to concentrate the protein if the concentration in the formulation is lower than the target value for the bulk sample.

3. The volume of the dialysate should be at least 200 times greater than that of the protein solution. If the dialysate volume needs to be reduced, increase the number of dialysis steps to achieve the same total dilution factor, where the total dilution factor is the product of each individual factor. For instance, a step with a dilution factor equal to 625 can be replaced by two steps each with a dilution factor equal to 25.
4. Degas the protein solution by drawing into a syringe and sealing the syringe tip with Parafilm. Create a vacuum at the tip of the syringe by gently pulling on the plunger while holding the syringe with the tip facing upward. Tap the plunger to dislodge any microscopic bubbles and let them float to the syringe tip. Gently remove the Parafilm and expel the gas to the atmosphere. Use the same procedure to degas 20 mL of the dialysate. For surface-active proteins, formation of any bubbles needs to be avoided. In this case, degas the sample by vacuum. Place the sample in a side-arm flask using a rubber stopper to seal the flask top. Make sure the sample container is open to the atmosphere and create a vacuum using a pump attached to the side-arm via tubing. Other methods of degassing include placing the sample in a sonicator followed by centrifugation.
5. When filtering protein solutions, better quality data is achieved by using a smaller pore size. Generally, a 0.02 μm pore-size filter can be used for proteins with molecular weights less than 20 kDa, while a 0.1 μm filter is used for larger proteins. A trial and error process might be required to determine the optimal pore size as proteins have different shapes and tendencies for interfacial adsorption, which causes membrane fouling and clogging.
6. When carrying out dilutions of bulk protein sample, use precise volumes so that each sample concentration can be calculated from the dilution factor and the bulk sample protein concentration. UV absorbance measurements should also be used to check each sample concentration especially when using small volumes where surface adsorption can lead to protein loss.
7. Before carrying out any experiment, the scattering profile from the pure solvent should be measured to check if the flow cell or cuvette is clean. Check that the measured correlation function appears as random noise rather than a detectable exponential decay function. Unless there are large molecular weight species in the solvent, the static light scattering signal, or count rate, is expected to remain constant. Changes to the solvent reading indicate the flow cell or cuvette is dirty, or there are large particles in the solvent, which is confirmed by having nonrandom noise in the correlation function.

8. Measurements on protein samples are generally carried out in either ascending or descending order of protein concentration. For cuvette or multiwell plate readers, each reading is done in triplicate as a minimum. For a flow through system, the measurement is commenced only after the static light scattering reading (or sample count rate) is time invariant.
9. The number of acquisitions and acquisition time needs to be set before each sample measurement. A correlation function is determined from averaging over each acquisition. For weakly scattering samples (e.g., at low protein concentration), longer acquisition times can be used to reduce noise by increased sampling, but lead to an increased probability of large particulates (dust, aggregates) entering the scattering volume, which skews the correlation function. If particles are problematic, reduce the acquisition time, but increase the number of acquisitions and only include acquisitions obtained for particle-free scattering.
10. Care must be taken to maintain a clean cuvette. Before each use, wash the cuvette thoroughly by rinsing with deionized water or mild acid followed by a volatile solvent such as ethanol or isopropanol. Dry the cuvette using compressed gas, either filtered air or nitrogen, or by using a cuvette dryer. Check the cuvette surface for any smudges and remove them by using dust-free lens tissue (for reduced volume cuvettes a magnification eyepiece may be required to properly inspect the cuvette). More heavily soiled cuvettes are soaked in a surfactant solution overnight such as 2% Hellmanex to remove material.
11. Using a pipette or syringe gently load the sample into the bottom of the cuvette while avoiding any bubble formation, which should be confirmed by visual observation. Place the cuvette into the instrument in the correct orientation.
12. Allow ample time for the sample to equilibrate at the desired temperature while checking that the count rate is higher than that for the pure solvent. If the count rate is fluctuating by greater than 20%, it may indicate the cuvette is not clean or that the sample has not been properly filtered.
13. Clean and remove any dust from the underside of the multiwell plate using dust-free lens tissue and volatile solvent followed by drying with compressed filtered air or nitrogen. The plate is kept covered as much as possible to reduce chances of contamination by dust.
14. Use a Pipetman or needle with syringe to load the required amount of sample into each well from the bottom up with the tip or needle not touching the plate. Centrifuge the plate at $800 \times g$ for 1–2 min to remove bubbles.

15. Place the plate in the sample chamber and provide an adequate time for temperature equilibration and then carry out the data acquisition. If a sample is evaporating the autocorrelation function is likely to increase because buffer is evaporating from the well, which increases the protein concentration.
16. Paraffin oil is used to prevent sample evaporation for small volume samples (e.g., less than 20 μL or the duration for scanning the plate is greater than a couple hours). Add an ample amount of oil to cover the sample. Recentrifuge the plate at 800 x *g* for 1–2 min. It is also recommended that the oil is filtered using a 0.22 μm pore size filter.
17. Inject the sample at a constant flowrate with a syringe pump connected to an in-line filter (with a 0.1 or 0.22 μm pore size) that precedes the light scattering detector. The sample injection volume should be between 0.5 and 1 mL to overcome any band broadening effects.
18. After injection, let the sample equilibrate in the flow cell for a minimum of 60 s. Before taking the measurement, the noise in the static light scattering reading should be reduced to an acceptable level and the reading should be at a steady-state. Excessive noise or a drifting static light scattering reading might be indicative of a dirty flow cell or possibly aggregation processes occurring in the sample cell.
19. If possible, use an on-line UV absorbance detector for simultaneous concentration determination. A variable path-length UV flow cell can be used for measuring concentrated protein samples that have absorbances too large for the Beer-Lambert law when using a 1 cm long pathlength flow cell. If not available, collect sample at outlet of detector for a batch UV absorbance reading. This step is necessary to ensure no protein was lost during the on-line filtration step.
20. Choose the maximum delay time for fitting the correlation function $\ln g_2(q, \tau)$ to the cumulant analysis. The cutoff should correspond to when the correlation function has decayed to approximately 5% of its initial value.
21. After the fitting has been carried out, check the quality of each sample reading. Quick-check readouts of data quality include the following:
    (a) *Polydispersity* (*P*): A polydispersity less than 0.1 indicates a monodispersed protein sample where the correlation function is predominantly weighted by the monomeric species. Greater values than 0.1 indicate significant contamination with higher molecular weight species.
    (b) *Baseline* (*B*): The fit value for the baseline should be equal to 1. A small amount of noise in the data is expected to cause deviations in the baseline on the order of 0.1%.

Any baseline reading greater than 1 indicates the presence of dust or large particles in the scattering volume, which will skew the fit values obtained using the cumulant analysis.

(c) *SOS*: The SOS is the sum of squares difference between the experimental correlation function and the best fit to the cumulant analysis, which provides a measure of the goodness of fit between the measured and theoretical data. There is no absolute SOS value that represents a good fit. Rather the SOS should only be compared between measurements of the same sample, where abnormally high values are indicative of either a dust particle interfering with the measurement or a dirty system.

(d) *Number of peaks and peak intensity*: Most instrumental software carries out a regularization analysis, which is applicable for characterizing multinodal populations with large size differences. The outputs of this analysis include the number of populations or peaks in the diffusion coefficient distribution and the intensity weighting of each peak. A single peak with a weighting of 100% by intensity indicates all high molecular weight impurities have been removed from the sample to undetectable levels. If multiple peaks occur in the analysis, the measured diffusion coefficients will be skewed from their true values.

22. Check that the theoretical correlation curve overlays well with the experimental curve. Deviations in the long time decay indicate presence of larger species, which will also be reflected by a larger polydispersity, a baseline not equal to 1.0, or the presence of a significant second peak in the regularization analysis. Deviations in the short time fitting might indicate the presence of short-time decay processes that arise in solutions containing cosolvent molecules due to their large size or self-association propensity. The latter effect becomes more pronounced at lower protein concentrations as the relative scattering signal of the solvent versus the protein increases.

23. All sample measurements are carried out in a minimum of triplicate. If one of the sample readings occurs outside of two standard deviations of the mean, the reading is discarded and the mean and standard deviation recalculated. Sample readings should also be discarded based on the criteria given in **Notes 2** and **3** above. If carrying out measurements using a cuvette, repeat measurements are required if data quality is low.

24. For systems with poor data quality, examine the fitting results of the individual data acquisitions for each sample reading.

Omit from the average any acquisitions with abnormally high values for polydispersity, SOS, and baseline readings.

25. The hydrodynamic size of the protein determined from D_0 provides an additional indicator for the quality of the data. The hydrodynamic size is independent of solvent conditions and temperature if the protein remains folded and solvent additives (or cosolvents) do not bind strongly to the protein. When comparing solution conditions that meet this criterion, an increase in the measured value of $R_{H,\,0}$ indicates the presence of small irreversibly formed aggregates or other impurities with sizes larger than the protein of interest. Conversely, a value of $R_{H,\,0}$ less than the expected value indicates the presence of species smaller than the protein.

References

1. Kirkwood JG, Goldberg RJ (1950) Light scattering arising from composition fluctuations in multi-component systems. J Chem Phys 18:54–57
2. Casassa EF, Eisenberg H (1964) Thermodynamic analysis of multicomponent solutions. Adv Protein Chem 19:287–395
3. Stockmayer WH (1950) Light scattering in multi-component systems. J Chem Phys 18:58–61
4. Vilker VL, Colton CK, Smith KA (1981) The osmotic-pressure of concentrated protein solutions – effect of concentration and pH in saline solutions of bovine serum-albumin. J Colloid Interface Sci 79:548–566
5. Binabaji E, Rao S, Zydney AL (2014) The osmotic pressure of highly concentrated monoclonal antibody solutions: effect of solution conditions. Biotechnol Bioeng 111:529–536
6. Ang S, Rowe AJ (2010) Evaluation of the information content of sedimentation equilibrium data in self-interacting systems. Macromol Biosci 10:798–807
7. Kuehner DE, Heyer C, Ramsch C, Fornefeld UM, Blanch HW, Prausnitz JM (1997) Interactions of lysozyme in concentrated electrolyte solutions from dynamic light-scattering measurements. Biophys J 73:3211–3224
8. Muschol M, Rosenberger F (1995) Interactions in undersaturated and supersaturated lysozyme solutions: static and dynamic light scattering results. J Chem Phys 103:10424–10432
9. Minton AP (2016) Recent applications of light scattering measurement in the biological and biopharmaceutical sciences. Anal Biochem 501:4–22
10. Connolly BD, Petry C, Yadav S, Demeule B, Ciaccio N, Moore JMR et al (2012) Weak interactions govern the viscosity of concentrated antibody solutions: high-throughput analysis using the diffusion interaction parameter. Biophys J 103:69–78
11. Lehermayr C, Mahler H-C, Maeder K, Fischer S (2011) Assessment of net charge and protein-protein interactions of different monoclonal antibodies. J Pharm Sci 100:2551–2562
12. Saito S, Hasegawa J, Kobayashi N, Kishi N, Uchiyama S, Fukui K (2012) Behavior of monoclonal antibodies: relation between the second virial coefficient (B-2) at low concentrations and aggregation propensity and viscosity at high concentrations. Pharm Res 29:397–410
13. Rosenbaum D, Zamora PC, Zukoski CF (1996) Phase behavior of small attractive colloidal particles. Phys Rev Lett 76:150–153
14. Rosenbaum DF, Kulkarni A, Ramakrishnan S, Zukoski CF (1999) Protein interactions and phase behavior: sensitivity to the form of the pair potential. J Chem Phys 111:9882–9890
15. Ahamed T, Esteban BNA, Ottens M, van Dedem GWK, van der Wielen LAM, Bisschops MAT et al (2007) Phase behavior of an intact monoclonal antibody. Biophys J 93:610–619
16. Curtis RA, Lue L (2006) A molecular approach to bioseparations: protein-protein and protein-salt interactions. Chem Eng Sci 61:907–923
17. Guo B, Kao S, McDonald H, Asanov A, Combs LL, Wilson WW (1999) Correlation of second virial coefficients and solubilities useful in protein crystal growth. J Cryst Growth 196:424–433

18. George A, Wilson WW (1994) Predicting protein crystallization from a dilute-solution property. Acta Crystallogr D 50:361–365
19. ten Wolde PR, Frenkel D (1997) Enhancement of protein crystal nucleation by critical density fluctuations. Science 277:1975–1978
20. Piazza R, Peyre V, Degiorgio V (1998) "Sticky hard spheres" model of proteins near crystallization: a test based on the osmotic compressibility of lysozyme solutions. Phys Rev E 58: R2733–R2736
21. Lekkerkerker HNW, Poon WCK, Pusey PN, Stroobants A, Warren PB (1992) Phase-behavior of colloid plus polymer mixtures. Europhys Lett 20:559–564
22. Brummitt RK, Nesta DP, Chang LQ, Chase SF, Laue TM, Roberts CJ (2011) Nonnative aggregation of an IgG1 antibody in acidic conditions: part 1. Unfolding, colloidal interactions, and formation of high-molecular-weight aggregates. J Pharm Sci 100:2087–2103
23. Sahin E, Grillo AO, Perkins MD, Roberts CJ (2010) Comparative effects of pH and ionic strength on protein-protein interactions, unfolding, and aggregation for IgG1 antibodies. J Pharm Sci 99:4830–4848
24. Chi EY, Krishnan S, Kendrick BS, Chang BS, Carpenter JF, Randolph TW (2003) Roles of conformational stability and colloidal stability in the aggregation of recombinant human granulocyte colony-stimulating factor. Protein Sci 12:903–913
25. Berne BJ, Pecora R (eds) (1976) Dynamic light scattering, with applications to chemistry, biology, and physics. John Wiley & Sons, New York, NY
26. McMillan WG, Mayer JE (1945) The statistical thermodynamics of multicomponent systems. J Chem Phys 13:276–305
27. Israelachvili JN (ed) (1992) Intermolecular and surface forces: With applications to colloidal and biological systems, 2nd edn. Academic, New York, NY
28. Gruenberger A, Lai P-K, Blanco MA, Roberts CJ (2013) Coarse-grained modeling of protein second osmotic virial coefficients: Sterics and short-ranged attractions. J Phys Chem B 117:763–770
29. Russel WB, Glendinning AB (1981) The effective diffusion-coefficient detected by dynamic light-scattering. J Chem Phys 74:948–952
30. Nagele G (1996) On the dynamics and structure of charge-stabilized suspensions. Phys Rep 272:216–372
31. Batchelor GK (1972) Sedimentation in a dilute dispersion of spheres. J Fluid Mech 52:245–268
32. Batchelor GK (1982) Sedimentation in a dilute polydisperse system of interacting spheres. 1. General-theory. J Fluid Mech 119:379–408
33. Felderhof BU (1978) Diffusion of interacting brownian particles. J Phys A Math Gen 11:929–937
34. Jeffrey DJ, Onishi Y (1984) Calculation of the resistance and mobility functions for 2 unequal rigid spheres in low-reynolds-number flow. J Fluid Mech 139:261–290
35. Roberts D, Keeling R, Tracka M, van der Walle CF, Uddin S, Warwicker J et al (2014) The role of electrostatics in protein-protein interactions of a monoclonal antibody. Mol Pharm 11:2475–2489
36. Arzensek D, Kuzman D, Podgornik R (2012) Colloidal interactions between monoclonal antibodies in aqueous solutions. J Colloid Interface Sci 384:207–216
37. Eberstein W, Georgalis Y, Saenger W (1994) Molecular interactions in crystallizing lysozyme solutions studied by photon-correlation spectroscopy. J Cryst Growth 143:71–78
38. Li W, Persson BA, Morin M, Behrens MA, Lund M, Oskolkova MZ (2015) Charge-induced patchy attractions between proteins. J Phys Chem B 119:503–508
39. Fusco D, Headd JJ, De Simone A, Wang J, Charbonneau P (2014) Characterizing protein crystal contacts and their role in crystallization: rubredoxin as a case study. Soft Matter 10:290–302

Chapter 2

Light Scattering to Quantify Protein–Protein Interactions at High Protein Concentrations

Mahlet A. Woldeyes, Cesar Calero-Rubio, Eric M. Furst, and Christopher J. Roberts

Abstract

Static and dynamic (laser) light scattering (SLS and DLS, respectively) can be used to measure the so-called weak or colloidal protein–protein interactions in solution from low to high protein concentrations (c_2). This chapter describes a methodology to measure protein–protein self-interactions using SLS and DLS, with illustrative examples for monoclonal antibody solutions from low to high protein concentrations ($c_2 \sim 1\text{–}10^2$ g/L).

Key words Protein–protein interactions, Static light scattering, Dynamic light scattering, Structure factor, Hydrodynamic factor

1 Introduction and Background

The solution behavior of proteins can be impacted by short and long-ranged "weak" protein–protein interactions (PPI). PPI have been shown to influence physical properties and processes such as opalescence, crystallization, liquid–liquid phase separation, aggregation rates and mechanisms, and elevated solution viscosity [1–7]. Typically and for practical reasons, PPI have been quantified experimentally using SLS and DLS at low protein concentrations (c_2)—where the subscript 2 denotes protein as the second component, water is component 1, and additional components are labeled 3, 4, etc. [1, 4, 8, 9]. While the same types of intermolecular forces exist at low and high c_2, the average distance between protein molecules is much smaller at high c_2, and the distance-dependence of different types of interactions can differ substantially [10]. This includes both direct interactions (e.g., dispersion forces, hydrogen bonding, steric repulsions) and solvent-averaged interactions (e.g., solvation forces, screened electrostatics). Consequently, it is not always clear whether the balance between short- and long-ranged

Jennifer J. McManus (ed.), *Protein Self-Assembly: Methods and Protocols*, Methods in Molecular Biology, vol. 2039,
https://doi.org/10.1007/978-1-4939-9678-0_2,

contributions to PPI at high c_2 can be predicted accurately from low-c_2 experimental behavior. Therefore, it is important to measure PPI at both high and low c_2. At high c_2, PPI can be quantified in terms of the Kirkwood–Buff integral (G_{22}) or the low-angle (zero-q) static structure factor ($S_{q\,=\,0}$) via static light scattering with a monochromatic source (i.e., laser scattering), SLS, or small-angle neutron/X-ray scattering [4, 9, 11–14]. Additionally, DLS can be used at high c_2 to measure the collective diffusion coefficient (D_C) [15]. D_C is influenced by contributions from thermodynamic as well as hydrodynamic interactions. The theoretical relations can be used to combine results from SLS and DLS to quantify the hydrodynamic interactions in terms of the hydrodynamic factor ($H_{q\,=\,0}$) [13, 16].

Batch SLS experiments are conducted at a fixed laser wavelength (λ) and constant temperature (usually between 20 and 25 °C). Protein particles scatter light due to the difference in refractive index from the buffer and fluctuations in local composition. The average scattered intensity is measured at a defined angle (e.g., 90° or 173° for backscattering) and used to calculate the excess Rayleigh ratio, represented as R^{ex}:

$$R^{\mathrm{ex}} = \frac{I_{\mathrm{sample}} - I_{\mathrm{buffer}}}{I_{\mathrm{toluene}} - I_{\mathrm{background}}} R_{\mathrm{toluene}} \times n_{\mathrm{solvent}}^2 \tag{1}$$

where I is the measured scattered light intensity of sample (I_{sample}), buffer (I_{buffer}), toluene (I_{toluene}), and background radiation ($I_{\mathrm{background}}$); R_{toluene} is the Rayleigh ratio of toluene at the measured temperature, and n_{solvent} is the refractive index of the solvent. R^{ex} for a protein solution as a function of c_2 can be used to estimate protein–protein interactions in the form of the protein–protein Kirkwood–Buff integral, G_{22}:

$$\frac{R^{\mathrm{ex}}}{K} = M_{\mathrm{w}} c_2 \left(\frac{M_{\mathrm{w,app}}}{M_{\mathrm{w}}} + G_{22} c_2 \right) \tag{2}$$

$$K = \frac{4\pi^2 n^2 \left(\frac{\mathrm{d}n}{\mathrm{d}c_2} \right)^2}{N_{\mathrm{A}} \lambda^4} \tag{3}$$

where $M_{\mathrm{w,app}}$ is the protein apparent molecular weight and M_{w} is the protein true molecular weight. K is a constant for a given protein and solution condition with a given experimental scattering configuration, where n is the solution refractive index, ($\mathrm{d}n/\mathrm{d}c_2$) is the change in refractive index of the solution as a function of c_2, N_{A} is Avogadro's number, and other symbols are as defined above [9]. The zero-q limit for the structure factor ($S_{q\,=\,0}$) can be obtained by dividing the right hand side of Eq. 2 by $c_2 M_{\mathrm{w}}$, with the canonical simplification that $M_{\mathrm{w,app}} \approx M_{\mathrm{w}}$ [9]. In this case, $S_{q\,=\,0}$ is equal to $1 + c_2 G_{22}$ and is dimensionless. Negative (positive) G_{22} values are equivalent to $S_{q\,=\,0}$ values below (above) 1, and correspond to net repulsive (attractive) interactions. Correspondingly, in dilute solutions, positive (negative) second osmotic virial coefficient (B_{22}) values indicate net repulsions (attractions) between protein molecules at low c_2. Note, $G_{22} \rightarrow -2B_{22}$ in the limit of dilute protein concentrations (i.e., c_2 below ~10 g/L and $|c_2 G_{22}| < 0.1$) [9]. As B_{22} is independent of c_2, B_{22} values can also be obtained by fitting experimental excess Rayleigh profiles to Eq. 2 for low-c_2 conditions, and equating $G_{22} = -2B_{22}$ for that low-c_2 limit.

In DLS, the time dependence of the fluctuations of scattered light is measured using a detector and autocorrelator. The resulting intensity autocorrelation function $g_2(t)$ is used to calculate the collective diffusion coefficient (D_C) and polydispersity (p_2) using the method of cumulants shown in Eq. 4 [17].

$$g_2(t) = B + \beta e^{-2tD_C q^2} \left(1 + \frac{\mu_2}{2} t^2\right)^2 \quad (4)$$

B corresponds to the average baseline intensity, and β is the amplitude of the autocorrelation function $g_2(t)$ as $t \rightarrow 0$. The magnitude of the scattering wave vector $q = \frac{4\pi n}{\lambda} \sin\left(\frac{\theta}{2}\right)$, where θ is the detector angle and n is the refractive index of the sample. The polydispersity can be calculated using $p_2 = \frac{\mu_2}{(D_C q^2)^2}$.

If one uses the analysis by Nägele [18], D_C is related to $H_{q=0}$ and $S_{q=0}$ via Eq. 5, where D_0 is the infinite-dilution or tracer diffusion coefficient.

$$D_C = \frac{D_o H_{q=0}}{S_{q=0}} \quad (5)$$

Therefore, by combining result from SLS and DLS, $H_{q=0}$ can also be quantified as a function of c_2.

2 Materials

All materials must be soluble in water, and resulting solutions should typically be transparent to the naked eye. As light scattering signals are typically dominated by the largest scattering species in solution, cosolutes larger than the protein of interest should be avoided if possible, otherwise their contributions to scattering signals will either convolute the data, or require an alternative experimental design to those illustrated below [19, 20]. Submicron filters (average pore size ~0.22 μm) and/or benchtop centrifugation are able to remove most large particulates, but one should avoid chemicals or materials that contain contaminants that would not be removed with such methods (*see* **Note 1**).

1. 20 mM sodium acetate solution (or desired buffer solutions and cosolutes).
2. 5 M sodium hydroxide solution (NaOH).
3. 0.45 μm pore size membrane filters (e.g., polyvinylidene fluoride, PVDF) (*see* **Note 2**).
4. 0.22 μm pore size low-protein binding (e.g., PVDF) syringe filters.
5. 0.22 μm pore size syringe filters (e.g., polytetrafluoroethylene, PTFE) for filtration of nonaqueous solutions.
6. Monoclonal antibody, denoted MAb1 (or desired Protein stock solution(s) or lyophilized protein powder(s)).
7. Dialysis membrane with a molecular-weight (MW) cutoff at least one half of the nominal protein MW (alternatively, diafiltration with an appropriate membrane can be used).
8. Centrifugal filter units (*see* **Note 3**).
9. Microcentrifuge tubes.
10. pH meter.
11. 10% w/w sodium dodecyl sulfate and/or 0.2% v/v Hellmanex® III (or equivalent) solutions.
12. >98% purity ethanol and/or acetone.
13. Lens paper.
14. Laser scattering instruments: For SLS—DAWN-HELEOS II Multi-Angle Light Scattering (MALS) instrument (Wyatt Technology, Santa Barbara, CA) and for DLS—Zetasizer (Malvern Instruments, Malvern, UK).
15. Data analysis software (e.g., Origin, MATLAB, Igor).

3 Methods

Carry out all procedures at room temperature, unless specified otherwise. The example numbers used below are based on a desired volume of 10 mL of initial protein stock solution, and can be scaled accordingly if larger or smaller starting volumes are desired. All steps should be performed wearing appropriate personal protective equipment. At a minimum, lab gloves (e.g., nitrile based) must be worn for all steps, both to protect the person performing the experiment, and to prevent contamination of samples and sample holders.

3.1 Dialysis

Before starting dialysis or other solvent exchange methods to attain the desired pH and cosolute concentrations at high protein concentrations, a few steps need to be taken if starting with a liquid protein stock solution with a relatively low protein concentration (e.g., c_2 below ~ 20–50 g/L). *See* **Notes 4–7** before proceeding to the following steps.

1. Prepare 2 L of 20 mM acetate at pH 4.7 (denoted as buffer solution; alternate pH and ionic strengths should be used if desired) using distilled, deionized water (typical resistivity ~18 MΩ cm) and use a 0.45 μm filter to filter the buffer solution.
2. Prepare MAb1 solution at ~20–50 g/L (*see* **Note 8**) and filter using a 0.22 μm syringe filter (*see* **Note 9**).
3. Prepare dialysis membrane tubing or cassette with appropriate molecular weight cut off (MWCO) (*see* **Note 10**) for MAb1 (*see* **Note 11**).
4. Transfer the MAb1 solution into the dialysis tubing or cassette, making sure to avoid/eliminate bubbles.
5. Place the dialysis tubing or cassette loaded with MAb1 solution into a 600 mL beaker filled with 500 mL of buffer solution.
6. Store the dialysis system and remaining buffer solution in refrigerated conditions (2–8 °C).
7. Exchange (swap out) the external buffer solution every 8–12 h for a total of four buffer exchanges of ~500 mL each. Leave ~100 mL of buffer to be used for subsequent dilutions (Subheading 3.2).
8. At the end of the buffer exchange time, remove the dialysis tubing or cassette from the beaker and transfer the MAb1 solution to a suitable syringe for filtration using a 0.22 μm syringe filter.
9. Transfer the filtered protein solution to an Eppendorf or equivalent test tube (depending on volume) and centrifuge at 3200–10,000 RCF for 10 min to eliminate residual bubbles and any precipitates or remaining insoluble particles that were not removed by filtration.

3.2 Concentrated Protein Solutions and Sample Preparation Using Gravimetric Dilutions

1. Before concentrating protein solutions, measure the concentration of the dialyzed protein sample using UV absorbance at 280 nm and the corresponding extinction coefficient (1.586 cm^2/mg in the present case) for MAb1.
2. From the measured concentration calculate the final volume of the concentrated MAb1 sample to attain the desired higher protein concentration (*see* **Note 12**).
3. Using the buffer solution, rinse the inside of a 10 kDa MWCO (adjust based on protein size) centrifugal filter tube and transfer the MAb1 solution to be concentrated (*see* **Note 13**).

4. Centrifuge the centrifugal filter tube at ~3000 RCF and 7–15 °C (*see* **Note 14**) until the desired volume is reached (*see* **Note 15**).
5. Using a fine-tip transfer pipette gently mix the concentrated MAb1 solution inside the centrifugal filter chamber to make it easier to pipette.
6. Transfer the solution to a microcentrifuge tube and gently tap and invert the tube a few times to mix the solution.
7. Centrifuge at ~10,000 RCF for 10 min to eliminate bubbles that might form from the previous step and to sediment any large particles/dust potentially introduced in prior steps.
8. Measure the pH of the concentrated MAb1 solution to ensure it is at pH 5.0 ± 0.05. *See* **Note 5** for steps to take if desired pH is not reached.
9. Readjust the pH of the buffer solution from pH 4.7 to pH 5.0 using sodium hydroxide solution.
10. Measure the concentration of the concentrated MAb1 solution using UV absorbance.
11. Use a spreadsheet to calculate the mass or volume of pH 5 buffer and concentrated MAb1 needed for each lower-concentration sample.
12. Gravimetric dilutions are performed using mass plus corrections for density increases with protein concentration (when available): final concentration = initial concentration × dilution factor.
13. Upon the addition of buffer/formulation solution, mix by gently inverting the microcentrifuge tube to avoid formation of concentration gradient of both protein and cosolutes.
14. Centrifuge all samples for 5–15 min at ~10,000 RCF immediately prior to loading into the light scattering cuvette or analogous sample holder.

3.3 Light Scattering Experiment

1. Turn on the light scattering instrument and corresponding software to be used (Fig. 1). Follow instrument instructions with regards to instrument warm up time, and safety procedures.
2. Wash the cuvette vigorously as described below (*see* **Note 16**).
3. Start by rinsing out the scattering cuvette with water. If the sample previously contained in the cuvette is not miscible with water, first wash the cuvette using acetone or ethanol before rinsing with water (*see* **Note 17**).
4. Wash both inside and outside of the cuvette using 10% w/w SDS or 0.2% v/v Hellmanex followed by thorough rinsing with water. (Optional: rinse the inside of the cuvette again with

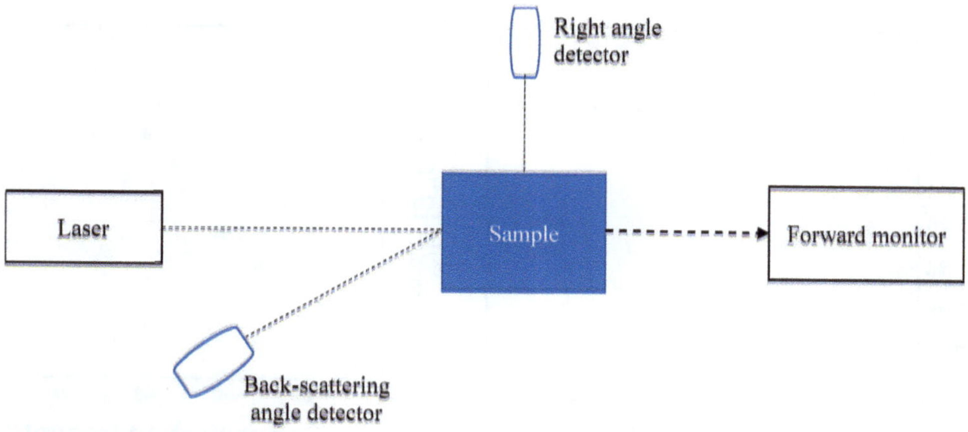

Fig. 1 Simplified schematic of a light scattering instrument. For instruments with DLS capability, an autocorrelator will either be connected to or replace the detector(s). An attenuator may also be located between the incident laser beam and the sample, or between the sample and the detector(s)

ethanol taking care to avoid residue formation. Avoid touching the scattering window(s) of the cuvette.)

5. Use dry and clean compressed air to completely dry the inside and outside of the cuvette.
6. Make sure the outside of the cuvette is clean; Use lens cleaning paper or task wipes (e.g., Kimwipes) with acetone/ethanol/isopropanol/lab-grade quartz cleaner to gently clean the outside surface and avoid leaving any residue on the outside of the cuvette after final cleaning.
7. Calibration with standard solution (e.g., toluene) when doing SLS measurements.
8. Follow instructions from the manufacturer regarding instrument/software calibration using a standard solution (usually toluene).
9. For DAWN-HELEOS II, using a glass syringe with PTFE based syringe filter, filter toluene into the clean cuvette and place the cuvette into the instrument. Check to make sure the scattering signal is clean and the temperature has stabilized (~5 min in common configurations). In the instrument software, open a calibration script and start the run. At the end of the run (e.g., 1 min) the software will output a calibration constant. Write down the calibration constant.
10. Wash the cuvette thoroughly after calibration and before adding an aqueous sample into the cuvette/sample holder.
11. To load the protein solution into the LS cuvette, using an automatic pipette, load 50–100 μL of sample into the cuvette or follow the volumes recommended by the cuvette/instrument manufacturer (*see* **Note 18**).

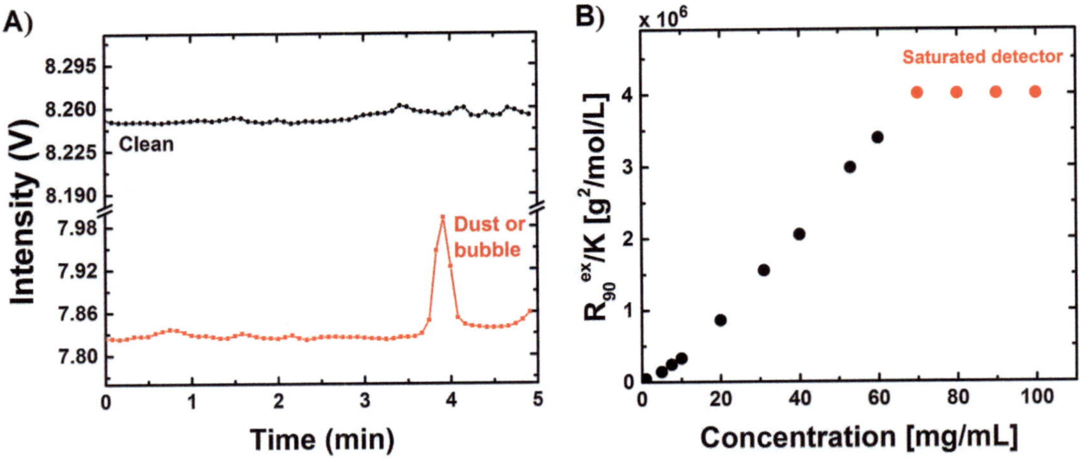

Fig. 2 Illustrative examples of (**a**) raw scatter intensity for clean sample (black circles) and a sample with dust or bubble (red rectangles). (**b**) Excess Rayleigh ratio as a function of protein concentration where detector saturation at the highest concentrations (red circles)

12. After loading the sample into cuvette, check for bubbles on internal cuvette surfaces and clean the cuvette external surface.
13. Place the loaded cuvette inside the LS instrument and allow the sample to equilibrate/reach the instrument set-point temperature. Typically, this is ~5–6 min for 20–25 °C measurements.
14. Check for detector saturation. For higher concentrations samples, and/or very large proteins, and for conditions with large net-attractive interactions, this may lead to sufficiently high scattering intensity. Therefore, it may be necessary to decrease the intensity of the incident laser. This is especially the case for instruments that do not have automatic attenuation. Detector saturation levels can be significantly different for SLS and DLS instruments.
15. If detector saturation is observed (Fig. 2b), decrease the intensity of the laser source either by decreasing the voltage or adding neutral density filters between the laser source and the sample. Be sure to perform a second calibration under the new conditions after the sample measurements are complete.
16. For DLS, it is essential that the instrument have automatic attenuation to prevent detector/autocorrelator saturations. Figure 3b illustrates the effect of laser power on the fitted diffusion coefficient if detector saturation is an issue.
17. Check for dust particles and/or bubbles based on scattering signal. Scattering intensity should be random with little to no significant spikes in signal. An illustrative scattering plot with clean data vs. poor data is shown in Fig. 2a (*see* **Note 19**).

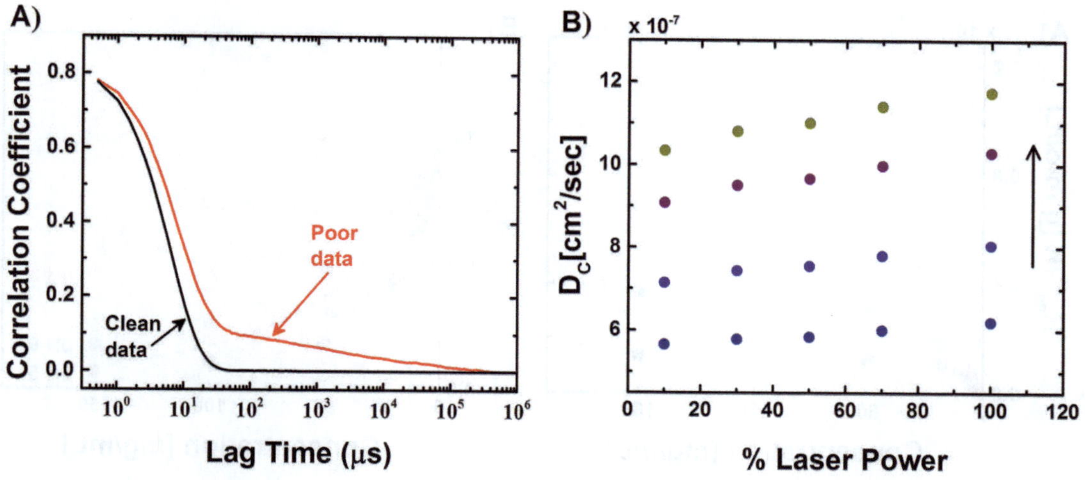

Fig. 3 Illustrative examples of (**a**) correlation coefficient for clean sample (black line) and a sample with dust or aggregate (red line). (**b**) Collective diffusion coefficient as a function of percent laser power at various protein concentrations. Arrow indicates increasing protein concentrations

18. DLS correlograms should be clean. The autocorrelation function should be horizontal at short lag times and show a single exponential decay to a baseline near zero at longer lag times (*see* **Note 20**).
19. It is recommended to collect data for a minimum of 5 min to assure high quality signal-to-noise ratio, as well as to confirm no artifacts from time-dependent aggregation events. For DLS, ten measurements with 10 s acquisition time repeated three times are recommended.
20. Export the data, if desired, to a data analysis package (Origin, Matlab, Igor, etc.) following software instructions.

3.4 SLS Data Analysis and Examples

1. For each protein concentration at a given solution condition, take a time-average of the scattering light intensity, excluding any obvious anomalies such as the "spike" shown in Fig. 2a that is likely due to dust or other contamination, unless the user has reason to expect those "spikes" are relevant for the sample.
2. Calibration correction using values attained from the standard solution and following the manufacturer recommended procedure. Final values should be given in the form of Rayleigh ratios as in Eq. 1.
3. Subtract measured average value(s) from protein-free solution to obtain excess Rayleigh ratios (R^{ex}) and divide the obtained R^{ex} values by the calibration/optical constant K as shown in Eqs. 2 and 3.

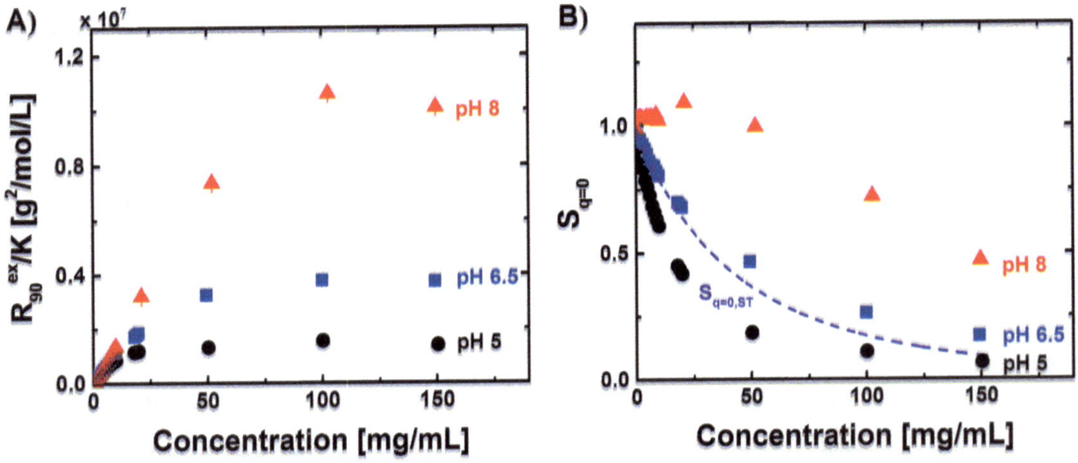

Fig. 4 (**a**) Excess Rayleigh scattering (R^{ex}/K) and (**b**) zero-q static structure factor ($S_{q=0}$) for MAb1 as a function of protein concentrations for pH 5 (black circles), pH 6.5 (blue rectangles) and pH 8 (red triangles). The purple dashed line corresponds to a reference steric-only coarse-grained MAb model (*see* Eq. 7 in ref. 23)

4. An illustrative example of Rayleigh scattering (R^{ex}/K) as a function of protein concentration for a monoclonal antibody (MAb1) is shown in Fig. 4a [21].
5. Protein apparent molecular weights and/or protein–protein interactions (via G_{22}) can be obtained by fitting Eq. 2 [9] against the experimental R^{ex}/K vs. c_2 data at low concentrations.
6. At high protein concentrations, protein–protein interaction can be quantified in terms of zero-q limit structure factor as shown in Fig. 4b, or G_{22} (via inversion of $S_{q=0} = 1 + c_2G_{22}$). The dashed line in Fig. 4b represents the steric-only contribution to $S_{q=0}$ (denoted $S_{q=0,\mathrm{ST}}$) that was attained from coarse-grained molecular simulations of a MAb model [22, 23]. $S_{q=0,\mathrm{ST}}$ is less than or equal to 1, and decreases monotonically with increasing c_2, which can be attributed to an increase in repulsive interactions due to molecular crowding at high c_2. The MAb in this example displays net-repulsive PPI relative to steric-only behavior at pH 5, as $S_{q=0}$ is less than $S_{q=0,\mathrm{ST}}$ at all concentrations above zero. At pH 8, $S_{q=0}$ is greater than $S_{q=0,\mathrm{ST}}$ at all tested concentrations, which is indicative of net-attractive PPI. At pH 6.5, the MAb has $S_{q=0}$ values that are close to $S_{q=0,\mathrm{ST}}$ as a function of c_2, indicating net-PPI that is not significantly different from steric repulsions [21].

3.5 DLS Data Analysis and Examples

1. If the experimental DLS correlation function displays a single decay, use the method of cumulants to determine the collective diffusion coefficient (D_C) and polydispersity index (p_2) from the intensity autocorrelation function ($g_2(t)$) as shown in Eq. 4 [17].

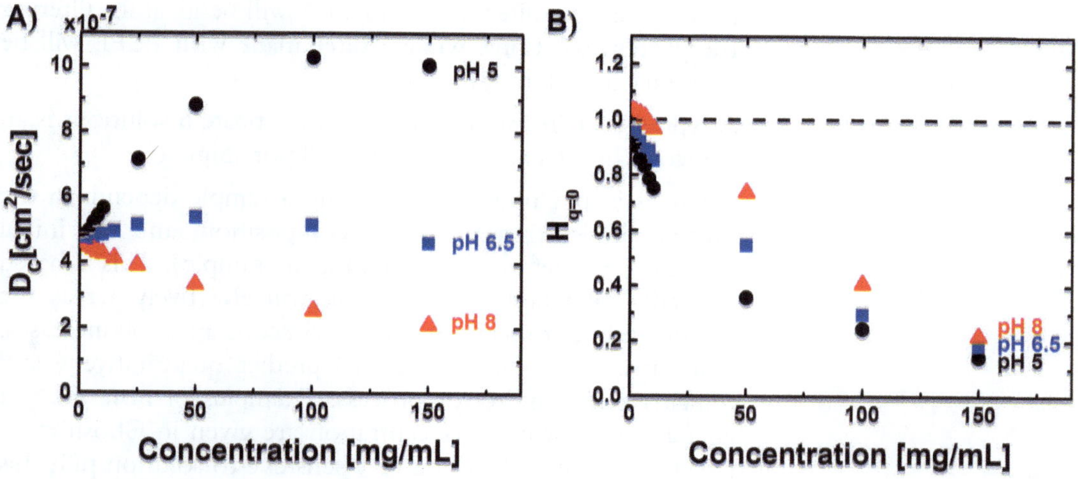

Fig. 5 (**a**) Collective diffusion coefficient (D_C) and (**b**) zero-q hydrodynamic factor ($H_{q=0}$) for MAb1 as a function of protein concentrations for pH 5 (black circles), pH 6.5 (blue rectangles), and pH 8 (red triangles)

2. The collective diffusion coefficient (D_C) as a function of protein concentration for MAb1 is shown in Fig. 5a [21]. D_C initially increases with increasing c_2 at pH 5 and pH 6.5, suggesting net-repulsive protein–protein interactions. D_C then decreases after a maximum value which is presumably due to competing effects of thermodynamic and hydrodynamic interactions as shown by the relation in Eq. 5. At pH 8, D_C decreases monotonically with increased c_2, consistent with net-attractive PPI.
3. As mentioned above, the collective diffusion coefficient has contributions from thermodynamic and hydrodynamic interactions (Eq. 5). By combining the results from SLS ($S_{q=0}$) and DLS (D_C and D_0) one can calculate the hydrodynamic factor ($H_{q=0}$) by inverting Eq. 5 to express $H_{q=0}$ in terms of D_C, $S_{q=0}$, and D_0. An example of $H_{q=0}$ versus protein concentration is shown in Fig. 5b. For MAb1, $H_{q=0}$ decreases with increasing c_2 at all pH values in this example.

4 Notes

1. Buffer, protein, and cosolute materials can be prepared/purchased as desired. However, the presence of large particles does considerably affect the quality of scattering data. Consequently, it is strongly encouraged to avoid sources that might contain particles (e.g., nano-sized particulate matter) that cannot be filtered with a 0.22 μm filter.
2. The choice of filter and syringe should be appropriate for the solvent and/or solution to be filtered. For example, low-

protein-binding filters such as PVDF will be ideal for filtering the protein solutions, while a filter made with PTFE will be necessary for filtering toluene.

3. A typical centrifugal filter unit used for protein solutions is an Amicon Ultra Centrifugal filter (Millipore Sigma).

4. Changes in pH upon concentrating a sample depend on the protein, ionic strength, buffer composition, and the initial solution pH (before concentrating the sample). This is based primarily on whether the protein will effectively act as the dominant buffer species as the protein concentration increases. Currently, this is not simple to predict quantitatively, and remains a trial-and-error process. Examples of how the pH changes with protein concentration are given in Ghosh et al. [4]. Given that PPI can be very sensitive to solution pH, this can be a key issue. In some cases, using higher concentrations of buffer might help mitigate this problem, although that necessitates working at higher ionic strength conditions that may not be desirable for some applications.

5. To account for possible pH shifts as the protein concentration increases (*see* **Note 4**): (a) Dialyze the protein solution at an appropriate pH at low protein concentration or (b) concentrate to the highest desired final protein concentration with buffer/formulation exchange steps using centrifugal filtration (*see* **Note 6**), but starting at lower (higher) solution pH than the desired final pH; trial-and-error to refine the correct starting pH value to achieve the desired final pH for a given final protein concentration (*see* ref. 4 for an example with a monoclonal antibody); achieve lower-concentration solutions by dilution with buffer matched to the desired final pH value, and this should match the pH value achieved after concentrating the initial (more dilute) protein stock solution.

6. Drawbacks: increased solution viscosity as c_2 increases; potential for incomplete protein material recovery (<90%); protein aggregation at high concentrations; need to iterate the starting pH to achieve the desired final pH and maximum protein concentration.

7. To account for possible changes in cosolute concentration (due to strong protein–cosolute interactions) after concentrating the protein, perform two additional buffer exchange steps at high protein concentration.

8. The starting concentration and volume is determined by how much protein is needed to perform the experiment. LS cuvettes or well plates often require volumes greater than 30 μL per sample (higher volumes are typically necessary for high-concentration samples because of losses due to handling with high-viscosity samples). For example, if one desires to do LS on

a series of samples with concentrations of 150, 125, 100, 75, 50, 25, and 10 mg/mL, a total of approximately 60 mg of protein will be required to have 100 μL of each sample. While scattering is inherently a nondestructive method, it is often desirable to run a set of samples immediately in series or in parallel, and therefore recovering the sample for dilution or reconcentrating is often not practical in realistic workflows.

9. During the dialysis steps, only the last filtration step is needed for high-quality data. However, the initial filtration step is encouraged for relatively unstable proteins and those sensitive to forming large particles/aggregates during dialysis.

10. A typical dialysis membrane for proteins with molecular weights greater than 25 kDa is a Spectra/Por 7, 10 kDa molecular weight cutoff (MWCO) dialysis membrane (Spectrum Laboratories). Slide-A-Lyzer 10K MWCO dialysis cassettes (Thermo Scientific) are also commonly used for dialysis.

11. Follow instructions to prepare the dialysis membrane tubing/cassette before transferring the protein solution into the membrane tubing. For example, some membranes require soaking in water for 15 min, while others may require a more extensive treatment, as many membranes are stored in solutions containing preservatives that will damage or jeopardize the stability of protein solutions.

12. It is important to a priori determine the mass of protein material that will be needed for all samples that will be measured for the experiment (*see* **Note 8**).

13. In some cases, particle shedding from the filter material has been reported during filtration steps. No robust solution is yet available, so the reader must be aware that excessive sample filtration might be detrimental to scattering data quality [24].

14. Temperatures significantly higher than refrigerated conditions (i.e., greater than ~15 °C) are recommended during centrifugation for shorter concentrating times with samples that exhibit a pronounced increase in solution viscosity with increased c_2. However, proteins that are extremely sensitive to temperature should be centrifuged at lower temperatures (2–8 °C) to avoid possible loss due to aggregation during this step.

15. The time it takes to reach the desired concentration, which can range from ~10 min to hours, is dependent on the viscosity of the solutions as well as the protein interactions. Empirically, solution conditions which lead to attractive PPI often take longer than those with repulsive PPI. However, as the point of the LS exercise is to determine those PPI, it is not typically possible to predict the necessary centrifugation time without

preliminary data. Therefore, it is recommended that trial runs for different centrifugation times are used to optimize as needed. Additionally, the user must select the temperature for this step based on potential solution viscosity and thermal stability of the protein sample.

16. Light scattering experiments are very sensitive to dust and particulates, due to the fact that the intensity of light scattering increases nonlinearly with the size or characteristic dimension (s) of the scattering species [25].Therefore, cleanliness during sample preparation and experimental setup (cleaning of pipette tips, tubes, cuvette, etc.) and proper filtering is essential. If available, cleaning and filling cuvettes/sample holders in a flow hood or laminar flow hood can help minimize contamination by dust [25].
17. Variations to the cleaning protocol can be used, and typically cuvette or instrument vendors have suggestions for best cleaning practices.
18. For high protein concentration samples, more sample volume may need to be loaded into the pipette tip than at low concentrations, as samples may be viscous.
19. The signal intensity must look random as a function of time. The presence of a trend in the scattering intensity as a function of time might highlight the presence of bubbles toward the walls of the cuvette or long ranged correlations due to an exceedingly strong protein–protein attractions or time-dependent aggregation or sample degradation (e.g., protein cleavage).
20. The correlation function should be a single exponential decay.

Acknowledgments

Support from the National Science Foundation (NSF GRF to M.A.W.), Amgen, and the National Institute of Health (NIH R01EB006006) are gratefully acknowledged.

References

1. Raut AS, Kalonia DS (2015) Opalescence in monoclonal antibody solutions and its correlation with intermolecular interactions in dilute and concentrated solutions. J Pharm Sci 104:1263–1274. https://doi.org/10.1002/jps.24326
2. Neergaard MS, Kalonia DS, Parshad H et al (2013) Viscosity of high concentration protein formulations of monoclonal antibodies of the IgG1 and IgG4 subclass – prediction of viscosity through protein–protein interaction measurements. Eur J Pharm Sci 49:400–410. https://doi.org/10.1016/j.ejps.2013.04.019
3. Connolly BD, Petry C, Yadav S et al (2012) Weak interactions govern the viscosity of concentrated antibody solutions: high-throughput analysis using the diffusion interaction parameter. Biophys J 103:69–78. https://doi.org/10.1016/j.bpj.2012.04.047

4. Ghosh R, Calero-Rubio C, Saluja A, Roberts CJ (2016) Relating protein-protein interactions and aggregation rates from low to high concentrations. J Pharm Sci 105:1086–1096. https://doi.org/10.1016/j.xphs.2016.01.004
5. George A, Wilson WW (1994) Predicting protein crystallization from a dilute solution property. Acta Crystallogr D Biol Crystallogr 50:361–365. https://doi.org/10.1107/S0907444994001216
6. Dumetz AC, Chockla AM, Kaler EW, Lenhoff AM (2008) Effects of pH on protein-protein interactions and implications for protein phase behavior. Biochim Biophys Acta 1784:600–610. https://doi.org/10.1016/j.bbapap.2007.12.016
7. Mason BD, Zhang L, Remmele RL, Zhang J (2011) Opalescence of an IgG2 monoclonal antibody solution as it relates to liquid-liquid phase separation. J Pharm Sci 100:4587–4596. https://doi.org/10.1002/jps.22650
8. Saluja A, Fesinmeyer RM, Hogan S et al (2010) Diffusion and sedimentation interaction parameters for measuring the second virial coefficient and their utility as predictors of protein aggregation. Biophys J 99:2657–2665. https://doi.org/10.1016/j.bpj.2010.08.020
9. Blanco MA, Sahin E, Li Y, Roberts CJ (2011) Reexamining protein-protein and protein-solvent interactions from Kirkwood-Buff analysis of light scattering in multi-component solutions. J Chem Phys 134:225103. https://doi.org/10.1063/1.3596726
10. Woldeyes MA, Calero-Rubio C, Furst EM, Roberts CJ (2017) Predicting protein interactions of concentrated globular protein solutions using colloidal models. J Phys Chem B 121:4756–4767. https://doi.org/10.1021/acs.jpcb.7b02183
11. Scherer TM, Liu J, Shire SJ, Minton AP (2010) Intermolecular interactions of IgG1 monoclonal antibodies at high concentrations characterized by light scattering. J Phys Chem B 114:12948–12957. https://doi.org/10.1021/jp1028646
12. Arzensek D, Kuzman D, Podgornik R et al (2015) Hofmeister effects in monoclonal antibody solution interactions. J Phys Chem B 119:10375–10389. https://doi.org/10.1021/acs.jpcb.5b02459
13. Fine BM, Lomakin A, Ogun OO, Benedek GB (1996) Static structure factor and collective diffusion of globular proteins in concentrated aqueous solution. J Chem Phys 104:326–335. https://doi.org/10.1063/1.470904
14. Stradner A, Sedgwick H, Cardinaux F et al (2004) Equilibrium cluster formation in concentrated protein solutions and colloids. Nature 432:492–495. https://doi.org/10.1038/nature03109
15. Berne BJ, Pecora R (2000) Dynamic light scattering with applications to chemistry, biology, and physics. Dover Publications, Mineola, NY
16. Blanco MA, Perevozchikova T, Martorana V et al (2014) Protein-protein interactions in dilute to concentrated solutions: α-chymotrypsinogen in acidic conditions. J Phys Chem B 118:5817–5831. https://doi.org/10.1021/jp412301h
17. Frisken BJ (2001) Revisiting the method of cumulants for the analysis of dynamic light-scattering data. Appl Opt 40:4087–4091
18. Nägele G (1996) On the dynamics and structure of charge-stabilized suspensions. Phys Rep 272:215–372. https://doi.org/10.1016/0370-1573(95)00078-X
19. Fernández C, Minton AP (2009) Static light scattering from concentrated protein solutions II: experimental test of theory for protein mixtures and weakly self-associating proteins. Biophys J 96:1992–1998. https://doi.org/10.1016/j.bpj.2008.11.054
20. Attri AK, Minton AP (2005) Composition gradient static light scattering: a new technique for rapid detection and quantitative characterization of reversible macromolecular hetero-associations in solution. Anal Biochem 346:132–138. https://doi.org/10.1016/j.ab.2005.08.013
21. Woldeyes MA, Qi W, Razinkov VI et al (2018) How well do low- and high-concentration protein interactions predict solution viscosities of monoclonal antibodies? J Pharm Sci. https://doi.org/10.1016/j.xphs.2018.07.007
22. Calero-Rubio C, Saluja A, Roberts CJ (2016) Coarse-grained antibody models for "weak" protein–protein interactions from low to high concentrations. J Phys Chem B 120:6592–6605. https://doi.org/10.1021/acs.jpcb.6b04907
23. Calero-Rubio C, Ghosh R, Saluja A, Roberts CJ (2018) Predicting protein-protein interactions of concentrated antibody solutions using dilute solution data and coarse-grained molecular models. J Pharm Sci 107:1269–1281. https://doi.org/10.1016/j.xphs.2017.12.015
24. Liu L, Randolph TW, Carpenter JF (2012) Particles shed from syringe filters and their effects on agitation-induced protein aggregation. J Pharm Sci 101:2952–2959. https://doi.org/10.1002/jps.23225
25. Schärtl W (2007) Light scattering from polymer solutions and nanoparticle dispersions. Springer, Berlin

Chapter 3

Quantitative Evaluation of Protein Solubility in Aqueous Solutions by PEG-Induced Liquid–Liquid Phase Separation

Ying Wang and Ramil F. Latypov

Abstract

This chapter describes an experimental method to quantitatively evaluate the solubility of proteins in aqueous solutions. Measurement of protein solubility can be challenging because low solubility can be manifested through various pathways (e.g., crystallization, aggregation, gelation, and liquid–liquid phase separation), some of which may occur over long periods of time. In the method described here, a nonionic polymer, polyethylene glycol (PEG), is added to a protein solution of interest to induce instantaneous formation of protein-rich liquid droplets. After incubation at a given temperature, the samples are centrifuged. The protein concentration in the supernatant is measured at various PEG concentrations to calculate an equilibrium binding free energy, which provides a measure of protein solubility. Based on the first principles of thermodynamics, this method is highly reproducible and applicable to various proteins and buffer conditions.

Key words Protein solubility, Colloidal stability, Aggregation, Crystallization, Liquid–liquid phase separation, PEG, Gelation, Precipitation

1 Introduction

Low protein solubility may negatively affect protein function and cause unwanted aggregation [1, 2]. Due to the attractive interprotein interactions, folded proteins can become insoluble through various pathways, including crystallization, liquid–liquid phase separation, colloidal aggregation, and gelation [3, 4]. In contrast to high temperature unfolding, these protein condensation phenomena usually occur at nondenaturing temperatures and relatively high protein concentrations. Because all condensation phenomena are driven by the inherent attractive interactions between protein molecules, they are closely related to each other as shown in the schematic phase diagram in Fig. 1 [5–7]. In principle, protein solubility can be evaluated by studying any one of these condensation phenomena. However, protein crystallization and aggregation

Jennifer J. McManus (ed.), *Protein Self-Assembly: Methods and Protocols*, Methods in Molecular Biology, vol. 2039,
https://doi.org/10.1007/978-1-4939-9678-0_3,

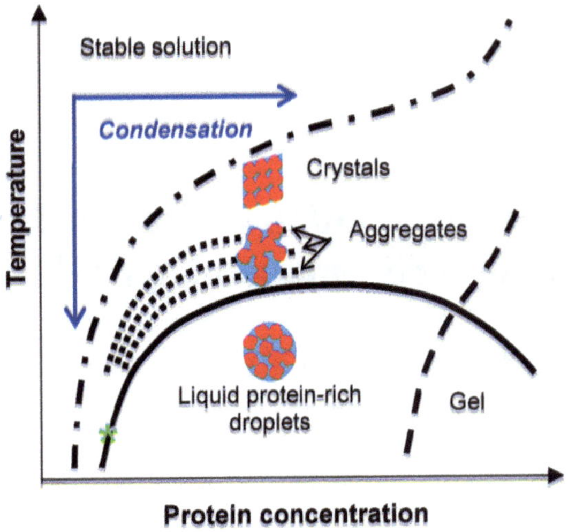

Fig. 1 A schematic phase diagram of a protein solution. The conditions for the various protein condensation phenomena are delineated by the phase boundaries in the phase diagram. Any point (e.g., the green star) on a phase boundary can be used to evaluate protein solubility. This method measures the points along the LLPS phase boundary indicated by the solid curve

can take many months, and protein crystals are not always available for solubility measurements.

In the method described herein, liquid–liquid phase separation (LLPS) in protein solutions is used to quickly gauge protein solubility. For most proteins, LLPS, marked by formation of protein-rich droplets (Fig. 2), cannot be directly observed at temperatures above the freezing point of the solution [8, 9]. To circumvent this problem, a nonionic polymer, polyethylene glycol (PEG) that is preferentially excluded from the immediate domain of the protein [10], can be added to protein solution to induce LLPS at temperatures above the freezing point [8, 9, 11]. In the presences of PEG, LLPS can be readily observed for many proteins and buffer conditions. In this method a sample is incubated for several hours and then centrifuged at a given temperature. The equilibrium protein concentration in the supernatant is measured at several initial PEG concentrations. PEG introduces an additional attractive interprotein interaction that is quantitatively described by a depletion interaction model [11, 12]. A binding free energy of the native interaction between protein molecules can be calculated by extrapolating the protein supernatant concentration to a no PEG condition [8, 9]. Low binding free energy indicates strong attractive interprotein interactions, low protein solubility, and therefore high probability of crystallization or aggregation [8, 9, 13]. With this method, protein solubility can be quantitatively evaluated in 1 day using less than 1 mg of protein. Also, a high throughput method can be developed to simultaneously screen solubility of various proteins in different buffer conditions.

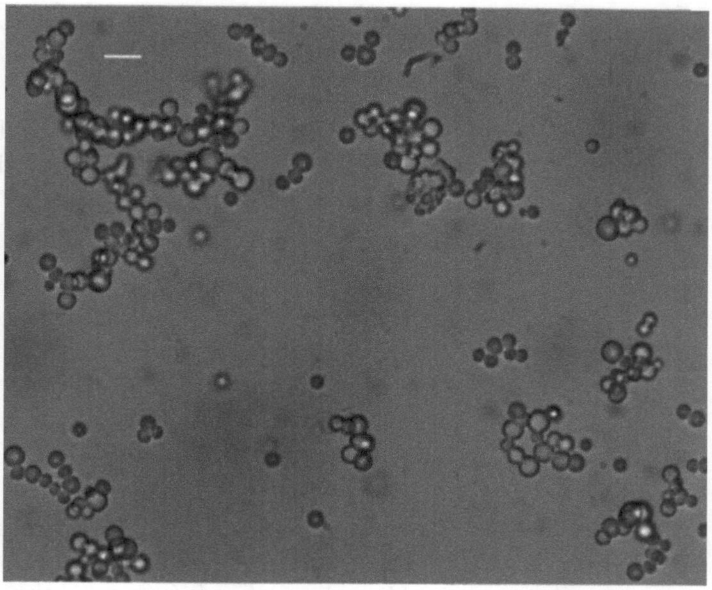

Fig. 2 Liquid–liquid phase separation in a 1 mg/mL IgG monoclonal antibody solution induced by the addition of 5.5% PEG3350. The scale of the white bar is 10 μm

2 Materials

1. Phosphate buffered saline (1× PBS), pH 7.2, or any other buffer of choice (*see* **Note 1**).
2. Polyethylene glycol MW 3350 (*see* **Note 2**).
3. Protein: an IgG monoclonal antibody or any other protein of interest (*see* **Note 3**).
4. Analytical balance.
5. Refrigerator, cold room, or a water bath calibrated for 2–8 °C.
6. Refrigerated microcentrifuge (e.g., Fisher Scientific accuSpin Micro 17R or equivalent).
7. UV-Vis spectrophotometer (Thermo Scientific NanoDrop™ 8000 or equivalent is useful when dealing with limited material).
8. ABBE Refractometer (ATAGO, Model DR-A1 or equivalent) (optional, *see* **Note 4**).
9. 0.5 mL clear polypropylene centrifugation tubes (Eppendorf or equivalent).

3 Methods

This section provides step by step instructions for executing PEG-LLPS assay of a monoclonal antibody in phosphate buffered saline (PBS) at pH 7.2. Same procedure is applicable for other proteins and buffers.

1. Prepare 1 g of 40% (w/w) PEG3350 stock solution by dissolving 0.4 g PEG3350 in 0.6 g of buffer by vortexing. Weigh both PEG and buffer (PBS hereafter in this example) using an analytical balance to calculate the accurate concentration of PEG. After vortexing, the concentrated PEG stock solution may contain air bubbles. Centrifuge at 3500 rcf for 2 min to remove bubbles (*see* **Note 5**).
2. Determine the concentration of the protein (IgG1 in this example) stock solution, c, by measuring the UV absorbance, A, at 280 nm and applying the Beer-Lambert law, $c = A/\varepsilon l$. Here, $\varepsilon = 1.5$ mL/mg cm is the extinction coefficient of the IgG1 at 280 nm, and $l = 1$ cm is the length of the light path. The protein concentration in the stock solution should be above 4 mg/mL. Prepare 0.5 mL of 2 mg/mL and 0.15 mL of 4 mg/mL protein solutions by diluting protein stock with PBS (*see* **Note 6**).
3. Determine the lowest PEG concentration needed for inducing LLPS in the protein solutions in the following way. Prepare 20 μL each of 5%, 10%, 20%, and 40% (w/w) PEG in clear 0.5 mL tubes by accurately mixing the 40% PEG stock and buffer (*see* **Note 7**). Add 20 μL of 2 mg/mL protein sample to each of the preprepared PEG solutions and immediately mix by vortexing for 3–5 s.
4. Incubate the samples at 4 °C for 15 min. Record the lowest final PEG concentration (2.5%, 5%, 10%, or 20%) at which the sample becomes cloudy due to LLPS (*see* **Note 8**). This step can be repeated using a narrower PEG concentration range to get a more accurate estimate of the minimum PEG concentration needed for inducing LLPS. For the antibody used here, 8% PEG3350 was found to be sufficient to induce LLPS.
5. Use the buffer and the PEG stock (40%) to prepare 200 μL of each of the three PEG solutions: (a) twice the minimum PEG concentration determined in **step 3** (e.g., $8 \times 2 = 16\%$), and (b, c) two samples at higher PEG concentrations with a 4% increment (in this case 20% and 24%).
6. Prepare the samples for solubility assay as follows. Use the 2 and 4 mg/mL protein solutions, buffer, and the three PEG solutions from **step 4** (i.e., 16%, 20%, and 24% PEG) to prepare seven samples shown in the Table 1. The recommended order

Table 1
The sample preparation chart for step 5

Sample list	Protein solution		Buffer	PEG solution			Final protein, mg/mL	Final PEG (w/w)
	2 mg/mL	4 mg/mL		16%	20%	24%		
1 (control)	50 μL	–	50 μL	–	–	–	1	0
2	50 μL	–	–	50 μL	–	–	1	0.08
3	–	50 μL	–	50 μL	–	–	2	0.08
4	50 μL	–	–	–	50 μL	–	1	0.10
5	–	50 μL	–	–	50 μL	–	2	0.10
6	50 μL	–	–	–	–	50 μL	1	0.12
7	–	50 μL	–	–	–	50 μL	2	0.12

of solution addition is protein solution, then buffer, and finally PEG. Mix each sample immediately after the addition of all three components by brief vortexing (3–5 s).

7. Incubate all samples including the control (0% PEG) at 4 °C overnight to reach equilibrium (*see* **Notes 9** and **10**).
8. Briefly centrifuge all seven samples at a maximum speed (10,000–17,000 × *g*) for 20–30 s in a refrigerated centrifuge set at the incubation temperature of 4 °C (*see* **Note 11**).
9. After centrifugation, a white precipitate is typically observed at the bottom of the test tube, whereas the supernatant is transparent. Immediately, place the samples back at 4 °C before proceeding to **step 10**.
10. Without disturbing the precipitate, slowly pipet 10 μL of the supernatant from the top center of the transparent solution to measure protein concentration. The collected supernatant can be kept at ambient temperature for concentration measurement, but the tube needs to be sealed to avoid drying. Return the rest of the sample to 4 °C (*see* **Note 12**).
11. Determine protein concentration, c_1, in the supernatant from **step 8** by measuring the UV absorbance at 280 nm and applying the Beer–Lambert law. The concentration of sample 1 (control without PEG) should remain the same as in the sample prepared originally. If there is significant reduction of the protein concentration in sample 1, this may suggest fast crystallization or aggregation (unusual in the case of antibodies). If this does occur, the incubation time should be shortened.

Table 2
The measured protein (IgG1 in this example) and PEG concentrations in supernatant, and the calculated values of ln c_1 and $\hat{\Pi}_2$ for the determination of the binding free energy

Supernatant	Protein concentration, c_1, (mg/mL)	PEG concentration, c_2, (w/w)	$\hat{\Pi}_2$	ln c_1
Sample 2	0.604	0.100	0.0788	−0.504
Sample 3	0.605	0.101	0.0802	−0.503
Sample 4	0.248	0.110	0.0935	−1.39
Sample 5	0.248	0.110	0.0935	−1.39
Sample 6	0.101	0.120	0.110	−2.29
Sample 7	0.118	0.119	0.108	−2.14

12. Check if the protein concentration of sample pairs with the same PEG content, that is, 2 and 3; 4 and 5; and 6 and 7, is within 20% error (corresponding to an acceptable small error in the calculated binding free energy). If this is confirmed, the samples are fully equilibrated. Then, record these protein concentration values, c_1, in Table 2 for the calculation in **step 15** (*see* **Note 13**).
13. Measure the refractive index (RI) of the supernatants to determine the PEG concentration, c_2, for samples 2–7 using a linear fit, shown in Fig. 3. Record the PEG concentrations, c_2, in Table 2 for the calculation in **step 14**. The use of PEG with a different molecular weight or in a different buffer may require performing a separate calibration experiment. The use of an experimentally measured PEG concentration in supernatant verified by RI improves the accuracy of the method (*see* **Note 4**).
14. Calculate the reduced osmotic pressure of PEG, $\hat{\Pi}_2$, for samples 2–7 from their PEG concentration, c_2, using Eq. 1.

$$\hat{\Pi}_2 = \frac{1000c_2}{M_2}\left[1 + 0.49\left(\frac{c_2}{c_2*}\right)^{1.25}\right] \quad (1)$$

c_2, is the PEG concentration of samples 2–7 in Table 2. M_2 is the molecular weight of PEG ($M_2 = 3350$ for PEG 3350). $c_2^* = \left(\frac{M_2-18}{44}\right)^{-0.8}/0.825$ is the dilute-semidilute crossover concentration of PEG. The calculated $\hat{\Pi}_2$ has a unit of mol/L. Fill Table 2 with calculated $\hat{\Pi}_2$ values. The values of $\hat{\Pi}_2$ for PEG 3350 in the concentration range 0–0.2 (i.e., 20%) can also be read from Fig. 4.
15. Use the supernatant protein concentration, c_1, of samples 2–7 in the units of mg/mL to calculate the natural logarithm lnc_1 and complete Table 2.

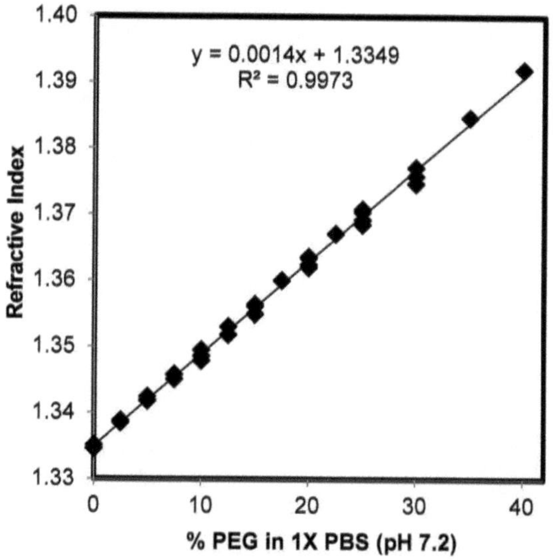

Fig. 3 PEG3350 (dissolved in PBS at pH 7.2) concentration (w/w) vs. RI

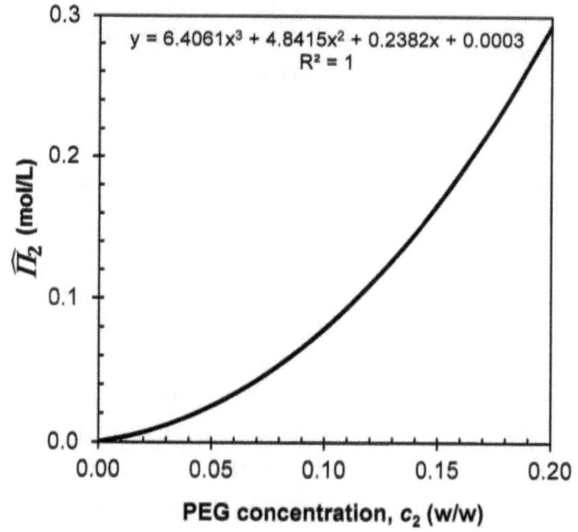

Fig. 4 The calculated $\hat{\Pi}_2$ of PEG3350 at various PEG concentration (w/w) in the range of 0–0.2 (20%)

16. Then, determine the binding free energy by linear regression of $\ln c_1$ vs. $\hat{\Pi}_2$ according to Eq. 2 (Fig. 5).

$$\ln c_1 = \sigma\hat{\Pi}_2 + \hat{\mu}_{\mathrm{cp}} \tag{2}$$

17. Two fitting parameters are obtained from the linear regression in Fig. 5. The absolute value of the negative slope, $\sigma = 57.9$ L/mol for this protein, is the molar depletion zone of the protein which is roughly proportional to the surface area of protein molecule. The intercept, $\hat{\mu}_{\mathrm{cp}} = 4.07$ for this protein in PBS, is

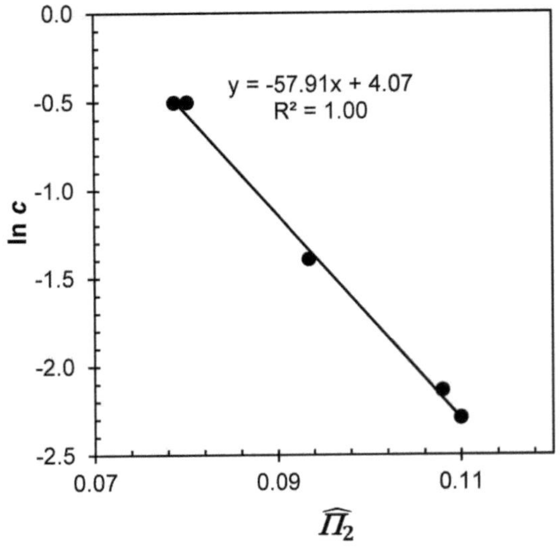

Fig. 5 Linear regression of ln *c* as a function of $\widehat{\Pi}_2$ for the samples 2–7

the binding free energy. $\hat{\mu}_{cp}$ provides a measure to compare the solubility of various proteins in different solutions. Larger positive $\hat{\mu}_{cp}$ indicates higher protein solubility and thereby lower risk of aggregation or precipitation (*see* **Note 14**).

4 Notes

1. If a different buffer is used in place of PBS, all PEG and protein solutions must be prepared in this buffer. In our experience various aqueous buffers can be used (e.g., phosphate, acetate, histidine, citrate, and Tris). Buffers containing organic solvents have not been tested; care should be taken when using such buffers that contain organic solvents, as this can potentially diminish PEG's ability to induce LLPS.
2. The recommended range of PEG molecular weights is from 1450 to 8000 Da. PEG molecules with a very low or very high molecular weight may not be able to effectively induce LLPS in protein solutions. Within the recommended molecular weight range, the higher the molecular weight of PEG, the stronger the depletion interaction it generates, that is, a lower percentage of PEG8000 would be needed to induce LLPS in a protein solution at the same temperature as compared to PEG1450.
3. This method is applicable to most proteins. However, two types of proteins are not suitable for this test. Type one is a protein that is very insoluble, in which case fast aggregation or crystallization (within a couple of hours) may preempt LLPS making this method inapplicable. Type two is a protein

of a low molecular weight, in which case PEG can be less effective in inducing LLPS. Therefore, monomeric peptides (MW < 10 kDa) are generally not suitable for this method. However, it works well in the case of peptides that form oligomers with high apparent molecular weights [14].

4. A refractometer should be used to measure the refractive index and thereby determine the PEG concentration in **step 9**. If the instrument is not available, the concentration of PEG solutions in **step 4** can be calculated by accurately recording weights of the PEG stock and buffer mixed during preparation. The final PEG concentrations in Table 1 shall then be used to fill in Table 2.
5. When using PEG with a higher molecular weight than PEG3350, a more dilute PEG stock solution can be prepared because less PEG is needed to induce LLPS (*see* **Note 2**). Also, it may become difficult to dissolve higher molecular weight PEG at a high concentration because of high viscosity. We found that brief heating of the solution in a glass vial on a hot plate for 3–5 s can facilitate PEG dissolution.
6. Protein concentrations higher than 2–4 mg/mL can be used. However, it is recommended that the final protein concentration in Table 1 does not exceed 5 mg/mL, for an accurate calculation of the binding free energy.
7. Because PEG solutions are very viscous, a positive displacement pipette should be used. Alternatively, great patience should be exercised during pipetting. We recommend waiting until the solution level in the pipette tip stops rising during aspiration, and to make sure that all solution within the tip is expelled during dispensation.
8. Proteins may exhibit greater solubility at pH values far from the isoelectric point. If LLPS is not observed with 20% PEG3350, both protein and PEG concentrations may be increased. Higher final PEG concentrations can be obtained, for example, by mixing 40% PEG3350 with 4 mg/mL protein in a 3:1 ratio. If no LLPS is observed even at the final concentrations of 5 mg/mL protein and 30% PEG3350, the protein has exceptionally high solubility in the given buffer and is unlikely to exhibit solubility issues over extended storage.
9. This protein solubility assay may also be performed at other incubation temperatures, such as room temperature, although higher PEG concentrations will be needed to induce LLPS at higher temperatures. Whatever the experimental temperature might be, the samples in **step 3** need to be incubated at the same temperature to determine the lowest required PEG concentration.

10. Protein samples can reach equilibrium within different incubation time ranging from minutes to 24 h. Equilibrium is confirmed as described in **step 12**. If equilibrium is reached quickly, shorter incubation may be used to avoid potential complications arising from crystallization or aggregation.
11. It is recommended that the centrifuge rotor is precooled to ensure the incubation temperature is maintained during centrifugation.
12. Care should be taken not to raise sample temperature during supernatant removal as this may resolubilize the pellet. During supernatant removal, a brief (~1 min) exposure to ambient temperature is allowed.
13. If the protein concentrations, c_1 of the samples 3, 5, and 7 are systematically higher than those of samples 2, 4, and 6, equilibrium has not been reached. In such a case, consider vortexing the leftover samples from **step 12** for 5 s to resuspend the precipitate. Allow further incubation and repeat **steps 7–12**.
14. $\hat{\mu}_{cp}$ is in the unit of RT, where R is the gas constant and T is the absolute temperature. Interested readers are referred to ref. 8 for derivation of Eq. 2, where a more detailed analysis specific to monoclonal IgGs was conducted and slightly different notations were used. In particular, $\hat{\Pi}_2$ used here is equal to $\frac{\Pi_2}{N_A kT}$ in the referenced publication; $\hat{\mu}_{cp}$ is equal to $\frac{-\varepsilon_B}{kT} + \ln\left(\frac{M_1}{\nu_0 N_A}\right)$; σ was denoted as $\Delta\nu$. In this chapter, $\hat{\mu}_{cp}$ is used to simplify data analysis. $\hat{\mu}_{cp}$ can be used to compare solubility of various proteins in different buffers, but its absolute value does not have rigorous physical meaning due to the dependency on the unit of protein concentration. The binding energy, ε_B, reported in reference [8] directly characterizes the energy of interprotein interaction. To calculate $\varepsilon_B = -kT\hat{\mu}_{cp} + kT\ln\left(\frac{M_1}{\nu_0 N_A}\right)$, the value of $\frac{M_1}{\nu_0 N_A} \approx 1100$ mg/mL can be used. This value was determined for IgGs as described in reference [8] and may be applied to other proteins as an approximation.

References

1. Uzunova VV, Pan W, Galkin O, Vekilov PG (2010) Free heme and the polymerization of sickle cell hemoglobin. Biophys J 99(6):1976–1985. https://doi.org/10.1016/j.bpj.2010.07.024
2. McManus JJ, Lomakin A, Ogun O, Pande A, Basan M, Pande J, Benedek GB (2007) Altered phase diagram due to a single point mutation in human gammaD-crystallin. Proc Natl Acad Sci U S A 104(43):16856–16861. https://doi.org/10.1073/pnas.0707412104
3. Dumetz AC, Chockla AM, Kaler EW, Lenhoff AM (2008) Protein phase behavior in aqueous solutions: crystallization, liquid-liquid phase separation, gels, and aggregates. Biophys J 94(2):570–583. https://doi.org/10.1529/biophysj.107.116152
4. Gunton JD, Shiryayev A, Pagan DL (2007) Protein condensation : kinetic pathways to crystallization and disease. Cambridge University Press, Cambridge

5. Asherie N (2004) Protein crystallization and phase diagrams. Methods 34(3):266–272. https://doi.org/10.1016/j.ymeth.2004.03.028
6. Asherie N, Lomakin A, Benedek GB (1996) Phase diagram of colloidal solutions. Phys Rev Lett 77(23):4832–4835. https://doi.org/10.1103/PhysRevLett.77.4832
7. Vekilov PG (2012) Phase diagrams and kinetics of phase transitions in protein solutions. J Phys Condens Matter 24(19):193101. https://doi.org/10.1088/0953-8984/24/19/193101
8. Wang Y, Latypov RF, Lomakin A, Meyer JA, Kerwin BA, Vunnum S, Benedek GB (2014) Quantitative evaluation of colloidal stability of antibody solutions using PEG-induced liquid-liquid phase separation. Mol Pharm 11 (5):1391–1402. https://doi.org/10.1021/mp400521b
9. Thompson RW Jr, Latypov RF, Wang Y, Lomakin A, Meyer JA, Vunnum S, Benedek GB (2016) Evaluation of effects of pH and ionic strength on colloidal stability of IgG solutions by PEG-induced liquid-liquid phase separation. J Chem Phys 145(18):185101. https://doi.org/10.1063/1.4966708
10. Bhat R, Timasheff SN (1992) Steric exclusion is the principal source of the preferential hydration of proteins in the presence of polyethylene glycols. Protein Sci 1(9):1133–1143
11. Annunziata O, Ogun O, Benedek GB (2003) Observation of liquid-liquid phase separation for eye lens gammaS-crystallin. Proc Natl Acad Sci U S A 100(3):970–974. https://doi.org/10.1073/pnas.242746499
12. Vivares D, Belloni L, Tardieu A, Bonnete F (2002) Catching the PEG-induced attractive interaction between proteins. Eur Phys J E Soft Matter 9(1):15–25. https://doi.org/10.1140/epje/i2002-10047-7
13. Banks DD, Latypov RF, Ketchem RR, Woodard J, Scavezze JL, Siska CC (2012) Razinkov VI native-state solubility and transfer free energy as predictive tools for selecting excipients to include in protein formulation development studies. J Pharm Sci 101 (8):2720–2732. https://doi.org/10.1002/jps.23219
14. Wang Y, Lomakin A, Kanai S, Alex R, Benedek GB (2017) Liquid-liquid phase separation in oligomeric peptide solutions. Langmuir 33 (31):7715–7721. https://doi.org/10.1021/acs.langmuir.7b01693

Chapter 4

Measuring Protein Solubility

Neer Asherie

Abstract

Protein solubility determines the conditions under which the protein will remain in solution. As a result, it is an important quantity in applications that involve concentrated protein solutions. Here I describe the solubility measurement of the protein thaumatin in the presence of tartrate ions as a function of temperature. This method can be used to measure the solubility of other proteins.

Key words Protein, Solubility, Thaumatin, Protein crystallization, Phase diagrams

1 Introduction

Protein solubility is the concentration of protein that is in equilibrium with a crystalline phase under a given set of conditions [1]. The solubility is a key thermodynamic quantity that provides insight into protein interactions [2]. It is also a fundamental parameter in many applications that require proteins to either remain soluble or form crystals. For example, the formulation and delivery of protein pharmaceuticals is often hampered by low protein solubility [3]. Progress in structural biology is subject to the opposite constraint: crystals for X-ray crystallography studies will only form if the solubility is exceeded [4].

The solubility of a protein will depend on the solubility conditions. The most common factors to be varied are the temperature, pH, and salt concentration (typically that of an additive, but the concentration of the buffer itself can also be varied) [5]. Temperature has the advantage of simplicity as it can be increased or decreased easily and reversibly [6]. The solubility curves as a function of temperature have been determined for many proteins, including lysozyme [7], α-amylase [6], bovine pancreatic trypsin inhibitor [6], γ-crystallins [8], glucose isomerase [9], human hemoglobin C [10], and IgG antibodies [11]. Here we use the intensely sweet protein thaumatin, which is readily crystallized in the presence of L-tartrate ions [12], to present a protocol for

Jennifer J. McManus (ed.), *Protein Self-Assembly: Methods and Protocols*, Methods in Molecular Biology, vol. 2039, https://doi.org/10.1007/978-1-4939-9678-0_4,

solubility measurements. Using this protocol we have obtained reproducible results for the solubility of thaumatin with both L- and D-tartrate [13].

2 Materials

Prepare all solutions using deionized water (resistivity of 18 MΩ cm at 25 °C) and analytical grade reagents. For reliable results, it is essential for both the protein and tartrate to be pure; contaminants in either the protein or the tartrate can significantly affect the solubility [14]. Prepare and store all reagents at room temperature (unless indicated otherwise). Filter all solutions with a 0.22 μm filter before use.

1. Pure monomeric thaumatin in buffer used for purification (*see* **Note 1**). Store at 4 °C.
2. Crystallization buffer: 10 mM sodium phosphate buffer (pH = 7.3; σ = 1.5 mS/cm) with 0.002% (m/v) sodium azide. Add 1.931 g of dibasic sodium phosphate (anhydrous), 0.883 g of monobasic sodium phosphate (monohydrate) and 0.040 g of sodium azide to 1900 ml. Stir until all solids have dissolved. Make up to 2 l with water and measure pH and conductivity. Store at 4 °C.
3. Tartrate solution (1 M). Weigh 1.501 g of L-tartaric acid and dissolve in 5 ml of the crystallization buffer. Adjust pH to 7.3 with NaOH as necessary (*see* **Note 2**) and make up to 10 ml with crystallization buffer.
4. Tartrate solution (0.5 M). Dilute 5 ml of the 1 M sodium L-tartrate solution with 5 ml of the crystallization buffer.

2.1 Equipment

1. Ultrafiltration stirred cell.
2. Centrifugal ultrafiltration device.
3. Disposable borosilicate culture tubes (6 × 50 mm).
4. Pyrex mixing beads (3 mm).
5. Temperature-controlled thermally isolated chamber.

3 Methods

3.1 Production of Crystals

Carry out all procedures at room temperature, unless otherwise specified.

1. Place approximately 150 mg of pure monomeric thaumatin solution in an ultrafiltration stirred cell with a 10 kDa membrane for diafiltration into the crystallization buffer (*see* **Note 3**). Filter the dialyzed solution with a 0.22 μm filter

(*see* **Note 4**). The final concentration should be approximately 12 mg/ml in 11 ml (typical protein losses are about 10% of the initial mass).

2. Concentrate the protein solution up to approximately 100 mg/ml in a centrifugal ultrafiltration device.
3. Collect clear supernatant (*see* **Note 5**) and measure its concentration (*see* **Note 6**).
4. Keep this high concentration protein solution on ice while the crystallization experiment is started (*see* **Note 7**).
5. Place 150 μl of protein solution in a disposable borosilicate culture tube and add 150 μl of crystallizing agent (1 M sodium L-tartrate solution). Stir briefly and gently on a vortexer (*see* **Note 8**).
6. Store the tartrate–thaumatin mixture at 4 °C and inspect periodically. If any turbidity or solid phase can be seen by eye, remove a 3 μl aliquot and check for crystals by bright field and polarized microscopy. Bipyramidal thaumatin crystals form within a few hours and sufficient crystals for solubility measurements form overnight (*see* **Note 9**).

3.2 Solubility Measurement

The solubility measurement described below is made by allowing the thaumatin crystals to dissolve into an undersaturated solution until equilibrium is reached. This approach is both rapid and reliable. To check the consistency of the results obtained this way, some solubility measurements can be made starting with supersaturated protein solutions (*see* ref. 15).

1. Choose the temperature for the solubility measurement. Here we will use 22 °C, which is assumed to be room temperature.
2. Remove the supernatant from the culture tube using a long-tipped pipette and measure its concentration (*see* **Note 10**). Leave the crystals behind.
3. Add 50 μl of equilibration solution (0.5 M sodium L-tartrate solution) to wash the crystals of any protein solution. Remove the supernatant and measure its concentration. Repeat this step as needed until the concentration of protein in the supernatant is sufficiently small (*see* **Note 11**).
4. Add 100 μl of equilibration solution (0.5 M sodium L-tartrate solution) and a 3 mm Pyrex mixing bead to the protein crystals.
5. Gently mix protein solution for 48 h (*see* Fig. 1).
6. Stop mixing and let the crystals sediment in the protein solution until the supernatant is clear (about 4 h).
7. Remove a 25 μl aliquot to measure concentration of protein in the supernatant (*see* **Notes 12** and **13**).
8. Resume mixing of protein solution for another 48 h.

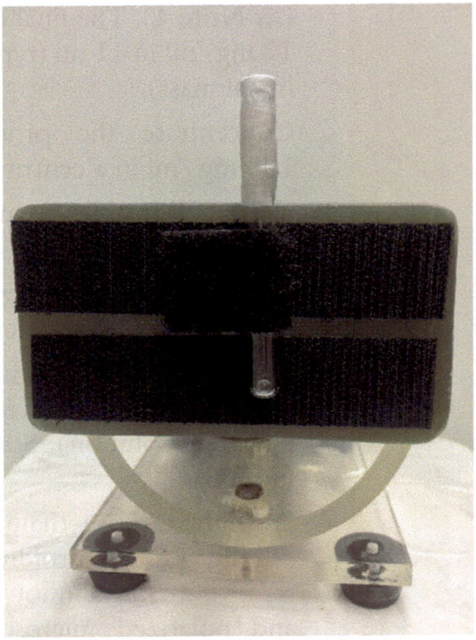

Fig. 1 Rotary mixer for solubility experiments. The tube containing the sample (crystals and solution not shown) and the glass bead are rotated at approximately 0.5 Hz to gently mix the crystals and the solution. The tube is attached to the mixer using Velcro. The whole device is placed in a thermally isolated, temperature-controlled chamber

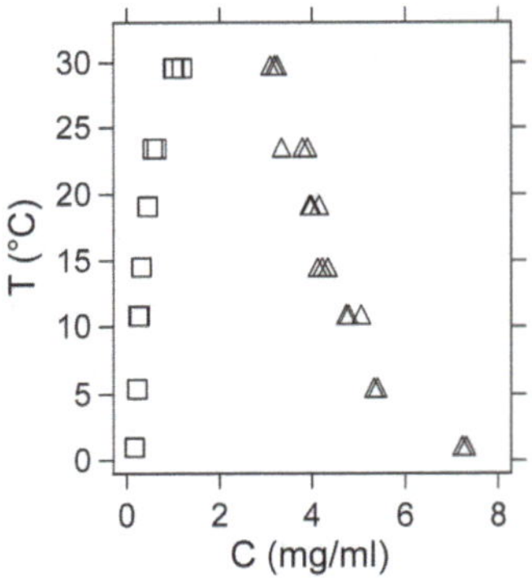

Fig. 2 The solubility of thaumatin. The solubility in 0.5 M sodium L-tartrate (squares) and 0.5 M sodium D-tartrate (triangles). All solutions contained 10 mM sodium phosphate (pH = 7.3) with 0.002% sodium azide. The D-tartrate results were obtained by the method described in the text. Reprinted with permission from ref. 13. Copyright 2008 American Chemical Society

9. Repeat **steps 6** and 7.
10. If the concentrations measured in **steps** 7 and **9** (i.e., after 48 and 96 h) are the same within experimental error, take the average and record this as the solubility measurement.

11. If the two differ, resume mixing and repeat **steps 6** and 7. If the 96 and 144 h measurements are the same, take the average and record this as the solubility measurement. Otherwise, repeat the solubility measurement with a new sample (*see* **Note 14**).
12. Once a solubility measurement has been made, start again at **step 1** by choosing a different temperature (*see* **Note 15**). Repeat until the solubility curve has been determined (*see* Fig. 2).

4 Notes

1. For this protein, collecting the monomer fraction after a single purification step with a preparatory scale size-exclusion chromatography column yields pure monomeric thaumatin. For details of the purification procedure, *see* ref. 14.
2. We use 5 M NaOH to raise the pH. To minimize the formation of the poorly soluble sodium hydrogen tartrate, the first amount of NaOH should be relatively large and the solution should be vigorously stirred. We find that adding 3 ml of 5 M NaOH to the initial 5 ml of tartaric acid results in the momentary formation of a small amount of white precipitate that quickly redissolves. The remaining NaOH (approximately another 300 μl) is added dropwise to reach the target pH of 7.3.
3. Our purification procedure produces thaumatin at 2.7 mg/ml in 0.275 M sodium acetate buffer (pH = 4.5), so we start with approximately 55 ml of protein. We add to it about 300 ml of crystallization buffer for the first diafiltration step and then repeat for an additional three steps. At each step we reduce the volume to approximately 20 ml and the refill the diafiltration cell up to 350 ml with the crystallization buffer.
4. Wet the filter with the crystallization buffer before filtering the protein solution. We find that a wet filter results in a smaller mass loss than using a dry filter.
5. Though some of the protein precipitated as it was concentrated, the precipitate adhered to the concentrator, making it possible to collect a clear solution of protein. The collection can be done directly from with a pipette or by inverting into the caps supplied by the manufacturer and centrifuging as instructed.
6. The extinction coefficient of thaumatin was taken to be $E^{0.1\%} = 1.27\ \mathrm{mg\ ml^{-1}\ cm^{-1}}$.
7. Keeping the high concentration solution on ice delays any further precipitation.

8. For simplicity, the protocol focuses on a single protein sample. However, to obtain reliable measurements, we typically make three protein samples at a time. Also, we run control samples with only protein (no precipitant) and only precipitant (no protein). These controls should remain clear throughout the experiment.
9. The solution conditions are such that only protein crystals should form. For a more detailed discussion about avoiding the formation of tartrate crystals (*see* ref. 14).
10. A useful check that all is going well: the concentration measured in this step should be smaller than the original concentration measured in Subheading 3.1, **step 3**.
11. The goal of the wash is to create an undersaturated solution by removing as much of the protein that is not in the crystals as possible. Care must be taken so as not to completely dissolve the crystals. For the solubility measurement to be reliable, there must always be crystals in the tube. Therefore, the equilibration solution should be at the temperature of the solubility measurement and the washes should be done with small volumes of solution and as rapidly as possible. We find that two washes typically suffice.
12. The supernatant solution must be clear so that no crystals are included in the aliquot for the protein concentration measurements.
13. This concentration measurement can usually be done directly in a spectrophotometer However, it is important to check that the precipitant and buffer do not absorb at 280 nm as this absorption can affect the accuracy of the measurement. For the thaumatin–tartrate mixtures (0.5 M tartrate), we find that the absorption at 280 nm is smaller than the experimental error for almost all of the data presented here. When high accuracy measurements are required, we separate the protein contribution to the absorbance using high-performance size-exclusion chromatography and determine the protein concentration from the area of the well-resolved protein peak (*see* ref. 14).
14. To overcome such problems, we measure the solubility for two or three protein samples simultaneously and then discard any inconsistent or inconclusive data.
15. For a protein with normal solubility—one that increases with temperature—it is simplest to start at a low temperature and then raise it. In this way, all solubility points will be determined through dissolution into undersaturated solutions. For a system with retrograde solubility—one that increases with temperature—it is easiest to start from a high temperature and then lower it. Of course, if the type of solubility is unknown, then the direction chosen will be an educated guess.

Acknowledgments

I gratefully acknowledge financial support from the National Science Foundation (DMR 0901260 and 1206416) and Yeshiva University.

References

1. McPherson A (1999) Crystallization of biological macromolecules. Cold Spring Harbor Laboratory Press, Cold Spring Harbor, NY
2. McManus JJ, Charbonneau P, Zaccarelli E, Asherie N (2016) The physics of protein self-assembly. Curr Opin Colloid Interface Sci 22:73–79
3. Trevino SR, Scholtz JM, Pace CN (2008) Measuring and increasing protein solubility. J Pharm Sci 97:4155–4166
4. Benvenuti M, Magnani S (2007) Nat Protoc 2:1633–1651
5. McPherson A, Gavira JA (2014) Introduction to protein crystallization. Acta Crystallogr F Struct Biol Commun 70(Pt 1):2–20
6. Astier J-P, Veesler S (2008) Using temperature to crystallize proteins: a mini-review. Cryst Growth Des 8:415–4219
7. Aldabeideh N, Jones MJ, Myerson AS, Ulrich J (2009) The solubility of orthorhombic lysozyme crystals obtained at high pH. Cryst Growth Des 9:3313–3317
8. Berland CR, Thurston GM, Kondo M, Broide ML, Pande J, Ogun O, Benedek GB (1992) Solid-liquid phase boundaries of lens protein solutions. Proc Natl Acad Sci U S A 89:1214–1218
9. Sleutel M, Willaert R, Gillespie C, Evrard C, Wyns L, Maes D (2009) Kinetics and thermodynamics of glucose isomerase crystallization. Cryst Growth Des 9:497–504
10. Feeling-Taylor AR, Banish RM, Hirsch RE, Vekilov PG (1999) Miniaturized scintillation technique for protein solubility determinations. Rev Sci Instrum 70:2845–2849
11. Rowe JB, Cancel RA, Evangelous TD, Flynn RP, Pechenov S, Subramony JA, Zhang J, Wang Y (2017) Metastability gap in the phase diagram of monoclonal IgG antibody. Biophys J 113:1750–1756
12. Ko TP, Day J, Greenwood A, McPherson A (1994) Structures of three crystal forms of the sweet protein thaumatin. Acta Cryst D 50:813–825
13. Asherie N, Ginsberg C, Blass S, Greenbaum A, Knafo S (2008) Solubility of thaumatin. Cryst Growth Des 8:1815–1817
14. Asherie N, Ginsberg C, Greenbaum A, Blass S, Knafo S (2008) Effect of protein purity and precipitant stereochemistry on the crystallization of thaumatin. Cryst Growth Des 8:4200–4207
15. Asherie N (2004) Protein crystallization and phase diagrams. Methods 34:266–272

Part II

Measuring Protein Self-Association, Aggregation, and Crystallization

Chapter 5

Integral *caa*$_3$-Cytochrome *c* Oxidase from *Thermus thermophilus*: Purification and Crystallization

Orla Slattery, Sabri Cherrak, and Tewfik Soulimane

Abstract

Cytochrome *c* oxidase is a respiratory enzyme catalyzing the energy-conserving reduction of molecular oxygen to water—a fundamental biological process of cell respiration. The first crystal structures of the type A cytochrome *c* oxidases, bovine heart and *Paracoccus denitrificans* cytochrome *c* oxidases, were published in 1995 and contributed immensely to the understanding of the enzyme's mechanism of action. The senior author's research focus was directed toward understanding the structure and function of the type B cytochrome *c* oxidases, *ba*$_3$-oxidase and type A2 *caa*$_3$-oxidase, both from the extreme thermophilic bacterium *Thermus thermophilus*. While the *ba*$_3$-oxidase structure was published in 2000 and functional characterization is well-documented in the literature, we recently successfully solved the structure of the *caa*$_3$-nature made enzyme-substrate complex. This chapter is dedicated to the purification and crystallization process of *caa*$_3$-cytochrome *c* oxidase.

Key words Cytochrome *c* oxidase, *caa*$_3$-Oxidase, Chromatography, Crystallization, Bioenergetics, *Thermus thermophilus*

1 Introduction

Cellular respiration is essential for life. Fundamental food molecules produced after digestion, such as glucose, are oxidized to carbon dioxide and water. In this process, the energy released is harnessed in the formation of ATP for use in the energy-expending processes carried out by the cell. Cytochrome *c* oxidase is the terminal enzyme of the respiratory chain belonging to the heme--copper oxidase (HCO) superfamily [1, 2]. The specific function of this enzyme is to catalyze the transfer of electrons from cytochrome *c* to oxygen, thereby reducing it to water, while also pumping protons across the membrane to create a charge gradient. According to vital residues within their proton pumping networks, cytochrome *c* oxidases are classified into groups A, B, and C.

Over the past two decades, a number of structures of mammalian and bacterial cytochrome *c* oxidases have been determined

Jennifer J. McManus (ed.), *Protein Self-Assembly: Methods and Protocols*, Methods in Molecular Biology, vol. 2039,
https://doi.org/10.1007/978-1-4939-9678-0_5,

[3–8] revealing key features of the structure/function relationship. In addition, a wealth of spectroscopic and site-directed mutagenesis data exists that has amplified knowledge of the enzyme's mechanism, although some conflicting theories are still controversially discussed.

One of the key organisms that have been studied in relation to its cytochrome *c* oxidases is the extreme thermophile, *Thermus thermophilus* HB8 (ATCC 27634). This gram-negative bacterium can be found living at temperatures in the region of 90 °C in hot springs of Izu in Japan [9]. It expresses two different cytochrome *c* oxidases, the ba_3-oxidase, which is expressed under low oxygen tension and the constitutively expressed caa_3-oxidase, the subject of this article. The ba_3-enzyme is the most divergent of the heme--copper oxidase superfamily, while the caa_3-type enzyme displays good sequence similarity to other members of the family, in particular, the purple bacterium *Rhodobacter sphaeroides* (44.8% sequence identity) and the soil organism *Paracoccus denitrificans* (44.5% sequence identity).

Cytochrome *c* oxidase consists of three core subunits found in all members of the HCO superfamily [10] with the exception of the ba_3-oxidase from *T. thermophilus* [5] which is a two-subunit enzyme. Subunits I and II/IIa are catalytically active. The exact function of subunit III in the cytochrome *c* oxidase family is not known, although it is believed to be important for structural stability and possibly to be involved in the assembly of the complex [11]. It has also been proposed to form the entrance to an oxygen channel leading to the active site [12]. Subunit I consists of twelve conserved transmembrane helices as well as the low-spin heme *a* and the binuclear center, heme a_3Cu_B, where oxygen reduction takes place. Subunit II holds the docking site for the enzyme's substrate, cytochrome *c*. It consists of two transmembrane helices that interact with subunit I and a ten-stranded β-barrel domain protruding into the extramembranal environment. This globular domain contains the Cu_A center, where electrons are received from cytochrome *c* [2].

According to the structures available, all but the ba_3-oxidase have additional subunits whose function is unknown. Bovine heart cytochrome *c* oxidase has ten additional subunits, making this the largest known enzyme of this kind. The enzymes from, *P. denitrificans* and *R. sphaeroides* both have one additional subunit, namely, subunit IV, a single membrane spanning helix, neither of which bears any sequence identity to the other or to subunit IV of the bovine heart cytochrome *c* oxidase.

Caa_3-cytochrome *c* oxidase from *T. thermophilus* is unique in that it consists of two fusion proteins linking subunits I with III (SUI/III) and subunit II with cytochrome *c* (SUIIc). The caa_3-oxidase was first described in 1980 as a two-subunit enzyme containing the metal centers characteristic of the heme–copper oxidase

[13]. Although maintaining high sequence homology to other cytochrome *c* oxidases, the arrangement of the subunits in this particular enzyme are slightly different. The 89 kDa subunit containing heme *a* and the binuclear center was identified as a fusion protein between the typical subunits I and III (SU I/III) [14] and most interestingly, the 39 kDa subunit II containing the Cu_A domain was found to be fused to a cytochrome *c* (SU IIc) [15]. Interestingly, this implies that the enzyme is coupled to its substrate, leading to the speculation that the soluble *T. thermophilus* cytochrome c_{552} electron carrier [16, 17] may not be the molecule responsible for electron transfer between the oxidase and the *T. thermophilus bc* complex; rather, the fused cytochrome *c* of SU IIc may act as the conduit molecule between the two complexes [18].

The crystal structure of the caa_3-oxidase was solved to 2.36 Å in 2012 [8] which confirmed the arrangement of SU I/III and SU IIc and additionally, revealed the presence of a fourth subunit (SU IV) comprising two transmembrane helices (Fig. 1). The structure also has uncovered many details about the entry of electrons into the protein complex, suggesting that electrons may be transferred via edge-to-edge interaction of the soluble electron transporter cytochrome c_{552} and cytochrome *c* of SU IIc. This theory goes against the earlier proposal regarding the formation of a *bc*-caa_3-oxidase "supercomplex" [18]. From there the

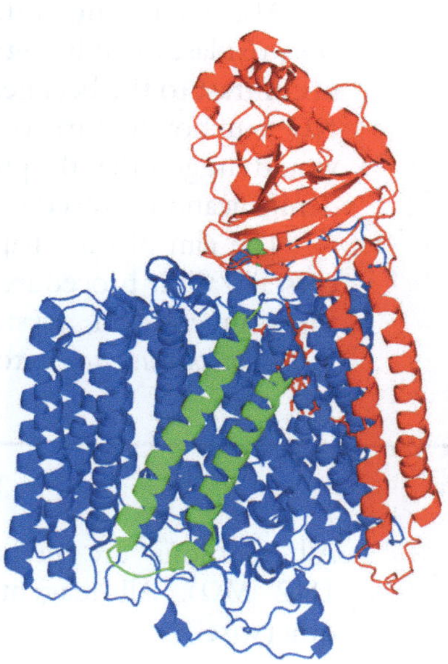

Fig. 1 The crystal structure of the caa_3-oxidase to 2.36 Å [8] showing the arrangement of the three subunits in ribbon form. SU I/III is colored blue, SU IIc is colored red, and SU IV is colored green and seen to be traversing the surface of SU I/III. Cu_A is depicted as a green sphere and hemes are colored red. This image was created using Pymol [19]

electrons are proposed to travel via the D-pyrrole and D-propionate of the SU II heme *c* on to a Phe and Cys residue and from there to the CuA center. The progress of electrons beyond this point is similar to other cytochrome oxidase enzymes. Briefly, the electron is subsequently passed to heme *a* in subunit I via two highly conserved Arg residues. From heme *a*, the electrons are passed to a conserved Phe residue and on to heme a_3 of the binuclear center. O_2 binding occurs once the binuclear center has been reduced by two electrons.

The caa_3-oxidase structure revealed evidence for the presence of the two main D and K proton pathways that are common to all Type A HCOs, due to the presence of an aspartic acid and a lysine residue at entrance to each respectively [20]. These pathways are required for protonation of molecular oxygen for water formation and for protons to be pumped across the membrane. These two pathways have been recognized in all cytochrome *c* oxidase structures, either through the identification of conserved key residues or as with the ba_3-oxidase, through comparable locations within the molecules [5]. By and large, these proton transfer pathways consist of hydrogen-bonded water molecules and protonatable polar amino acid side chains [21]. In the caa_3-oxidase, the terminus of the D-pathway was found to have a proton gating site consisting of a Tyr-Ser motif that differs from the canonical glutamate found in other Type A HCOs, hence the caa_3-oxidase is termed a Type A2 HCO [8, 20].

Also interesting is the fact that both *T. thermophilus* ba_3 and caa_3-oxidases, exhibit significant NO reductase activity [22]. This is in contrast to the beef heart oxidase which has been shown to bind NO but not to turn over the substrate to N_2O. This evidence would suggest that the processes of aerobic respiration and bacterial denitrification evolved from common ancestry.

The aim of this chapter is to summarize the purification and crystallization procedure that led to the growth of well-ordered three-dimensional crystals suitable for X-ray analysis of the *T. thermophilus* caa_3-cytochome *c* oxidase [8].

2 Materials

All solutions are prepared using ultrapure water (resistivity 18.2 MΩ). All reagents are analytical grade, unless specified otherwise.

2.1 Cell Paste

1. *T. thermophilus* HB8 biomass can be obtained from 100 L fermentation carried out according to a well-established protocol [23, 24] in a stainless steel jar fermenter under 0.5 volume air per volume medium per minute at 70 °C using the following media: EGTA (Titriplex IX) 0.04 M, Na_2SO_4 0.8 M,

$MgSO_4 \cdot 7H_2O$ 0.41 M, NaCl 9 mM, KCl 1.02 mM, $CaCl_2$ 0.04 mM, K_2HPO_4 2.8 mM, KH_2PO4 2.2 mM, $NaHCO_3$ 5–10 mM, Tris 40 mM, Fe-citrate 3.5 μM, $MnSO_4 \cdot H_2O$ 9.2 μM, $ZnSO_4 \cdot 7H_2O$ 1.9 μM, H_2SO_4 2.1 μM, $CuSO_4 \cdot 5H_2O$ 1 μM, $Na_2MoO_4 \cdot 2H_2O$ 0.1 μM, $CoCl_2$ 0.2 μM, D-biotin 8 μM, monosodium glutamate 50 mM, glucose 16.7 mM, yeast extract 5 kg, peptone 5 kg, antifoaming 5 mL, pH to 7.5.

2. Cells should be harvested in the early to middle exponential growth phase and stored at −80 °C.

2.2 Cell Lysis

1. 400 mg lysozyme per 100 g of biomass.
2. 1 mg DNase I per 100 g of biomass.
3. 1.25 mL of 1 M $MgCl_2$ solution per 100 g of biomass.

2.3 Buffers

All detergents used in the buffers should be high purity where possible (>99%).

1. Lysis Buffer 1: 0.1 M Tris–HCl, pH 7.6, 0.2 M NaCl.
2. Lysis Buffer 2: 0.1 M Tris–HCl, pH 7.6, 0.1 M NaCl.
3. Wash Buffer 1: 0.1 M Tris–HCl, pH 7.6, 0.1 M NaCl, 0.1% (v/v) Triton X-100.
4. Solubilization Buffer: 0.1 M Tris–HCl pH 7.6, 5% (v/v) Triton X-100, 0.1 M NaCl.
5. Equilibration Buffer 1: 0.01 M Tris–HCl, pH 7.6, 0.1% (v/v) Triton X-100.
6. Elution Buffer 1: 0.01 M Tris–HCl pH 7.6, 0.1% (v/v) Triton X-100, 0.2 M NaCl.
7. Elution Buffer 2: 0.01 M Tris–HCl, pH 7.6, 0.1% (v/v) Triton X-100, 1 M NaCl.
8. Equilibration Buffer 2: 0.01 M Tris–HCl pH 7.7, 0.05% (w/v) *n*-dodecyl-β-D-maltoside (DDM).
9. Elution Buffer 3: 0.01 M Tris–HCl, pH 7.6, 0.05% (w/v) DDM, 1 M NaCl.
10. Size Exclusion Buffer: 0.05 M Tris–HCl pH 7.6, 0.05% (w/v) DDM.
11. Equilibration Buffer 3: 0.01 M sodium phosphate pH 6.8 containing 0.05% (w/v) DDM.
12. Final Storage Buffer: 0.01 M Tris–HCl pH 7.6, 0.2% (w/v) *n*-decyl-β-D-maltoside (DM), 0.15 M NaCl.

2.4 Chromatography

Except for the first chromatography step on DEAE Biogel Agarose, all chromatography steps can be carried out using a fast protein liquid chromatography (FPLC) system. The reagents and equipment required are as follows:

1. DEAE Biogel Agarose.
2. Fractogel EMD TMAE.
3. Size exclusion chromatography: a prepacked, gel filtration column (i.d. 16 mm, column height 600 mm) containing 120 mL of gel filtration resin capable of separation of biomolecules with the range of approximately 10–600 kDa.
4. Hydroxyapatite "High Resolution" ion exchange resin.
5. Fraction collector.
6. Empty glass chromatography columns of three different sizes: (A) internal diameter (i.d.) 50 mm, column height 1000 mm; (B) i.d. 26 mm, column height 200 mm; (C) i.d. 16 mm, column height 200 mm.

2.5 Instrumentation

1. UV–visible spectrophotometer.
2. 10 mm path-length quartz cuvette
3. Conductivity meter.
4. Ultracentrifuge with a fixed angle rotor.
5. Benchtop refrigerated centrifuge with a fixed angle conical rotor to fit 15 and 50 mL tubes.

2.6 Dialysis and Concentration

1. 30 kDa cutoff dialysis membrane
2. Centrifugal concentrator with 50 kDa cutoff (max 2 mL volume).
3. Centrifugal concentrator with 50 kDa cutoff (max 7–8 mL volume).

2.7 SDS-PAGE

1. 2× Laemmli Sample Buffer.
2. 4–15% Precast Gels.
3. 10× Tris/Glycine/SDS Running Buffer.
4. Electrophoresis system for mini-gels.
5. Coomassie Brilliant Blue Staining Solution.
6. Coomassie Brilliant Blue Destaining Solution.

2.8 Crystallization

1. 7.7 MAG (Avanti Polar Lipids Inc., Alabama, USA).
2. Dual syringe mixing device (*described in refs.* 25, 26).
3. Cubic mesophase, 96-well glass sandwich plate (*prepared as per refs.* 26, 27).
4. Tungsten carbide glass cutter (any art supply shop).
5. Premounted loops for cryocrystallography (50–100 mm in diameter).
6. Precipitant solutions: 14–21% (v/v) PEG 400, 0.1 M NaCl, 0–0.1 M Li_2SO_4, and 0.1 M sodium citrate pH 4.5–5.0.

3 Methods

3.1 Solubilization of Thermus thermophilus Membranes

1. Mix 100 g of *T. thermophilus* biomass with 500 mL of Lysis Buffer 1 at 20 °C with stirring to break open the outer membrane of the cells (*see* **Note 1**).
2. Centrifuge the resuspended biomass at 154,383 × *g* at 4 °C for 20 min in an ultracentrifuge.
3. Resuspend the pelleted spheroblasts produced in **step 1** in 500 mL of Lysis Buffer 2 with the addition of 400 mg lysozyme (to lyse the inner membrane) in the presence of 1 mg of DNase I and 1.25 mL of 1 M $MgCl_2$ and stir for at least 20 min at 20 °C.
4. Recentrifuge the lysate as in **step 1** (*see* **Note 2**).
5. Wash the pelleted inner and outer membranes four times with 500 mL of Lysis Buffer 2. A final wash should be carried out with 500 mL of Wash Buffer. Between each wash, the membranes should be centrifuged as described in **step 1**.
6. Solubilize the membranes in 500 mL of Solubilization Buffer and stir for a minimum period of 3 h at 4 °C to ensure complete solubilization (*see* **Note 3**).
7. Following centrifugation as in **step 1**, to remove any unsolublized material, dilute the supernatant (500 mL) to 5 L with ultrapure water (*see* **Note 4**).

3.2 Column 1: Anion Exchange—DEAE Biogel Agarose

1. Fill an empty glass column (i.d. 100 mm, column height 200 mm) with 500 mL of DEAE Biogel Agarose and equilibrate at 4 °C (in a cold room) with 10 column volumes of Equilibration Buffer 1 (*see* **Note 5**).
2. Add the diluted solubilized membranes (5 L) using gravity flow (*see* **Note 6**). Upon successful binding, a brown band is observed at the top of the column (Fig. 2).
3. Wash the column with a minimum of 2 L of Equilibration Buffer 1 at a flow rate of 3 mL/min.
4. Prepare a 4 L linear elution gradient consisting of 2 L of Equilibration Buffer and 2 L of Elution Buffer using a homemade gradient mixer (Fig. 3), consisting of two pieces of rubber tubing and a disposable plastic syringe.
5. Apply the gradient to the column at a flow rate of 3 mL/min. The initiation of fraction collection should begin as soon as the gradient is applied and 12 mL fractions should be collected using a fraction collector. Elution of caa_3-oxidase will start when the concentration of NaCl reaches 0.1 M.

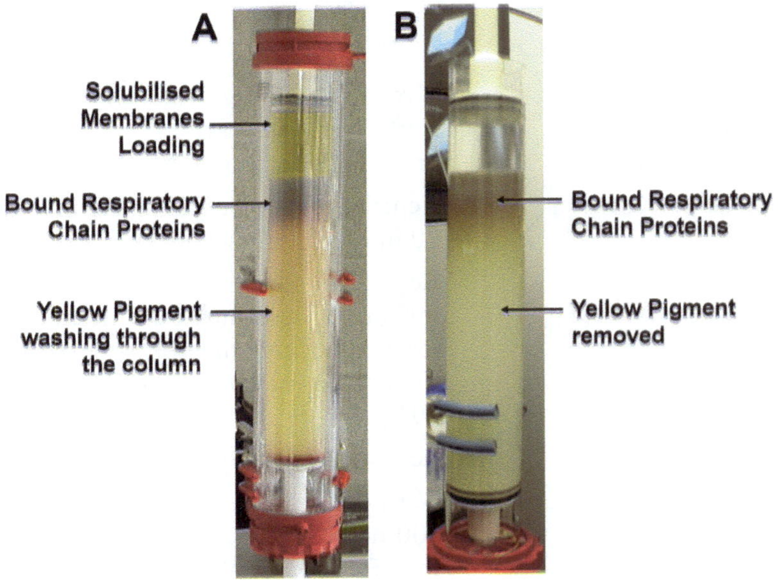

Fig. 2 Loading column 1: (**a**) Bound proteins create a strong brown/red colored band at the top of the column. The contaminant pigment along with cytochrome *b* is visible flowing through the lower half of the column. (**b**) Washing of the bound respiratory chain proteins with Equilibration Buffer 1. All the solubilized membrane has been loaded at this point and the yellow pigment has been washed out of the column

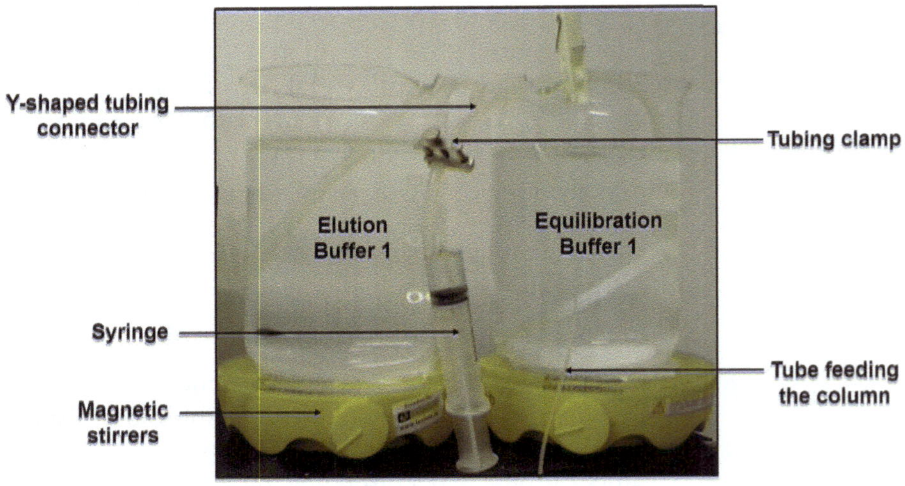

Fig. 3 Gradient-Maker: the beaker on the right side of the image is filled with 2 L of Equilibration Buffer and the beaker in the left-hand side of the images is filled with 2 L of Elution Buffer 1. The gradient maker consists of two pieces of rubber tubing connected via a Y-shaped connector, attached to a 50 mL syringe. A magnetic stirrer ensures continuous mixing of the buffer as the Elution Buffer flows into the Equilibration Buffer. Tubing leading to Column 1 can be seen emerging from the beaker on the right-hand side

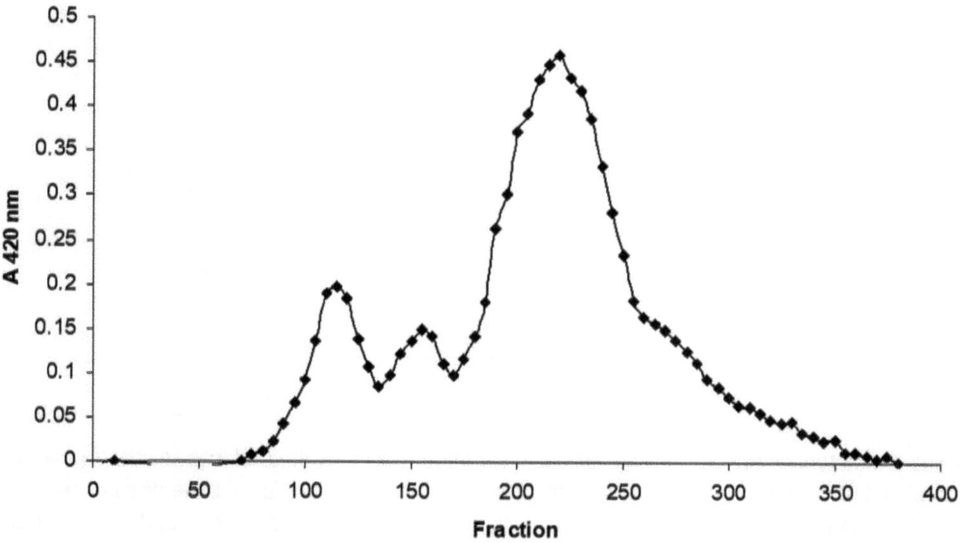

Fig. 4 Column 1 (DEAE Biogel Agarose) elution profile at 420 nm: caa_3-oxidase and other respiratory chain complexes from *T. thermophilus* as they elute from Column 1. Peaks 1 and 2 represent fractions containing the ba_3-oxidase and caa_3-oxidase, respectively, and are easily distinguishable. Peaks 3 and 4 represent a mixture of various cytochromes but in particular the *bc*-complex, its dissociated cytochromes $c_{554/549}$ and b_{562} as well as succinate-ubiquinone reductase

6. Record the absorbance of all fractions at 420 nm in a 10 mm path-length optical glass cuvette using a UV-visible spectrophotometer.
7. Create a chromatogram for the purification by plotting absorbance at 420 nm versus fraction number using a data analysis package (*see* Fig. 4 *for an example*).
8. Measure the reduced-minus-oxidized spectrum of peak fractions to identify the protein(s) present as follows: add the protein sample to 10 mm path-length optical glass cuvette and reduce the sample by adding a few grains of sodium dithionate. Invert the cuvette gently to dissolve the dithionate. Record the absorbance spectrum between 650 and 400 nm using nonreduced (oxidized) sample as a reference (*refer to* Fig. 5 *for an example spectrum*).
9. Pool fractions containing caa_3-oxidase. The final volume of this pool is usually ~1 L.
10. Dialyze the pooled caa_3-oxidase at room temperature (usually 12 h) using a 30 kDa cutoff dialysis membrane in a 10 L bath of Equilibration Buffer 1. Ensure the conductivity of the total volume of caa_3-oxidase is below 2 mS/cm after dialysis so that it is suitable for application on a subsequent column.

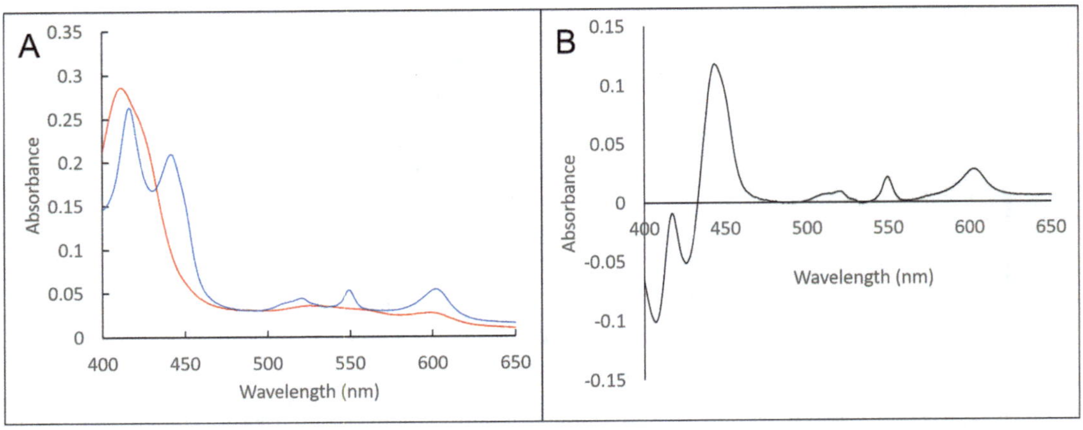

Fig. 5 The characteristic spectra of caa_3-oxidase: (**a**) the air oxidized state of caa_3-oxidase (—) with the Soret band at 411 nm and a shoulder at 420 nm. The sodium dithionite reduced state of caa_3-oxidase (—), with the Soret bands at 414 nm and 440 nm for the *c* and *a* hemes respectively, and the alpha bands at 547 nm and 602 nm for the *c* and *a* hemes respectively. (**b**) The characteristic reduced-minus-oxidized (—) spectrum indicating the purity of the caa_3-oxidase, with the Soret bands at 417 nm and 444 nm for the *c* and a hemes respectively. The alpha bands at 549 nm and 603 nm respectively are clearly visible

3.3 Column 2: Anion Exchange—Fractogel EMD TMAE

1. Fill an empty glass column (i.d. 26 mm, column height 200 mm) with ~30 mL of Fractogel EMD TMAE anion exchange material and equilibrate at room temperature (20–22 °C) with 10 column volumes of Equilibration Buffer 2.
2. Apply the dialyzed caa_3-oxidase sample at a flow rate of 4 mL/min. Upon successful binding, a brown band is observed at the top of the column.
3. Wash the column with ~150 mL of Equilibration Buffer 2 at a flow rate of 4 mL/min to ensure complete detergent exchange. The A_{280} absorbance output from the FPLC monitor should stabilize at <0.02 AU after the washing process, indicating that the Triton X-100 has been removed (*see* **Note** 7).
4. Elute the protein with a 0–0.3 M linear gradient of NaCl in Equilibration Buffer 2 over a period of 1 h at a flow rate of 4 mL/min. Use Elution Buffer 3 as the "high salt" buffer for the gradient.
5. Collect 10 mL fractions as soon as the gradient is initiated. The caa_3-oxidase usually elutes at 0.12 M NaCl (refer to Fig. 6).
6. As with Column 1, analyze peak fractions by measuring their reduced-minus-oxidized spectrum and pool caa_3-oxidase containing fractions. The total collected volume is usually ~100 mL.
7. Initially concentrate the pooled caa_3-oxidase containing fractions to a minimum volume of 2 mL using a large 50 kDa cutoff centrifugal concentrator (7–8 mL max volume) at a

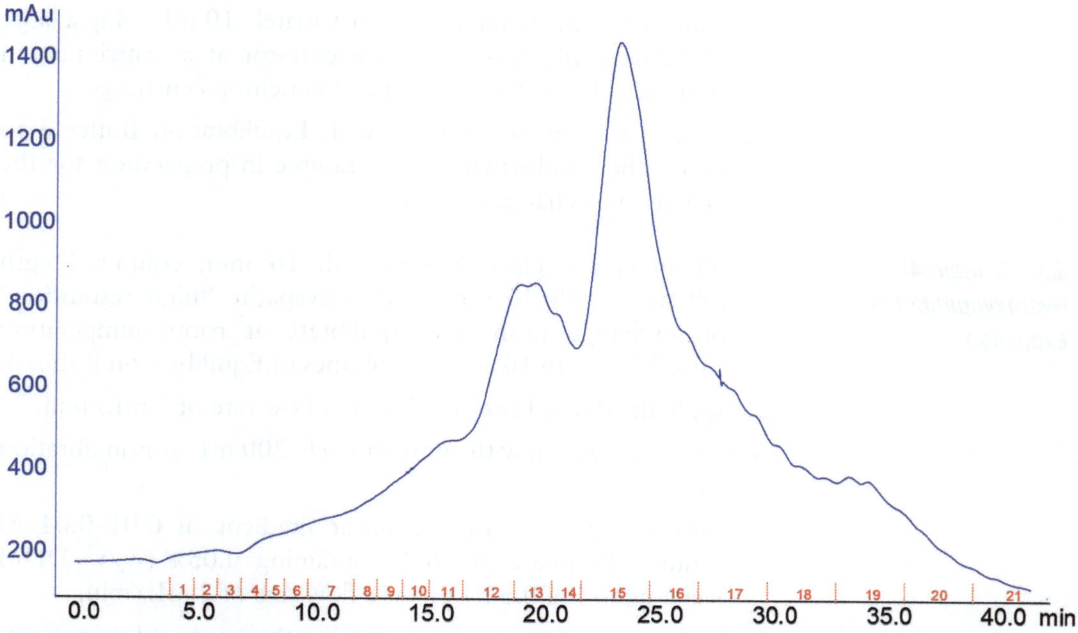

Fig. 6 Column 2 (EMD TMAE) elution profile: absorbance at 280 nm is shown in blue. Fractions are numbered along the *x*-axis. Remnants of ba_3-oxidase eluted in the first peak followed by the more dominant caa_3-oxidase peak. Fraction underneath this second peak were pooled as these were considered to contain the most spectroscopically pure caa_3-oxidase as seen in the reduced-minus-oxidized spectrum of pooled fractions (refer to Fig. 5). The final shoulder of the elution profile contained cytochrome $c_{549/554}$

centrifugation speed of 1500 × *g* at 4 °C using a benchtop centrifuge.

8. Further concentrate to ~500 μL with a small 50 kDa cutoff centrifugal concentration (2 mL max volume) at a centrifugation speed of 5000 × *g* at 4 °C, also in a benchtop centrifuge.

3.4 Column 3: Size Exclusion

1. Inject an aliquot of 500 μL of concentrated caa_3-oxidase using a 2.0 mL loop onto a prepacked, gel filtration column (i.d. 16 mm, column height 600 mm) containing 120 mL of gel filtration resin capable of separation of biomolecules with the range of approximately 10–600 kDa. The column should be preequilibrated at room temperature (20–22 °C) with Size Exclusion Buffer (*see* **Note 8**).
2. Elute the protein using the same buffer at a flow rate of 1 mL/min. The caa_3-oxidase generally elutes after 50–60 min.
3. As previously, analyze peak fractions by measuring their reduced-minus-oxidized spectrum and pool caa_3-oxidase containing fractions. The total volume of this pool is usually ~36 mL.

4. Concentrate the volume to approximately 10 mL using a large 50 kDa cut-off centrifugal concentrator at a centrifugation speed of 1500 × *g* at 4 °C using a benchtop centrifuge.
5. Dilute this pool to 100 mL with Equilibration Buffer 3 to reduce the conductivity of the sample in preparation for the next anion exchange column.

3.5 Column 4: Hydroxyapatite Ion Exchange

1. Fill an empty glass column (i.d. 16 mm, column length 200 mm) with 20 mL of hydroxyapatite "high resolution" ion exchange resin and equilibrate at room temperature (20–22 °C) with 10 column volumes of Equilibration Buffer 3.
2. Apply the diluted caa_3-oxidase at a flow rate of 2 mL/min.
3. Wash the column with approximately 200 mL of Equilibration Buffer 3.
4. Elute the protein using a linear gradient of 0.01–0.04 M sodium phosphate pH 6.8, containing 0.05% (w/v) DDM over a period of 0.5 h, also at a flow rate of 2 mL/min.
5. Analyze peak fractions by measuring their reduced-minus-oxidized spectrum and pool caa_3-oxidase containing fractions. The total volume collected is usually ~50 mL.
6. Dilute the sample tenfold with Equilibration Buffer 2 in preparation for another round of anion exchange chromatography.

3.6 Column 5: Anion Exchange (Repeat)—Fractogel EMD TMAE

1. Fill an empty glass column (i.d. 16 mm, column height 200 mm) with ~15 mL of Fractogel EMD TMAE anion exchange material and equilibrate at room temperature (20–22 °C) with 10 column volumes of Equilibration Buffer 2 (*see* **Note 9**).
2. Load the diluted caa_3-oxidase sample onto the column at a flow rate of 4 mL/min.
3. Wash the sample with ~100 mL Equilibration Buffer 2 at a flow rate of 4 mL/min.
4. Elute the protein with a 0–0.3 M linear gradient of NaCl in Equilibration Buffer 2 over a period of 1 h at a flow rate of 4 mL/min. Use Elution Buffer 3 as the "high salt" buffer for the gradient.
5. Collect 6 mL fractions as soon as the gradient is initiated. The caa_3-oxidase usually elutes at 0.12 M NaCl.
6. Analyze peak fractions by measuring their reduced-minus-oxidized spectrum and pool caa_3-oxidase containing fractions. The total volume collected is usually ~22 mL.
7. Initially concentrate this pooled sample to a minimum volume of 2 mL using a large 50 kDa cutoff centrifugal concentrator.

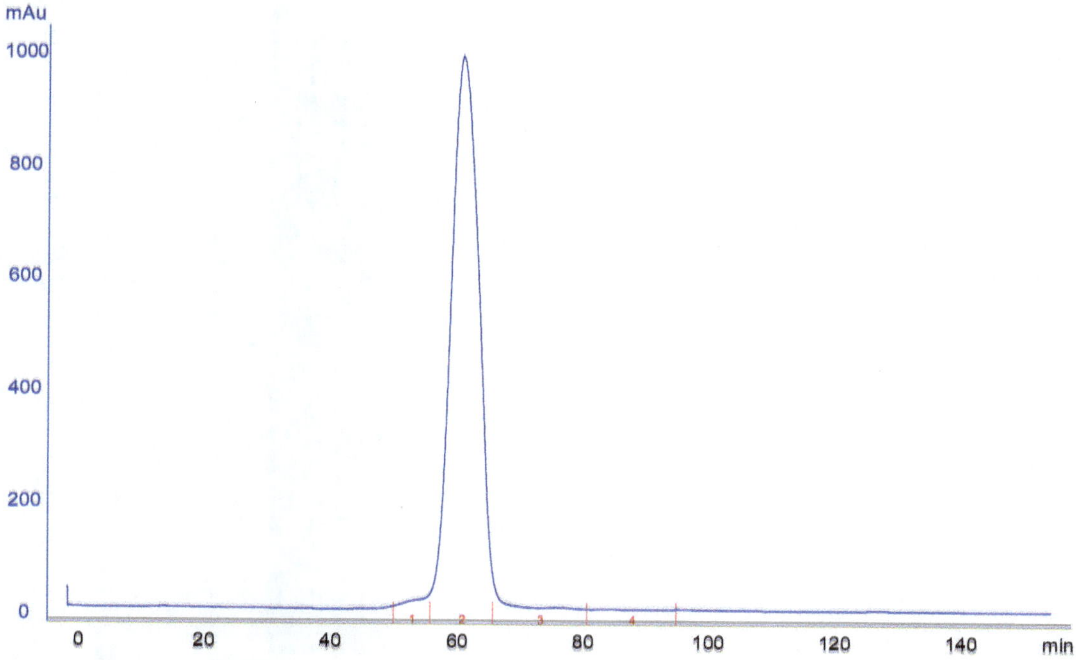

Fig. 7 Final size exclusion elution profile: absorbance at 280 nm is shown in blue with fractions numbered along the *x*-axis. The caa_3-oxidase eluted after approximately 60 min. This elution profile shows an example of a Gaussian peak indicating homogeneity of the caa_3-oxidase sample necessary for crystallization

8. Further concentrate the sample to ~500 μL with a small 50 kDa cutoff centrifugal concentrator in preparation for gel filtration chromatography.

3.7 Column 6: Size Exclusion (Repeat) and Detergent Exchange

1. Using a 2.0 mL loop, inject a 500 μL aliquot of concentrated caa_3-oxidase onto an a prepacked, gel filtration column (i.d. 16 mm, column height 600 mm) containing 120 mL of gel filtration resin capable of separation of biomolecules with the range of approximately 10–600 kDa. The column should be preequilibrated at room temperature (20–22 °C) with Final Storage Buffer (*see* **Note 10**).
2. Elute the protein at a flow rate of 1 mL/min using the same buffer. As previously noted, the caa_3-oxidase generally elutes after 50–60 min (refer to Fig. 7).
3. Following gel filtration, the caa_3-oxidase containing fractions (total volume approximately 12 mL) should be pooled and initially concentrated to approximately 2 mL using a large 50 kDa cutoff centrifugal concentrator and further concentrated to a final concentration of 10 mg/mL using a small 50 kDa cutoff centrifugal concentrator.

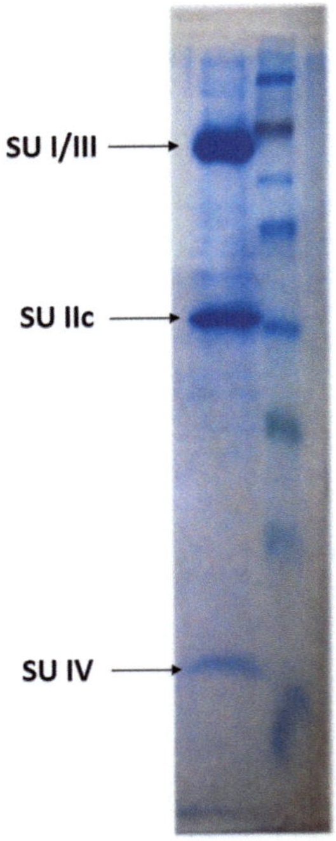

Fig. 8 SDS-PAGE of the purified *caa*$_3$-oxidase: the Coomassie-stained gel depicts the 3 subunits SU I/III (89 kDa), SU IIC (39 kDa), and SU IV (7 kDa) in the left-hand lane. The paucity of other bands indicates the high purity of the oxidase

3.8 Final Analysis and Storage

1. Aliquot the protein into 20 μL aliquots in 500 μL microcentrifuge tubes and flash-freeze using liquid nitrogen (*see* **Note 11**).
2. Take an aliquot of purified *caa*$_3$-oxidase for both spectroscopic and SDS-PAGE analysis (Fig. 8). Briefly, dilute 10 μL of purified protein to 2 mL with Final Storage Buffer and use to record the final reduced-minus-oxidized spectrum of the purified *caa*$_3$-oxidase. Dilute a further 10 μL with 2×-Laemmli Sample Buffer and load 5–10 μL of this sample loaded onto a precast 4–15% SDS-PAGE mini-gel.
3. Run the gel in 1× Laemmli Running Buffer at 100 V for 20 min, followed by 150 V for 1 h at room temperature (20–22 °C).
4. Stain the gel with Coomassie Brilliant Blue Staining Solution for 1 h and de-stain with Coomassie Brilliant Blue Destaining Solution.

3.9 Concentration Determination

1. Calculate the concentration of *caa*$_3$-oxidase from the heme *a* to protein ratio deduced from the reduced-minus-oxidized spectrum. The heme *a* concentration is determined using the reduced-minus-oxidized absorption peak at 604 nm (Fig. 5b)

Table 1
Example yield of caa_3-oxidase at key points of the purification

Column 1: anion exchange (mg)	Column 2: anion exchange (mg)	Column 3: gel filtration (mg)	Column 5: repeat anion exchange (mg)	Column 6: final gel filtration (mg)
16.5	15.5	10	6.4	5.6

with an extinction coefficient of 12,000 M^{-1} cm^{-1} [28]. Subsequently, the heme *a* concentration can be used to determine the protein concentration using the heme *a* to protein ratio of 2:1. An example of the enzyme yield in milligrams at key points of purification is outlined in Table 1.

3.10 Crystallization

1. Add sodium ascorbate to the purified caa_3-oxidase solution (10 mg/mL) to a final concentration of 0.2 μM.
2. Using a dual syringe mixing device [25] mix the solution with 7.7 MAG in a 1:1 ratio by weight to reconstitute the cubic mesophase [26].
3. With a syringe held vertically, load 50 nL of the cubic mesophase to each well in a 96-well glass sandwich plate (*see* **Note 12**).
4. Add 0.8 μL of the precipitant solution to each well.
5. Place a coverslip squarely over the wells to cover them uniformly.
6. Incubate the plates in a temperature-controlled chamber at 20 °C. Crystals should appear 5 days post-setup (Fig. 9).
7. To harvest the crystals, open the well of interest using a tungsten carbide glass cutter and remove crystals using a 50–100 mm premounted loop. Crystals can be cryo-cooled directly in liquid nitrogen.

4 Notes

1. The pH of all buffers was measured at room temperature (20–22 °C) only and the pH was measured before the addition of NaCl.
2. The supernatant can be retained after the first centrifugation for the purification of cytochrome c_{552}.
3. Usually left overnight in a cold room. After solubilization, the suspension will clarify noticeably.
4. The purpose of the dilution is to reduce the conductivity to less than 2 mS/cm, to decrease the detergent concentration to a final concentration of 0.5% (v/v) and to decrease the salt

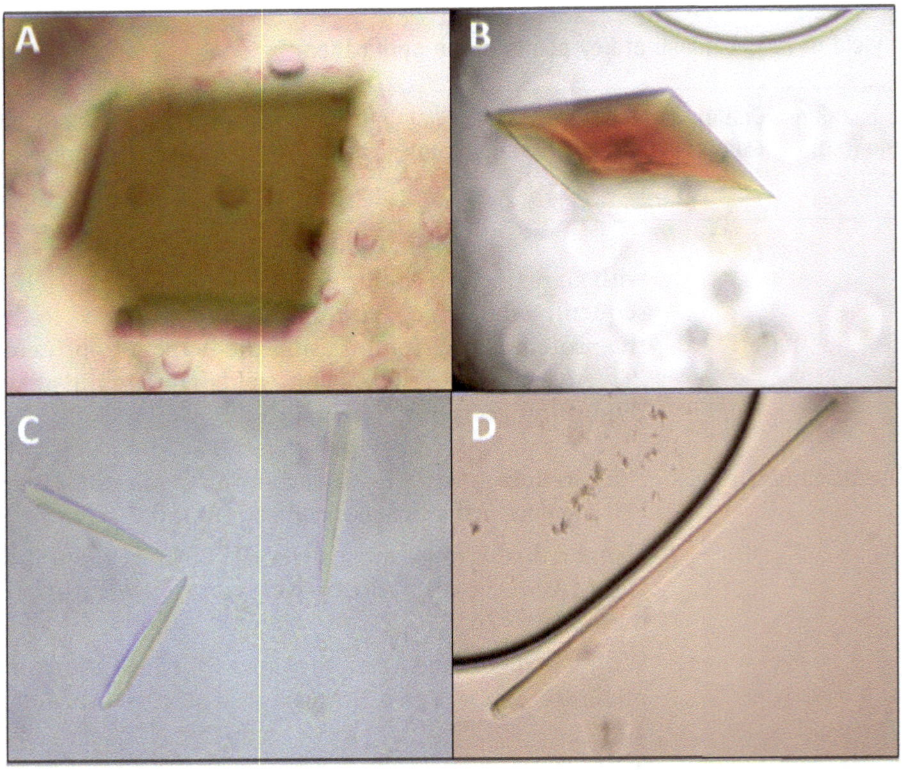

Fig. 9 *caa_3*-Oxidase crystals: grown in cubic mesophase in the presence of optimized precipitation screens

concentration to 10 mM NaCl in preparation for column chromatography.

5. The purpose of this first column is to crudely separate all the respiratory chain proteins present in the inner membrane of *T. thermophilus* and to remove a yellow pigment present in the membranes that can interfere with the spectroscopic analysis of the proteins. The best separation is achieved with the column recommended in the protocol. However, commercially available columns with an i.d. of 100 mm and a height of 200 mm are difficult to obtain and often must be custom made. Therefore, it is possible to carry out the purification with a glass column of i.d. 50 mm and a height of 1000 mm, which is more readily available commercially.
6. If application is not possible immediately, the material can be stored for up to 24 h at 4 °C.
7. The purpose of this column is to remove any remaining cytochrome *b*, *ba_3*-oxidase and some cytochrome $c_{549/554}$ from the *caa_3* sample and is also the first stage of the detergent exchange from Triton X-100 to DDM. Cytochrome *b* does not bind to this column and can be separated from the *caa_3* pool during the loading and washing process. The *ba_3*-oxidase elutes at a lower

NaCl concentration than the caa_3-oxidase. With sample bound to the column, detergent exchange from Triton X-100 to DDM is possible. The exchange of Triton X-100 can be controlled by following its absorption at 280 nm using a UV-detector. The intrinsic UV absorption of Triton X-100 decreases as the detergent is exchanged.

8. The purpose of this step is to remove cytochrome $c_{558/549}$ from the caa_3 sample. Cytochrome $c_{558/549}$ is significantly smaller than caa_3-oxidase, being only 26 kDa [29, 30]; therefore, it separates readily from the caa_3 on the size exclusion column.
9. At this stage of purification, the main contaminants of caa_3-oxidase, namely, numerous cytochrome *b*'s, ba_3-oxidase, and cytochrome $c_{558/549}$, are usually purged from the sample. However, in some instances, the *bc* complex (Complex III) remains as a contaminant. This complex usually separates into its main component parts, cytochrome b_{562} and cytochrome $c_{558/549}$ during the purification and is removed from the caa_3-sample as described, during column purifications 2 and 3. However, during some preparations, the complex remains intact and proves more difficult to remove. This final anion exchange column usually separates this complex from the pure caa_3-oxidase as it elutes at a higher salt concentration.
10. The purpose of this step is to provide a homogeneous caa_3-oxidase sample for crystallization as well as to exchange the detergent from DDM to DM which is the optimal detergent for crystallization of the protein.
11. Frozen samples are stored at −80 °C. The reduced-minus-oxidized can be used as a measure of stability. In contrast to bovine heart cytochrome oxidase, *T. thermophilus* cytochrome *c* oxidases can withstand numerous freeze–thaw steps [31]; however, the protein should be aliquoted to avoid this.
12. The tip of the needle should be no more than a few hundred micrometers above the base of the well to ensure proper delivery. If the tip is too far away the mesophase will usually curl up and away from the base; if it is too close the mesophase will remain stuck to the needle and will not be delivered in to the well. Reproducible delivery is easily achieved with practice.

References

1. Ferguson-Miller S, Babcock GT (1996) Heme/copper terminal oxidases. Chem Rev 96:2889–2908
2. Michel H, Behr J, Harrenga A, Kannt A (1998) Cytochrome *C* oxidase: structure and spectroscopy. Annu Rev Biophys Biomol Struct 27:329–356. https://doi.org/10.1146/annurev.biophys.27.1.329
3. Tsukihara T, Aoyama H, Yamashita E et al (1996) The whole structure of the 13-subunit oxidized cytochrome c oxidase at 2.8 A. Science 272:1136–1144

4. Iwata S, Ostermeier C, Ludwig B, Michel H (1995) Structure at 2.8 A resolution of cytochrome c oxidase from Paracoccus denitrificans. Nature 376:660–669. https://doi.org/10.1038/376660a0
5. Soulimane T, Buse G, Bourenkov GP et al (2000) Structure and mechanism of the aberrant ba(3)-cytochrome c oxidase from thermus thermophilus. EMBO J 19:1766–1776. https://doi.org/10.1093/emboj/19.8.1766
6. Svensson-Ek M, Abramson J, Larsson G et al (2002) The X-ray crystal structures of wild-type and EQ(I-286) mutant cytochrome c oxidases from Rhodobacter sphaeroides. J Mol Biol 321:329–339
7. Buschmann S, Warkentin E, Xie H et al (2010) The structure of cbb3 cytochrome oxidase provides insights into proton pumping. Science 329:327–330. https://doi.org/10.1126/science.1187303
8. Lyons JA, Aragão D, Slattery O et al (2012) Structural insights into electron transfer in caa3-type cytochrome oxidase. Nature 487:514–518. https://doi.org/10.1038/nature11182
9. OSHIMA T, IMAHORI K (1974) Description of Thermus thermophilus (Yoshida and Oshima) comb. nov., a nonsporulating thermophilic bacterium from a Japanese thermal spa. Int J Syst Bacteriol 24:102–112. https://doi.org/10.1099/00207713-24-1-102
10. Rich PR (2003) The molecular machinery of Keilin's respiratory chain. Biochem Soc Trans 31:1095–1105
11. Haltia T, Finel M, Harms N et al (1989) Deletion of the gene for subunit III leads to defective assembly of bacterial cytochrome oxidase. EMBO J 8:3571–3579
12. Riistama S, Puustinen A, García-Horsman A et al (1996) Channelling of dioxygen into the respiratory enzyme. Biochim Biophys Acta 1275:1–4
13. Fee JA, Choc MG, Findling KL et al (1980) Properties of a copper-containing cytochrome c1aa3 complex: a terminal oxidase of the extreme thermophile Thermus thermophilus HB8. Proc Natl Acad Sci U S A 77:147–151
14. Mather MW, Springer P, Hensel S et al (1993) Cytochrome oxidase genes from Thermus thermophilus. Nucleotide sequence of the fused gene and analysis of the deduced primary structures for subunits I and III of cytochrome caa3. J Biol Chem 268:5395–5408
15. Mather MW, Springer P, Fee JA (1991) Cytochrome oxidase genes from Thermus thermophilus. Nucleotide sequence and analysis of the deduced primary structure of subunit IIc of cytochrome caa3. J Biol Chem 266:5025–5035
16. Than ME, Hof P, Huber R et al (1997) Thermus thermophilus cytochrome-c552: a new highly thermostable cytochrome-c structure obtained by MAD phasing. J Mol Biol 271:629–644. https://doi.org/10.1006/jmbi.1997.1181
17. Lecomte S, Hilleriteau C, Forgerit JP et al (2001) Structural changes of cytochrome c (552) from Thermus thermophilus adsorbed on anionic and hydrophobic surfaces probed by FTIR and 2D-FTIR spectroscopy. Chembiochem 2:180–189
18. Janzon J, Ludwig B, Malatesta F (2007) Electron transfer kinetics of soluble fragments indicate a direct interaction between complex III and the caa3 oxidase in Thermus thermophilus. IUBMB Life 59:563–569. https://doi.org/10.1080/15216540701242482
19. Schrodinger L PyMOL Molecular Graphics System, Version 2.0. Schrödinger, LLC, New York, NY
20. Sousa PMF, Videira M a M, Bohn A et al (2012) The aerobic respiratory chain of *Escherichia coli*: from genes to supercomplexes. Microbiology 158:2408–2418. https://doi.org/10.1099/mic.0.056531-0
21. Brzezinski P (2004) Redox-driven membrane-bound proton pumps. Trends Biochem Sci 29:380–387. https://doi.org/10.1016/j.tibs.2004.05.008
22. Giuffrè A, Stubauer G, Sarti P et al (1999) The heme-copper oxidases of Thermus thermophilus catalyze the reduction of nitric oxide: evolutionary implications. Proc Natl Acad Sci U S A 96:14718–14723
23. Castenholz RW (1969) Thermophilic blue-green algae and the thermal environment. Bacteriol Rev 33:476–504
24. Keightley JA, Zimmermann BH, Mather MW et al (1995) Molecular genetic and protein chemical characterization of the cytochrome ba3 from Thermus thermophilus HB8. J Biol Chem 270:20345–20358
25. Cheng A, Hummel B, Qiu H, Caffrey M (1998) A simple mechanical mixer for small viscous lipid-containing samples. Chem Phys Lipids 95:11–21
26. Caffrey M, Cherezov V (2009) Crystallizing membrane proteins using lipidic mesophases. Nat Protoc 4:706–731. https://doi.org/10.1038/nprot.2009.31
27. Cherezov V, Caffrey M, IUCr (2003) Nano-volume plates with excellent optical properties for fast, inexpensive crystallization screening of membrane proteins. J Appl Crystallogr

36:1372–1377. https://doi.org/10.1107/S002188980301906X

28. van Gelder BF (1966) On cytochrome c oxidase. I. The extinction coefficients of cytochrome a and cytochrome a3. Biochim Biophys Acta 118:36–46
29. Mooser D, Maneg O, Corvey C et al (2005) A four-subunit cytochrome bc(1) complex complements the respiratory chain of Thermus thermophilus. Biochim Biophys Acta 1708:262–274. https://doi.org/10.1016/j.bbabio.2005.03.008
30. Mooser D, Maneg O, MacMillan F et al (2006) The menaquinol-oxidizing cytochrome bc complex from Thermus thermophilus: protein domains and subunits. Biochim Biophys Acta 1757:1084–1095. https://doi.org/10.1016/j.bbabio.2006.05.033
31. Soulimane T (1993) Activity and preparation for crystallisation of various cytochrome oxidase preparations. Rheinish-Westfalishen Technishen Hochshule, Aachen

Chapter 6

Aggregation Profiling of *C9orf72* Dipeptide Repeat Proteins Transgenically Expressed in *Drosophila melanogaster* Using an Analytical Ultracentrifuge Equipped with Fluorescence Detection

Bashkim Kokona, Nicole R. Cunningham, Jeanne M. Quinn, and Robert Fairman

Abstract

The recent development of a fluorescence detection system for the analytical ultracentrifuge has allowed for the characterization of protein size and aggregation in complex mixtures. Protocols are described here to analyze protein aggregation seen in various human neurodegenerative diseases as they are presented in transgenic animal model systems. Proper preparation of crude extracts in appropriate sample buffers is critical for success in analyzing protein aggregation using sedimentation velocity methods. Furthermore, recent advances in sedimentation velocity analysis have led to data collection using single multispeed experiments, which may be analyzed using a wide distribution analysis approach. In this chapter, we describe the use of these new sedimentation velocity methods for faster determination of a wider range of sizes. In Chapter 7 of this book, we describe how agarose gel electrophoresis can be used to complement the analytical ultracentrifugation work, often as a prelude to careful biophysical analysis to help screen conditions in order to improve the success of sedimentation velocity experiments.

Key words Analytical ultracentrifugation, Sedimentation velocity, Protein aggregation, Neurodegeneration, Amyotrophic lateral sclerosis, Frontotemporal dementia, *Drosophila melanogaster*

1 Introduction

There is an emerging need for fluorescence methods to study protein assemblies as they exist in vivo or in the context of the aqueous milieu that represents the macromolecules present in the cell. One new fluorescence approach that has been used to study the size of protein assemblies is the analytical ultracentrifuge equipped with fluorescence detection [1, 2]. This instrument has allowed for the analysis of fluorescently labeled protein aggregates in crude extracts prepared from complex cellular and even animal model systems [3–6]. This approach is particularly valuable since it

Jennifer J. McManus (ed.), *Protein Self-Assembly: Methods and Protocols*, Methods in Molecular Biology, vol. 2039, https://doi.org/10.1007/978-1-4939-9678-0_6,

does not require the use of strong denaturants or matrices for fractionation, representing a true solution method in which sedimentation can be analyzed by appropriate mass transport equations. We have employed multispeed methods using wide distribution analysis (MSM-WDA) to collect and analyze data over a wide range of polymer sizes using the sedimentation velocity approach [7, 8], providing aggregation profiles in the 1 S–1000 S size range. We have used such methods to study polyglutamine and huntingtin aggregation in both fly and worm transgenic models [3, 4], and more recently, have used this approach to study aggregation in mutations in the *c9orf72* gene that are the principal familial causal agent for amyotrophic lateral sclerosis (ALS) and frontotemporal dementia (FTD) [9, 10]. Here, we describe the unique challenges in preparing samples from *D. melanogaster* [11, 12] for sedimentation velocity characterization of the dipeptide repeat proteins (DPRs) produced by this mutant gene. This work builds on previous published protocols developed for sedimentation velocity measurements using fluorescence detection [13].

2 Materials

1. 10–15 third instar larva expressing *C9orf72*-derived DPRs fused to GFP or eGFP.
2. 4-(2-Hydroxyethyl)-1-piperazineethanesulfonic acid.
3. Phenylmethylsulfonyl fluoride.
4. Ethylene glycol tetraacetic acid.
5. Dithiothreitol.
6. Protease inhibitor tablets.
7. 1 mL glass homogenizer for tissue disruption.
8. Liquid nitrogen/ or a dry/ice ethanol bath.
9. Coomassie Plus Protein Assay Reagent.
10. Analytical balance.
11. Lysis buffer prepared as follows:
    - 100 mM 4-(2-hydroxyethyl)-1-piperazineethanesulfonic acid (HEPES), pH 7.3.
    - 2 mM phenylmethylsulfonyl fluoride (PMSF) using a 100 mM stock prepared in isopropanol.
    - 2 mM ethylene glycol tetraacetic acid (EGTA).
    - 2 mM dithiothreitol (DTT); may be prepared as 1 M stock solutions and frozen at −20 °C.
    - 2× Protease Inhibitor.

12. An-60 Ti Rotor, 4-Place or 8-Place using Epon-charcoal two-- sectored centerpieces.
13. Beckman ProteomeLab XL-A or XL-I analytical ultracentrifuge equipped with an AVIV Biomedical fluorescence detection system.
14. Heavy mineral oil (FC43 oil, Fluorinert).
15. Analysis Software, SedAnal v.6.80x64 [14]. SedAnal is a sedimentation analysis software originally created by Peter Sherwood and Walter Stafford, Boston Biomedical Research Institute, Watertown MA, USA, which can be downloaded for free from http://www.sedanal.org/. A more detailed description of the multispeed method (MSM) data collection approach and the wide distribution analysis method (WDA) for polydisperse solutions has been described [8]. The WDA method is a variation of the dc/dt approach of analyzing sedimentation boundaries [7, 15]. This method is able to handle multispeed data and eliminates both the time- and radial-independent noise. The multiple speed protocol can accommodate molecules with sedimentation coefficients ranging from 1.0 S to 250,000 S. The final relation from which the distribution function $g(s^*)$ is a function of itself, thus the relation is evaluated by iteration:

$$g(s^*) = \left(\frac{\partial c}{\partial s*}\right)_r + \left(2\omega^2 \int_{s*=0}^{s=s^*} g(s^*)\mathrm{d}s^*\right)\left(\frac{\partial t}{\partial s*}\right)_r$$

The value of $\left(\frac{\partial t}{\partial s*}\right)_r$ can be obtained by implicit differentiation of $s* \ (t_i, r_j) = (1/\omega^2 t_i) \times \ln\,(r_j/r_{men})$ equal to $(-t/s^*)$. Since the range of s^* values is large, the resulting analysis is plotted as $s\,g(s^*)$ vs. $\ln(s^*)$. A 2% smoothing is applied and the s^* grid of 0.01 is chosen.
16. Analysis Software, Sednterp v. 1.09 [16]. Sednterp is used to calculate solvent density and viscosity.
17. Analysis software, Sedfit v. 14.4d [17].

3 Methods

3.1 Mechanical Disruption of Larvae

The key to profiling a wide range of protein aggregation, from monomer to oligomer to inclusion-sized particles is to avoid detergents or solvent systems that might result in disrupting high molecular weight protein aggregates. Since aggregates approaching the size of inclusion bodies can spin out with even low centrifugal forces, we do not advise any precentrifugation steps for preparing lysates. Instead, we allow cell debris to gravity settle on ice.

1. Resuspend 10–15 larvae in 300 μL prechilled 2× lysis buffer.

2. For more efficient lysis, flash-freeze the samples three times using liquid nitrogen followed by thawing on ice.
3. Homogenize the samples in a prechilled glass tissue homogenizer by 50 twists, store it on ice for 5 min then follow up with an additional ten twists. Samples should be kept on ice at all times to avoid heating of the samples.
4. Transfer lysate into a 1.5 mL microfuge tube and store it on ice for 45 min to sediment large particulates by gravity. Since proteins are known to form sizeable aggregates avoid spinning the lysates. Then transfer 150–200 μL of the supernatant into a fresh microfuge tube and label it appropriately. It is valuable to estimate and record the volume of supernatant that is transferred, as this helps with calculations to determine how much sample one has for analytical ultracentrifugation, and other assays.
5. Protein concentrations can be determined using a Coomassie Plus Protein Assay Reagent Kit prior to making aliquots, as this way samples will not undergo multiple freeze–thaw cycles.
6. Set aside 20–40 μg total protein into a new microfuge tube for other assays, while the rest is used for sedimentation velocity studies.
7. Snap-freeze samples in liquid nitrogen and store them at −80 °C until required.

3.2 Collecting Sedimentation Velocity Data

In order to capture a complete sedimentation profile of aggregates in solution, prior approaches have involved collecting separate sets of sedimentation velocity experiments at different rotor speeds starting with low-speed experiments at 3000 rpm and higher speeds up to 50,000 rpm (Fig. 1a). Low-speed experiments will resolve aggregates equivalent in mass to that of the ribosome (3.2×10^6 Da) and larger, while high-speed experiments will likely capture GFP tagged monomers ($>2.7 \times 10^3$Da) and low-mass oligomers in the range of 10^4 Da. Given the wide range of particle size distributions commonly observed for protein aggregation in various transgenic animal model systems expressing proteins involved in neurodegeneration, and limitations of individual speed experiments, the multispeed method can be applied successfully here. In the multispeed method, the speed is varied during the run starting with low speed so that large particles in the range 10^6 Da can be observed, then increased to a speed in which the smallest particles of interest ($<10^3$ Da) have cleared the meniscus (Fig. 1b). Data collected using MSM are analyzed using WDA [15], eliminating both the time independent and radial independent noise, as implemented in Sedanal. Most notable prior applications of MSM have been on known mixtures [7, 8]. Results from WDA can be complemented by aggregate analysis using

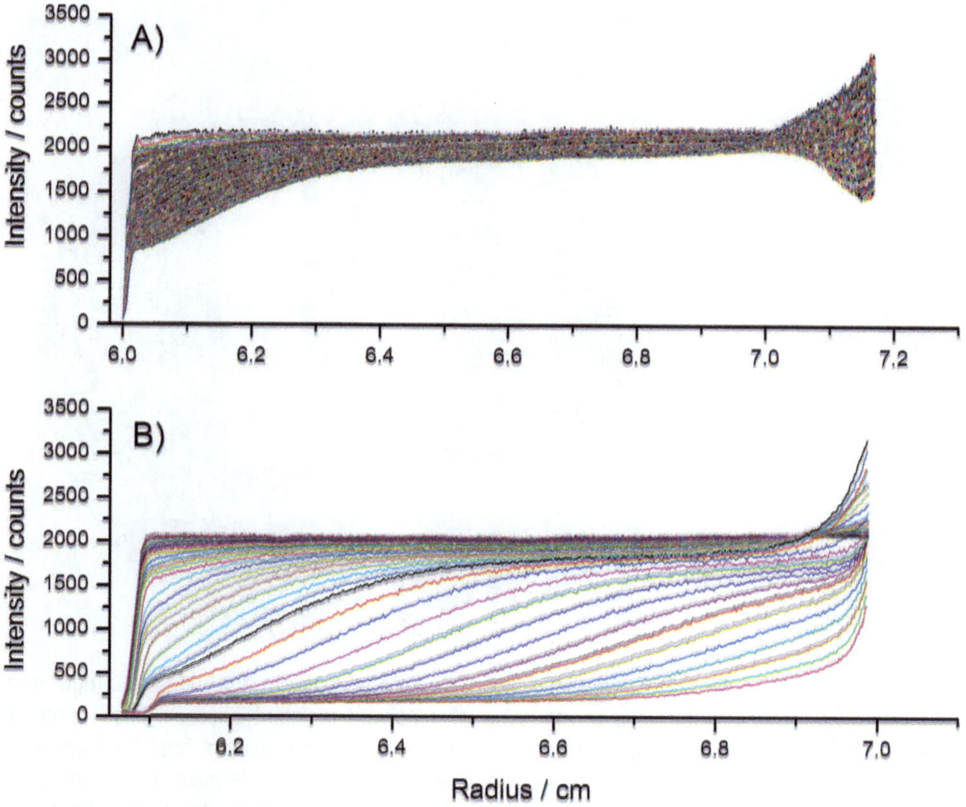

Fig. 1 Sedimentation boundaries for data collected at a single speed versus multiple speeds. (**a**) 250 scans collected at 20,000 rpm. (**b**) 750 scans collected at speeds of 3000, 6000, 10,000, 20,000, 30,000, and 50,000 rpm, showing only every tenth scan for clarity. The lysate is generated from the progeny of a cross of transgenic flies harboring Hsp70 or an Htt46-eGFP fusion construct, and is loaded at 0.5 mg/mL total protein concentration. [3]

semidenaturant detergent agarose gel electrophoresis (SDD-AGE), a method capable of detecting polymers in the 10^3–10^6 Da range (*see* Chapter 7).

1. Using gel-loading tips, layer 20–30 μL of Fluorinert FC43 oil into each compartment of a two-sectored charcoal-filled Epon centerpiece (*see* **Note 1** and Fig. 2). Failure to use FC43 Fluorinert oil will result in a truncation of the signal below 7.1 cm near the base of the cell, appearing as a downward slope, thus missing large particles sedimenting at the beginning of the run.
2. Thaw samples from the −80 °C freezer on ice. Dilute samples to 0.5–0.25 mg/mL total protein concentration and 1× lysis buffer. Concentrations above this level can lead to significant sedimentation artifacts.

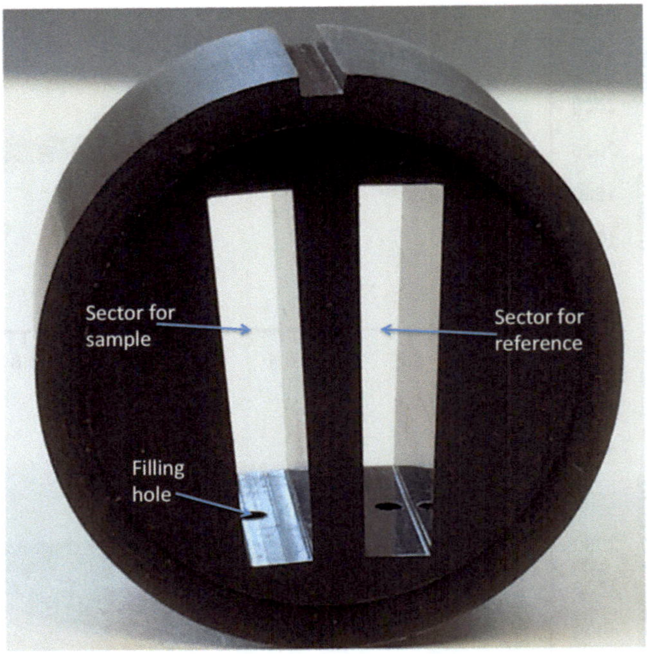

Fig. 2 Centerpiece used to hold sample in the ultracentrifuge rotor. The centerpiece is made up of a solvent-resistant charcoal-filled Epon material, manufactured to contain two sectors. These sectors hold the sample solution and the reference solution, as labeled. The centerpiece is a component of an assembly that includes quartz or sapphire windows bracketing the centerpiece, all held in place in an aluminum housing. The centerpiece has filling holes that can be accessed for sample and reference solution addition after the parts are assembled in the housing

3. Using gel-loading tips, carefully layer 350 μL of sample on top of the FC43 in each sector of the cell and avoid loading bubbles (*see* **Note 2**).
4. Place two red housing cap gaskets and cover them with housing plugs. Hand-tighten the plugs with a flat head screw driver.
5. Before loading the rotor make sure the counterbalance and sample cell are within 0.5 g using an analytical balance.
6. Insert the cells into the rotor and use an alignment tool to make sure it is all the way down inside the rotor.
7. Looking underneath the rotor, align the scribe mark on the rotor to that of the counterbalance or the working cell.
8. Install the Aviv FDS unit and allow the vacuum to fall below 50 μm before turning on the laser.
9. Allow the temperature to equilibrate at 20 °C (*see* **Note 3**).
10. After switching on the laser, start the AOS software (*see* **Note 1**) and begin spinning the rotor at 3000 rpm.

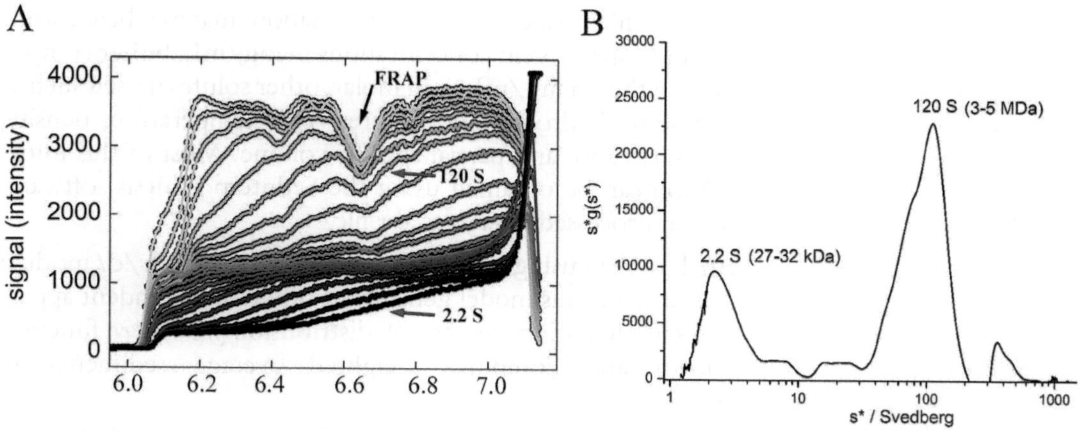

Fig. 3 Sedimentation data for GFP-$(GA)_{50}$ lysates. (**a**) Data were collected at speeds of 3000, 10,000, 20,000, 30,000, and 50,000 rpm. The background fluorescence intensity is about 200 counts. (**b**) $s^* g(s^*)$ distribution plotted against s^* (Svedberg range: 0.9 S–1200 S). The sedimentation boundaries associated with each peak are identified in 2A with red arrows. FRAP-like sedimentation profile is identified using a black arrow

11. Once the laser is locked and the magnet angle of the rotor has been established, set the fluorescence signal of the samples by adjusting the gain between 400 and 4000 counts, with the upper limit established to avoid saturation of the detection system. In our experience, we have been able to gain meaningful information even when intensities have been below 400 counts (*see* **Note 4**).
12. The rotor speed and running times can be individually tailored depending on the mixture, however, in our experience, it is best to set the speeds to 3000, 10,000, 20,000, 30,000, and 50,000 rpm, allowing the capture of particles in the 10^2–10^6 Da range. Make sure to spin the rotor long enough for the meniscus to clear.
13. Figure 3a shows the sedimentation boundaries of a 0.5 mg/mL GFP-$(GA)_{50}$ extract collected at speeds ranging from 3000 to 50,000 rpm. Even with the use of FC43 Fluorinert oil some attenuation of the signal is observed (*see* **Note 5**). Figure 3b shows the dc/dt transformed data from Fig. 3a using WDA (*see* **Note 6**). The peak centered at 2.2 S represents the GFP-$(GA)_{50}$ monomer, while the other peak centered at 120 S represents large aggregates, as observed in SDD-AGE experiments.
14. Analysis of the multispeed method (MSM) experimental data is done using SedAnal.

3.3 Analyzing Sedimentation Velocity Data Using MSM-WDA

1. Preprocess sedimentation velocity data: read scan files or cell data files, locate the base and meniscus, and specify the range to fit, typically the complete range is included from 6.1 to 7.2 cm and an s^* grid of 0.01 is chosen. Save reprocessed data as a .abr file.

2. Create a file with experiment information, material being studied and total protein concentrations in mg/mL, buffer components either in mg/mL or in molar, other solutes if used such as guanidine hydrochloride or urea, pH, temperature, density, viscosity ratio, and partial specific volume. Most of this information can be obtained using the Sednterp analysis software. Save reprocessed data as a .abr file.
3. Analyze data using the WDA model part of the d*c*/d*t* module in SedAnal. This model generates a model-independent apparent sedimentation coefficient distribution $g(s^*)$ (*see* function above) and computes weighted average sedimentation coefficient.
4. Select all radial points for WDA by selecting the default (blue color).
5. A 2% smoothing percentage is applied to the distribution and distribution data are saved as a text file.
6. Data are plotted in Origin and graphs are exported as a tiff.

4 Notes

1. The rotor used for the XLA is an An-Ti60 rotor with velocity type cells containing double-sector charcoal-filled Epon centerpieces with quartz windows. For calibration, a cell containing fluorescein is used either at position 4 or 8. Operating control software used is Advanced Operating Software (AOS) from Aviv Biomedical.
2. Samples with similar total protein concentrations are examined for consistent comparison between samples. The smallest volume for experiments with FDS should be no less than 350 μL.
3. We perform our experiments at 20 °C. The measurements of densities and viscosities are calculated using Sednterp.
4. The baseline of fluorescence scans is always offset from zero due to a small "dark count" signal in our experiments at about 200 counts in intensity.
5. Blinking and FRAP-like sedimentation are observed on occasion [18]. FRAP-like sedimenting molecules initially located in the trough will diffuse and exchange with fluorescent molecules from outside, thus diminishing the trough. Also, an upward-sloping plateau is observed due to small errors in the tracking laser beam caused by shifts of the focal point during the scan. The impact of these on accuracy can be mitigated by correcting for time-independent (TI) noise in the analysis software, Sedfit; however, in SedAnal both time and radially independent noise are eliminated.

6. Sedimentation analysis of biologically complex solutions is difficult due to problems with nonideality and protein–protein and protein–nucleic acid interactions. Nevertheless, our studies of crude lysates have shown that the system is robust when compared to previous studies of the GFP purified recombinant protein in either buffer or serum provided that total protein concentrations are kept at or below 0.5 mg/mL [1, 2]. Deviation of *s*-values as a result of nonideality or interactions with other proteins is unlikely to have a significant impact on the broader conclusions on the degree of heterogeneity and wide distribution of aggregates.

References

1. Kingsbury JS, Laue TM (2011) Fluorescence-detected sedimentation in dilute and highly concentrated solutions. Methods Enzymol 492:283–304
2. Kroe RR, Laue TM (2009) NUTS and BOLTS: applications of fluorescence-detected sedimentation. Anal Biochem 390:1–13
3. Kim SA, D'Acunto VF, Kokona B, Hofmann J, Cunningham NR, Bistline EM et al (2017) Sedimentation velocity analysis with fluorescence detection of mutant huntingtin exon 1 aggregation in Drosophila melanogaster and Caenorhabditis elegans. Biochemistry 56:4676–4688
4. Kokona B, May CA, Cunningham NR, Richmond L, Garcia FJ, Durante JC et al (2015) Studying polyglutamine aggregation in Caenorhabditis elegans using an analytical ultracentrifuge equipped with fluorescence detection. Protein Sci 25:605–617
5. Olshina MA, Angley LM, Ramdzan YM, Tang J, Bailey MF, Hill AF et al (2010) Tracking mutant huntingtin aggregation kinetics in cells reveals three major populations that include an invariant oligomer pool. J Biol Chem 285:21807–21816
6. Xi W, Wang X, Laue TM, Denis CL (2016) Multiple discrete soluble aggregates influence polyglutamine toxicity in a Huntington's disease model system. Sci Rep 6:34916
7. Runge MS, Laue TM, Yphantis DA, Lifsics MR, Saito A, Altin M et al (1981) ATP-induced formation of an associated complex between microtubules and neurofilaments. Proc Natl Acad Sci U S A 78:1431–1435
8. Stafford WF, Braswell EH (2004) Sedimentation velocity, multi-speed method for analyzing polydisperse solutions. Biophys Chem 108:273–279
9. DeJesus-Hernandez M, Mackenzie IR, Boeve BF, Boxer AL, Baker M, Rutherford NJ et al (2011) Expanded GGGGCC hexanucleotide repeat in noncoding region of C9ORF72 causes chromosome 9p-linked FTD and ALS. Neuron 72:245–256
10. Renton AE, Majounie E, Waite A, Simon-Sanchez J, Rollinson S, Gibbs JR et al (2011) A hexanucleotide repeat expansion in C9ORF72 is the cause of chromosome 9p21-linked ALS-FTD. Neuron 72:257–268
11. Freibaum BD, Lu Y, Lopez-Gonzalez R, Kim NC, Almeida S, Lee KH et al (2015) GGGGCC repeat expansion in C9orf72 compromises nucleocytoplasmic transport. Nature 525:129–133
12. Haeusler AR, Donnelly CJ, Periz G, Simko EA, Shaw PG, Kim MS et al (2014) C9orf72 nucleotide repeat structures initiate molecular cascades of disease. Nature 507:195–200
13. Polling S, Hatters DM, Mok YF (2013) Size analysis of polyglutamine protein aggregates using fluorescence detection in an analytical ultracentrifuge. Methods Mol Biol 1017:59–71
14. Stafford WF, Sherwood PJ (2004) Analysis of heterologous interacting systems by sedimentation velocity: curve fitting algorithms for estimation of sedimentation coefficients, equilibrium and kinetic constants. Biophys Chem 108:231–243
15. Stafford WF 3rd (1992) Boundary analysis in sedimentation transport experiments: a procedure for obtaining sedimentation coefficient distributions using the time derivative of the concentration profile. Anal Biochem 203:295–301
16. Schuck P, Zhao H (2011) Editorial for the special issue of methods "modern analytical ultracentrifugation". Methods 54:1–3

17. Dam J, Velikovsky CA, Mariuzza RA, Urbanke C, Schuck P (2005) Sedimentation velocity analysis of heterogeneous protein-protein interactions: Lamm equation modeling and sedimentation coefficient distributions c(s). Biophys J 89:619–634

18. Zhao H, Fu Y, Glasser C, Andrade Alba EJ, Mayer ML, Patterson G et al (2016) Monochromatic multicomponent fluorescence sedimentation velocity for the study of high-affinity protein interactions. eLife 5:pii: e17812

Chapter 7

Size Analysis of *C9orf72* Dipeptide Repeat Proteins Expressed in *Drosophila melanogaster* Using Semidenaturing Detergent Agarose Gel Electrophoresis

Nicole R. Cunningham, Bashkim Kokona, Jeanne M. Quinn, and Robert Fairman

Abstract

This chapter supplements Chapter 6 on sample preparation and analysis using an analytical ultracentrifuge with fluorescence detection. In this related chapter, we describe how semidenaturing detergent agarose gel electrophoresis can be used to complement the analytical ultracentrifugation work, often as a prelude to careful biophysical analysis to help screen conditions to improve the success of sedimentation velocity experiments. We describe preparation of crude lysates made using *Drosophila melanogaster* and provide a protocol giving detailed instructions for successful fractionation of protein aggregates using SDD-AGE. While limited in resolving power, this method can identify fractionation in three pools based on sample migration in the gel: that of a monomer or limiting small oligomer species; intermediate aggregation pools, which are typically heterogeneous, represented as high retention smears; and large-scale aggregation, found caught up in the wells.

Key words Semidenaturing detergent agarose gel electrophoresis, Western blotting, Protein aggregation, Neurodegeneration, Amyotrophic lateral sclerosis, Frontotemporal dementia, *Drosophila melanogaster*

1 Introduction

The involvement of the *C9orf72* gene products in amyotrophic lateral sclerosis (ALS) and frontotemporal dementia (FTD) is described in Chapter 6 [1, 2]. Due to problems with repeat-associated non-AUG (RAN) translation, five independent dipeptide repeat (DPR) polypeptides are produced, and they aggregate to varying degrees, which can be explored applying either sedimentation velocity methods using an analytical ultracentrifuge equipped with fluorescence detection (Chapter 6) or semidenaturing detergent agarose gel electrophoresis (SDD-AGE), described in this chapter. The five DPRs that are found in human disease include

Jennifer J. McManus (ed.), *Protein Self-Assembly: Methods and Protocols*, Methods in Molecular Biology, vol. 2039, https://doi.org/10.1007/978-1-4939-9678-0_7,

polyGA, polyGP, polyPA, polyGR, and polyPR, and have also been studied in transgenic *Drosophila melanogaster* [3, 4].

SDD-AGE was first described by Kryndushkin et al. in studies of SDS-resistant protein aggregates formed by prions in yeast [5]. The method was then expanded as a general tool to screen for aggregation of amyloid and amyloid-like aggregates in yeast [6]. One caveat described in this latter work is the importance of using protease inhibitors since the concentration of SDS used (typically 0.1%) will not necessarily be sufficiently potent to denature all proteases present in a cell. Our protocol for sample preparation (*see* Chapter 6) includes such inhibitors. Other similar electrophoresis methods have been described, such as native agarose gel electrophoresis (NAGE), which avoids the use of SDS and its potential for disrupting partially solvated or noncovalent aggregates. NAGE has been applied to the study of polyglutamine construct aggregation for modeling Huntington's disease, using lysates derived from transgenic expression in *Caenorhabditis elegans* [7]. The disadvantage of this approach is that migration of material through the gel will depend to some extent on the p*I* of the polypeptide chains.

We have applied the SDD-AGE method to the study of the aggregation of huntingtin protein fragments in both *Drosophila melanogaster* and *Caenorhabditis elegans* transgenic animal model systems [8, 9] and show its use here to study aggregation in *C9orf72* DPRs.

2 Materials

1. Crude Lysate Extracts stored at −80 °C (aliquoted to minimize freeze-thaw) prepared as described in Chapter 6.
2. High molecular weight prestained protein standard. Store at −20 °C.
3. 2× SDS sample buffer (5× can also be used; store at −20 °C) prepared with:
   - 250 mM Tris–HCl, pH 6.8.
   - 2% (w/v) SDS.
   - 0.2% (w/v) bromophenol blue.
   - 20% (v/v) glycerol.
   - 10% β-mercaptoethanol or 200 mM dithiothreitol (DTT).
4. Agarose powder. Store at room temperature.
5. 1× Tris–acetate with ethylenediaminetetraacetic acid (TAE). Store at room temperature.
6. 1× Tris–acetate with ethylenediaminetetraacetic acid (TAE) + 0.1% SDS. Store at room temperature.

7. Tris-buffered saline (TBS). Store at room temperature.
8. Tween 20 or Tween 80. Store at room temperature.
9. 1× phosphate buffered saline (PBS). Store at room temperature
10. PBS–Tween: 0.1% (v/v) Tween 20 or Tween 80 in phosphate buffered saline. Store at room temperature.
11. Blotto: 5% nonfat powdered milk in PBS–Tween. Should be used within 24 h for best results. Store at 4 °C.
12. PBS–Tween +1% bovine serum albumin (BSA). Store at 4 °C.
13. SDS-PAGE running buffer (for 1 L; store at room temperature):
    - 3 g Tris base.
    - 14.4 g glycine.
    - 1 g SDS.
14. Electroblotting transfer buffer: SDS-PAGE running buffer +10% methanol. Store at room temperature.
15. Microwave oven.
16. Horizontal gel electrophoresis apparatus (typically used for agarose gels).
17. Electrophoresis power supply.
18. Nitrocellulose membrane or polyvinylidene difluoride (PVDF) membrane (*see* **Note 1**).
19. Blotting paper (*see* **Note 2**).
20. Transfer setup (Fig. 1):
    - Tray.
    - 24 pieces of dry blotting paper.
    - Four pieces of wet blotting paper.
    - Nitrocellulose (or PVDF) membrane.
    - Agarose gel.
    - Wet wick of blotting paper.
    - Shallow container.
    - 500 mL bottle of water
    - Two reservoirs of 1× TBS.
21. Polyethylene sealable bags: may be purchased in 6 × 8″ sizes but resealable plastic sleeves for paper can be used as well.
22. Plastic heat sealer.
23. Anti-green fluorescent protein (GFP) antibody. Aliquot and store at −20 °C.

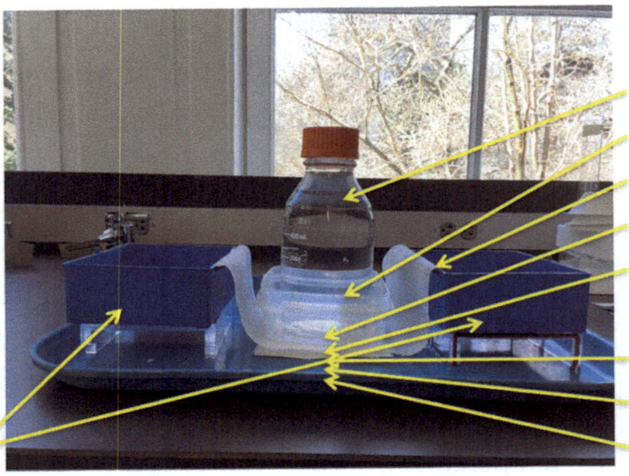

Fig. 1 SDD-AGE transfer setup. Two containers filled with 1× TBS on each side of the setup help keep the blotting paper wet overnight. The agarose gel is sandwiched between a nitrocellulose membrane and three small sheets of wet blotting paper. The nitrocellulose membrane sits on top of 25 wet blotting papers. A 500 mL bottle filled with liquid is placed on top of the setup

24. Anti-rabbit secondary antibody conjugated to horseradish peroxidase (HRP). Store at −20 °C.
25. Supersignal West Femto Maximum Sensitivity Substrate. Store at room temperature or 4 °C for long-term storage.
26. Chemiluminescence detecting imaging unit.
27. High energy autoradiography pen.
28. Heating dry bath equipped with blocks that can hold 1.5 mL microfuge tubes.
29. Vertical gel electrophoresis apparatus (typically used for polyacrylamide gels).
30. Electroblotting apparatus.

3 Methods

3.1 Larval Lysate Preparation

This protocol is identical to that described in Chapter 6.

3.2 Agarose Gel Electrophoresis (See Note 3)

1. Prepare a 1.5% (w/v) agarose gel with 0.1% SDS in 1× TAE by adding an appropriate amount of agarose powder to 1× TAE. Most small gel electrophoresis units typically use a 50 mL bed of agarose, requiring 0.75 g of agarose. Dissolve the agarose by heating using repeated 30-s microwave sessions until the agarose is fully dissolved.
2. Allow the agarose liquid to cool to about 55 °C (able to touch with gloved hand), then add the 20% SDS solution to achieve a

final concentration of 0.1% SDS (*see* **Note 4**). Swirl gently to mix.

3. Assemble the gel casting system with a comb using 10–15 teeth, which will allow for 20–30 μL of sample to be loaded per well, with the range depending on the thickness of the gel. Pour the gel and allow it to solidify for at least 40 min.
4. Thaw the larval lysate extracts on ice and keep on ice unless indicated otherwise (*see* **Note 5**).
5. Combine the appropriate volume of larval lysates with an equal volume of 2× SDS sample buffer and mix by pipetting up and down a couple of times (*see* **Note 6**).
6. Incubate the samples at room temperature for 5 min. DO NOT BOIL SDD-AGE samples. Do not use samples that have been previously boiled.
7. When the agarose gel has finished solidifying, remove the comb to reveal the wells and move the gel onto the horizontal gel electrophoresis unit. Add the 1× TAE containing 0.1% SDS as the running buffer. Samples should run toward the red electrode (positive electrode). A sufficient volume of 1× TAE buffer should be added to the electrophoresis tank to completely submerge the gel.
8. Load 1–5 μg of protein from the larval lysate (previously combined with SDS sample buffer) in each well. It is important to run a high molecular weight prestained protein ladder in order to visualize aggregates in one well next to your samples.
9. Plug the horizontal gel apparatus into the power supply. Run the gel at a low voltage (45 V) for 3–5 h at 4 °C in a temperature controlled cold room. The dye front should travel at least two thirds of the distance of the gel. After running is complete, gently remove agarose gel from apparatus and place in a small container with 1× TBS (*see* **Note 7**).

3.3 Transfer to Membrane and Development of Bands

1. Cut a piece of nitrocellulose (or PVDF; *see* **Note 1**) membrane slightly larger than the size of the agarose gel and place in 1× TBS to soak for a few minutes. Do not let the gel or nitrocellulose dry—keep them in 1× TBS buffer. Cut a total of 28 pieces of blotting paper that are slightly larger than the agarose gel (*see* **Note 2**).
2. Cut another piece of blotting paper that is the same width as the other pieces of blotting paper but is much longer in length, about 18–24″ (46–61 cm).
3. Place a large flat tray on the bench. Place 24 pieces of dry blotting paper in the middle of the tray. Wet one piece of blotting paper in 1× TBS and place this on top of the 24 pieces of dry blotting paper.

4. Place the nitrocellulose membrane on top of the piece of wet blotting paper. Immediately place the agarose gel on top of the nitrocellulose paper (*see* **Note 8**). A transfer pipette can be used to wet the area between the membrane and the agarose gel with 1× TBS. Ensure there are no bubbles present between agarose gel and nitrocellulose membrane during transfer. Bubbles will distort protein transfer or leave a portion of the membrane blank.
5. Place three sheets of wet blotting paper on top of the agarose gel. A piece of blotting paper the width of the stack and 18–24″ (46–61 cm) in length should be wetted in 1× TBS and placed on top of the stack. The long length of this blotting paper allows the ends to be submerged in two trays containing 1× TBS on either side of the stack. This serves as a wick to keep the gel and membrane moist during the transfer.
6. Finally, place a small tray on top of the entire stack and weight it down with a 500 mL bottle filled with water. The setup should now look like the apparatus in Fig. 1.
7. The protein is transferred to nitrocellulose using capillary action overnight at room temperature.
8. Make sure that the transfer was successful by looking for the protein ladder on the membrane. The dye front runs out of the gel and the ladder is usually compressed into a smear. Take a picture of the membrane, as the prestained high molecular weight markers will no longer be visible after the blotting procedure described below.
9. After transfer, the nitrocellulose should be placed immediately in 15–30 mL Blotto for blocking. Block for 30 min in a small container (the top of a P1000 pipette tip box is ideal) on a rotating or rocking platform.
10. Rinse with PBS–Tween; then follow with 2 × 5 min washes with PBS–Tween on a rotating or rocking platform.
11. Prepare a dilution of the appropriate primary antibody in 5 mL PBS–Tween +1% BSA. This dilution can be stored for a day or two at 4 °C prior to use. For the anti-GFP antibody (rabbit primary antibody in Fig. 2), a 1:8000 dilution factor was used.
12. Place the membrane in a resealable plastic bag. Three of the sides should be sealed with a heat sealer to minimize the amount of primary antibody solution that is used. Add 5 mL or less of the antibody solution in PBS–Tween +1% BSA.
13. Remove air bubbles by gently rolling a 10 mL pipette along the outside of the bag toward the open end. Seal the remaining open end with a heat sealer.
14. Place the plastic bag on a rocking platform overnight at 4 °C. This incubation can also be done at room temperature for 4 h, but is not recommended.

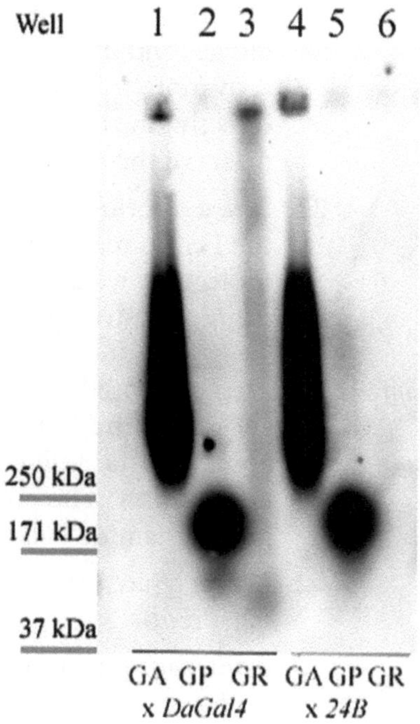

Fig. 2 SDD-AGE blot showing aggregation profiles for polyGA, polyGP, and polyGR DPRs as expressed in third instar fly larvae using the Da-Gal4 promoter (pan-tissue) or the 24B-Gal4 promoter (muscle-specific). polyGA shows a significant fraction of large scale aggregation (in the wells) and intermediate aggregation (smear in the center of gel). polyGP is largely monomeric. polyGR expression is significantly lower, but showing protein partitioning largely between monomeric or large aggregate states, with little intermediate aggregation

15. Remove the membrane from the plastic bag and block the membrane with Blotto for 30 min to 1 h on a rocking or rotating platform.
16. Wash membrane 3 × 5 min each with PBS–Tween in a small container on a rocking or rotating platform.
17. Dilute the anti-rabbit secondary antibody in Blotto according to the manufacturer's recommended specifications. Make sure your secondary antibody is conjugated to HRP. Then follow steps nine and ten above to insert the membrane and antibody solution into a plastic bag (*see* **Note 9**). Rock the bag at room temperature for at least 30 min.
18. Remove the membrane from the bag and rinse the membrane with PBS–Tween. Follow up with 4 × 5-min washes using a small container on a rocking or rotating platform.
19. Follow the manufacturer's protocol for preparing the solutions for activating the chemiluminescence through the action of the HRP, or horseradish peroxidase enzyme, linked to the secondary antibody.

20. Handle the membrane with tweezers, wearing latex or nitrile gloves, and insert the membrane into a plastic sleeve. Add the activation solutions directly to the membrane by pipetting directly onto the membrane. Then close the plastic sheet to preserve the solution.
21. Image capture should proceed immediately upon addition of the Femto reagents as the signal starts fading fairly rapidly. The molecular weight standards can be highlighted on the blot using an autoradiography pen (Fig. 2; *see* **Notes 10** and **11**).

3.4 Western Blotting Analysis of Protein Expression (See Notes 3 and 11)

1. Elements of the protocol here are similar to that described above for the SDD-AGE protocol, and appropriate sections above will be referenced to avoid unnecessary duplication of information. The protocol provided here assumes that precast polyacrylamide gels are purchased for protein electrophoresis.
2. Larval lysate preparation should be carried out as described in Chapter 6.
3. After thawing on ice, dilute the larval lysates in SDS sample buffer.
4. Set a heating dry bath to 95 °C, and when it comes to temperature, heat the larval lysates for 5 min in preparation for SDS-PAGE gel electrophoresis.
5. The percentage of acrylamide in the SDS-PAGE gel used is dependent on the size of the protein being evaluated. For the *C9orf72* DPR-GFP fusions used in this study, at about 32 kDa size, either a 4–20% gradient gel or a 12% resolving gel, using a 5–6% stacking gel, was used. A 10 or 15-well format is recommended, using 1.5 mm thick gels.
6. Insert the precast gel into appropriate vertical gel electrophoresis unit, fill the upper and lower chambers with running buffer, and remove the comb in preparation for loading samples.
7. Load approximately 35 μL of each of the samples, plus 10 μL protein molecular weight markers into the wells.
8. Perform the electrophoresis at 4 °C for 45 min to 1 h at 120 V.
9. Pour out the running buffer, remove the precast gel, and extract the gel from the plastic according to the manufacturer's instructions.
10. Cut a piece of nitrocellulose or PVDF membrane (*see* **Note 1**) to a size slightly larger than the gel. It is a good idea to mark a corner with a cut so the gel can be appropriately oriented later. Soak the membrane and the gel in electroblotting transfer buffer.

11. The assembly of the membrane and gel into the transfer cassette will depend on the type of transfer apparatus used. The instructions here are for a typical setup. All components should be prewet with electroblotting transfer buffer and assembled in the following order inside the transfer cassette:
    - Sponge (provided by the manufacturer).
    - Blotting paper.
    - Membrane (blot).
    - Gel.
    - Blotting paper.
    - Sponge.
12. While assembling the sandwich, place the blotting paper over the membrane (and gel) and roll out any bubbles with a 5 mL pipette. This removes bubbles between the membrane and the gel.
13. The assembled cassette sandwich can now be inserted into the electroblotting apparatus and filled with transfer buffer, making sure that the entire cassette is immersed in the solution.
14. Fill to blotting line with transfer buffer.
15. Transfer at 4 °C using 400 mAmps for 1–2 h or 40 mAmps overnight.
16. After the transfer is complete, remove the membrane from the transfer cassette, and follow instructions 5–16 in the SDD-AGE protocol (*see* Subheading 3.3, Fig. 3). The used transfer buffer should be discarded safely as organic waste.

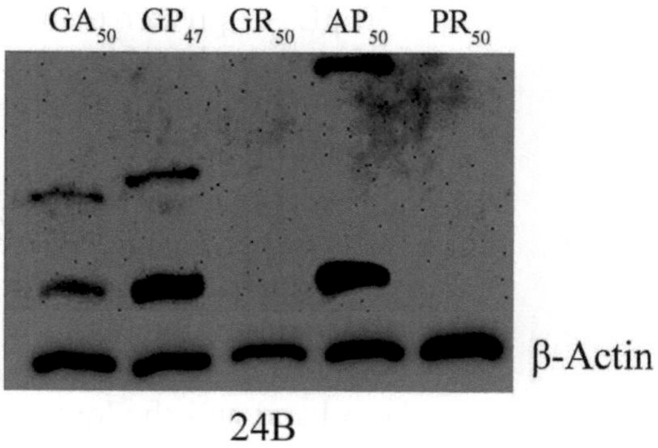

Fig. 3 Western blot showing protein levels for polyGA, polyGP, polyGR, polyAP, and polyPR fused to GFP, in crude extracts derived from third instar fly larvae. The total protein concentration in the extracts is 0.5 mg/mL. At this concentration, no polyGR or polyPR are detected since muscle-specific expression (24B is a 24B-Gal4 muscle-specific promoter) was too low to observe. The primary antibody used here was a rabbit anti-GFP antibody. β-Actin was used as a loading control

4 Notes

1. Nitrocellulose or PVDF membranes should only be handled by tweezers to avoid introducing foreign protein. It is best to use latex or nitrile gloves whenever handling these membranes. If using PVDF membrane, it MUST be activated first by soaking it for at least a minute in methanol and then rinsed in Milli-Q water prior to use.
2. Filter paper should be used when wet blotting paper is described. Standard laboratory paper towels may be used for dry blotting paper in order to save on cost.
3. We strongly recommend analyzing all samples for SDD-AGE by Western blotting or dot blotting, using the same antibody staining protocols, to confirm expression and to measure relative band intensities. These methods allow for straightforward trouble-shooting of band intensities (or lack thereof) in blots prepared from SDD-AGE experiments.
4. Certain protein aggregates are even more resistant to SDS. Increasing the concentration of SDS in agarose or acrylamide gels up to 1% or higher may be necessary.
5. Freeze-thawing of samples multiple times will likely change the composition of your sample (i.e., potentially induce further aggregation). It is important that appropriate aliquots are created after concentrations are determined using a standard Bradford assay when the lysate was prepared. Do not centrifuge or spin crude extracts at any point. Spinning may sediment possible aggregates you want to detect.
6. If the volume of the cell lysate plus SDS sample buffer exceeds 30 μL (or greater than what the agarose gel well will hold), compensate by adding less sample buffer. Do not reduce sample buffer amount too much or you will lose the semidenaturing effect.
7. In order to get similar results every time, it is important to keep agarose percentage, SDS concentrations, and voltage constant from run to run. Any change in those conditions will likely give different results. Also, the ladder does not separate well in the SDD-AGE gel. The best estimates one can make are where the top band and the lower band run. The ladder should not be used to resolve small oligomers, like dimers and trimers, from one another. In addition, the protein standards in the molecular weight ladder may become smeared during the running of the agarose gel. This is normal. Instead of a precise band, the ladder will appear stretched and may even fade during blotting.
8. Do not trim lanes off the agarose gel because aggregates may be in/near the loading well.

9. If cost is not an obstacle, use more secondary antibody solution for the blotting protocol. Replace the plastic bag with a small container, requiring the use of 15–30 mL of solution.
10. On occasion, the prestained molecular weight markers can lose their color through the various washes of the membrane. To avoid losing this information, cut very small slits at the edge of the membrane to indicate the locations of each prestained marker along the axis of electrophoresis.
11. Unlike a Western blot, which routinely has clear and distinct bands, SDD-AGE results in a blot that appears more smeared and stretched. To interpret the results, it can be helpful to run a control sample that is the same as one of your samples but is denatured by a 5-min incubation at 100 °C. This can help you determine where a protein monomer would run on the gel relative to the protein aggregates. SDD-AGE data should be interpreted only relative to other samples on the same gel. One type of aggregate is smaller or larger based on how far it migrated down the gel. Smaller aggregates run further, while larger aggregates do not migrate as far. If the conditions allow it, sometimes protein monomers are visible or even fragment at the lower part of the gel. Aggregates can greatly vary in size. In fact, some aggregates are so large that they do not fully penetrate the gel and can be detected in or near the well.

References

1. DeJesus-Hernandez M, Mackenzie IR, Boeve BF, Boxer AL, Baker M, Rutherford NJ et al (2011) Expanded GGGGCC hexanucleotide repeat in noncoding region of C9ORF72 causes chromosome 9p-linked FTD and ALS. Neuron 72:245–256
2. Renton AE, Majounie E, Waite A, Simon-Sanchez J, Rollinson S, Gibbs JR et al (2011) A hexanucleotide repeat expansion in C9ORF72 is the cause of chromosome 9p21-linked ALS-FTD. Neuron 72:257–268
3. Freibaum BD, Lu Y, Lopez-Gonzalez R, Kim NC, Almeida S, Lee KH et al (2015) GGGGCC repeat expansion in C9orf72 compromises nucleocytoplasmic transport. Nature 525:129–133
4. Haeusler AR, Donnelly CJ, Periz G, Simko EA, Shaw PG, Kim MS et al (2014) C9orf72 nucleotide repeat structures initiate molecular cascades of disease. Nature 507:195–200
5. Kryndushkin DS, Alexandrov IM, Ter-Avanesyan MD, Kushnirov VV (2003) Yeast (PSI+) prion aggregates are formed by small Sup35 polymers fragmented by Hsp104. J Biol Chem 278:49636–49643
6. Halfmann R, Lindquist S (2008) Screening for amyloid aggregation by semi-denaturing detergent-agarose gel electrophoresis. J Vis Exp 17:838. https://doi.org/10.3791/838
7. Holmberg M, Nollen EA (2013) Analyzing modifiers of protein aggregation in *C. elegans* by native agarose gel electrophoresis. Methods Mol Biol 1017:193–199
8. Kim SA, D'Acunto VF, Kokona B, Hofmann J, Cunningham NR, Bistline EM et al (2017) Sedimentation velocity analysis with fluorescence detection of mutant huntingtin exon 1 aggregation in Drosophila melanogaster and *Caenorhabditis elegans*. Biochemistry 56:4676–4688
9. Kokona B, May CA, Cunningham NR, Richmond L, Garcia FJ, Durante JC et al (2016) Studying polyglutamine aggregation in *Caenorhabditis elegans* using an analytical ultracentrifuge equipped with fluorescence detection. Protein Sci 25:605–617

Chapter 8

The Use of High Performance Liquid Chromatography for the Characterization of the Unfolding and Aggregation of Dairy Proteins

Sophie Jeanne Gaspard and André Brodkorb

Abstract

High-performance liquid chromatography (HPLC) is routinely used to identify and characterize proteins. HPLC can help to understand protein aggregation processes in dairy products, which are induced by common industrial processing steps such as heat treatment. In this chapter, three complementary chromatographic methods are described, which are based on the principles of size exclusion and reversed-phase chromatography. These methods are used to determine the degree of denaturation and aggregation of proteins, and estimate the molecular weight of these aggregates.

Key words HPLC, Proteins, Denaturation, Aggregation, Reversed-phase chromatography, Size exclusion chromatography

1 Introduction

High-performance liquid chromatography, abbreviated HPLC, is a routine technique developed in the 1960s to purify and analyze polar molecules with a high molecular weight in less than 1 h [1]. Thanks to significant improvements in chromatography matrices and the packing of columns, HPLC can now be used as a tool to analyze peptides, proteins, and biopolymers with great accuracy and reproducibility. The characterization of proteins, in particular protein unfolding and aggregation, is of great importance in the field of biochemistry, but also widely used in food and biomaterial sciences. This chapter describes the use of HPLC in dairy chemistry, in particular for the characterization of the state of dairy proteins (native, unfolded, aggregated) due to common process-induced changes during food production (heat treatment, high pressure, concentration, dehydration, change in acidity and ionic strength, etc.). Chromatographic separation is based on the size (gel permeation or size exclusion chromatography, SEC-HPLC) or

Jennifer J. McManus (ed.), *Protein Self-Assembly: Methods and Protocols*, Methods in Molecular Biology, vol. 2039, https://doi.org/10.1007/978-1-4939-9678-0_8, © Springer Science+Business Media, LLC, part of Springer Nature 2019

polarity (reversed-phase chromatography, RP-HPLC) of the protein material. By combining these methods, a detailed characterization of the extent of protein denaturation and aggregation is possible [2–4]. The chromatography results contribute to the overall kinetic and structural understanding of heat-induced changes in the structure of dairy proteins, which is of high scientific and industrial interest. A summary of other methods for the quantification of dairy proteins can be found elsewhere [5].

Three complementary HPLC methods are described in this chapter:

Method 1: RP-HPLC method for the total quantification of dairy proteins (caseins and whey proteins, native and aggregated), based on a method by Visser et al. [6]. It allows the quantification of individual proteins, including those in aggregates. Sample treatment involves the disruption of the intermolecular disulfide bonds and noncovalent interactions by β-mercaptoethanol and urea [7].

Method 2: RP-HPLC method for the quantification of native whey proteins based on a method by Beyer et al. [8, 9]. Sample treatment involves the isoelectric precipitation and removal of denatured whey proteins. The degree of protein denaturation can be calculated from the difference between the total amount of proteins and that of the native proteins. This is only applicable for whey proteins, which can unfold and denature because of their globular structure.

Method 3: SEC-HPLC method for the estimation of the molecular weight of the proteins and protein aggregates. The method is suitable for molecular weight ranges between approximately 10^4 and 5×10^5 Da, depending on the choice of chromatography column.

2 Materials

All solutions should be prepared using ultrapure water, such as Milli-Q® water and analytical or, if available, HPLC-grade reagents. All solutions containing acetonitrile (ACN) or trifluoroacetic acid (TFA) must be prepared in a fume hood using the correct PPE (lab coat, lab goggles, and appropriate gloves).

(a) *Preparation of mobile phase A (for method 1), 10% (v/v) acetonitrile (ACN) + 0.1% (v/v) trifluoroacetic acid (TFA) in water.* Carefully pour 200 mL of ACN into a 2 L volumetric flask. Add around 1500 mL of water and 2 mL of TFA (*see* **Note 1**). Invert the solution to thoroughly mix the organic phase and water. Fill up to 2 L with water. Rinse a filtration vessel and a 2 L glass bottle with a small amount of the filtered mobile phase and return it to the unfiltered buffer. Vacuum-filter the mobile phase with a 0.45 μm pore size hydrophilic filter (e.g., Durapore hydrophilic PVDF membrane filter type, Merck Millipore). Store at room temperature (*see* **Note 2**).

(b) *Preparation of mobile phase B (for method 1 and 2), 90% (v/v) ACN + 0.1% (v/v) trifluoroacetic acid (TFA) in water.* First add 200 mL of water to a 2 L volumetric flask (*see* **Note 3**). Carefully add 2 mL of TFA and slowly add ACN up to 10 cm below the fill line. Invert the solution to mix thoroughly. Wait 20 min to fully equilibrate and fill up to the 2 L mark (*see* **Note 4**). Rinse a filtration vessel and a 2 L glass bottle with a small amount of the filtered mobile phase and return it to the unfiltered buffer. Vacuum-filter the mobile phase with a 0.45 μm pore size hydrophobic filter (e.g., Durapore hydrophobic PVDF membrane filter type, Merck Millipore). Store at room temperature (*see* **Note 2**).

(c) *Preparation of mobile phase A (for method 2), 0.1% (v/v) trifluoroacetic acid (TFA) in water.* Mix approximately 1800 mL of water and 2 mL of TFA (*see* **Note 1**). Make up to 2 L with water. Invert the solution to thoroughly mix the water and TFA. Rinse a filtration vessel and a 2 L glass bottle with a small amount of the filtered mobile phase and return it to the unfiltered buffer. Vacuum-filter the mobile phase with a 0.45 μm pore size hydrophilic filter (e.g., Durapore hydrophilic PVDF membrane filter type, Merck Millipore). Store at room temperature (*see* **Note 5**).

(d) *Preparation of the mobile phase (for method 3), 20 mM sodium phosphate, pH 7.0.* Prepare 1 L of 20 mM monobasic sodium phosphate (NaH_2PO_4) and 1 L of 20 mM dibasic sodium phosphate (Na_2HPO_4). Add solid sodium azide to reach a concentration of 0.05% (w/v) in both solutions to inhibit undesirable microbial growth. Add 900 mL of 20 mM dibasic sodium phosphate to a 2 L beaker and stir continuously. Slowly add 20 mM monobasic sodium phosphate until pH 7.0 is reached. Rinse a filtration vessel and a 2 L glass bottle with a small amount of the filtered mobile phase and return it to the unfiltered buffer. Vacuum-filter the mobile phase through a 0.45 μm pore size hydrophilic filter (e.g., Durapore hydrophilic PVDF membrane filter type, Merck Millipore). Store at room temperature (*see* **Note 6**).

(e) *0.1 M Sodium acetate/acetic acid buffer pH 4.6.* Prepare 0.1 M of sodium acetate with water in 500 mL volumetric flask, in the fume hood. Prepare 0.1 M acetic acid with water in a 500 mL volumetric flask, in the fume hood. Transfer 400 mL of 0.1 M acetic acid solution to a 1 L beaker and slowly add 0.1 M sodium acetate until pH 4.6 is reached. Store at room temperature (*see* **Note 7**).

Table 1
Protein standards for SEC-HPLC of protein aggregates on a TSK Gel G2000SW$_{XL}$ and a TSK Gel G3000SW$_{XL}$ in series (Tosoh Bioscience GmbH, Griesheim, Germany)

Protein	Molecular weight (Da)
Blue dextran	>2,000,000
Thyroglobulin	669,000
Ferritin	440,000
Aldolase	158,000
Bovine serum albumin	66,267
β-Lactoglobulin	18,362
α-Lactalbumin	14,174

Proteins can be purchased as a high molecular weight kit (GE Healthcare, Little Chalfont, UK) in addition to bovine serum albumin, β-lactoglobulin, and α-lactalbumin (Sigma-Aldrich, St. Louis, MO, USA)

(f) *Denaturing sample buffer: 7 M urea + 20 mM bis-tris propane, pH 7.5.* Weigh 42 g of urea ($M_w = 60.06$ g/mol) and 0.56 g of bis-tris propane (1,3-bispropane, $M_w = 282.33$ g/mol) in a glass beaker with 80 mL of water (*see* **Note 8**). Stir and heat gently to aid dissolution. Adjust the pH to 7.5 using 0.1 M HCl or NaOH. Transfer to a 100 mL volumetric flask and rinse the transfer funnel with a small amount of water. Add water to 100 mL. Invert several times to mix thoroughly (*see* **Note 9**).

(g) *Molecular weight standards.* The molecular weight standards (*see* Table 1) are prepared in water.

3 Methods

3.1 Method 1: Quantification of Dairy Proteins Using RP-HPLC

3.1.1 Sample Preparation for Dairy Protein Quantification

Caseins exist in milk as large, colloidal particles (casein micelles, mean diameter ≈150 nm) suspended in the aqueous milk serum, the latter containing whey proteins. Caseins associate via noncovalent interactions [10]. In contrast to this, native whey proteins are in monomeric or dimeric form. Upon heating, whey proteins and caseins can associate via covalent disulfide bonds and other noncovalent interactions. For chromatographic separation, the proteins need to be dissociated and fully denatured prior injection onto the column. The noncovalent interactions and disulfide bonds can be disrupted by pretreating samples with urea and β-mercaptoethanol. In this method, the samples are mixed with the denaturing sample buffer in a ratio of 1:20 (*see* **Note 10**).

1. In order to reach the desired final concentration of 0.2% (w/v) of proteins (*see* **Note 11**), standardize the protein sample to 3.5–4% (w/v) protein. The protein standards, native whey proteins, and caseins are prepared in water.
2. Transfer the volume of sample buffer needed for a sample–buffer volume ratio of 1:20, to a polypropylene tube and add 50 μL of β-mercaptoethanol for every 10 mL of sample buffer.
3. Add 200 μL of each sample to 3.8 mL of urea and β-mercaptoethanol mixture. Vortex the samples. Leave at room temperature for 1 h and invert every 15 min (*see* **Note 12**).
4. Filter the samples through a 0.22 μm low protein binding and hydrophilic syringe filter (e.g., PVDF membrane filter type) into the HPLC vials. Fill to the neck.

3.1.2 HPLC System

The method requires an HPLC separation module with a UV/visible detector and the corresponding software for data analysis.

The results were obtained here using a Poroshell 300SB-C18 column measuring 2.1 × 75 mm from Agilent (Santa Clara, CA, USA).

One chromatographic run takes 35 min per sample at a flow rate of 0.5 mL/min. The injection volume is 5 μL. The column temperature is set at 35 °C.

3.1.3 HPLC Run and Analysis of the Elution Profiles

1. Equilibrate the column with 2–5 column volumes of a mobile phase mixture of 74% mobile phase A and 26% of mobile phase B at a flow rate of 0.5 mL/min. The absorbance is recorded at 214 and 280 nm (*see* **Note 13**). After equilibration, the absorbance should be constant and changes in absorption close to $\pm 10^{-5}$ AU; extend the equilibration if necessary.
2. Set up the HPLC instrument method to run the gradient detailed in Table 2, at a flow rate of 0.5 mL/min and at a column temperature of 35 °C.
3. Inject 5 μL of a blank (water or mobile phase) at the beginning and end of each set of samples to verify a clean baseline. Inject 5 μL of the samples and the standards. The order for the injection should follow an increasing protein concentration to reduce the risk of cross-contamination.
4. Compare the elution time of the standards to the elution time of the unknown proteins to identify the peaks. Use the software functions to integrate the individual peaks and deduce the protein content of each protein from a calibration curve for each protein standard. Anticipated elution profiles of caseins and whey proteins at 214 nm are shown in Fig. 1.

Table 2
Gradient of elution for the separation of caseins and whey proteins on a Poroshell 300SB-C18 column (Agilent, Santa Clara, CA, USA)

Time (min)	% A	% B
0.0	74	26
10.0	63	37
23.0	55	45
26.0	0	100
29.5	0	100
32.5	74	26
35.0	74	26

Solvent A: 10% (v/v) ACN in 0.1% (v/v) TFA; solvent B: 90% (v/v) ACN in 0.1% (v/v) TFA

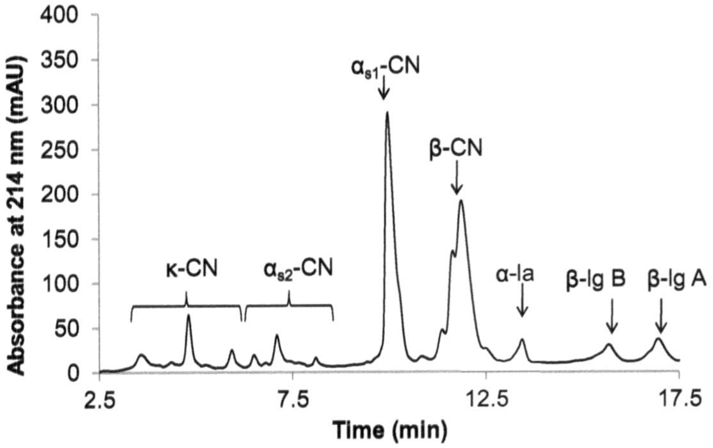

Fig. 1 Anticipated results (method 1) of elution of κ-casein (κ-CN), α_{s2}-casein (α_{s2}-CN), α_{s1}-casein (α_{s1}-CN), β-casein (β-CN), α-lactalbumin (α-la), β lacto-globulin (β-lg A and B) from skim milk on a Poroshell 300SB-C18 column (Agilent, Santa Clara, CA, USA) at flow rate of 0.5 mL/min. The mobile phase A was 10% (v/v) acetonitrile (ACN) + 0.1% (v/v) trifluoroacetic acid (TFA) in water and the mobile phase B was 90% (v/v) ACN + 0.1% (v/v) TFA in water

3.2 Method 2: Quantification of Whey Protein Denaturation by RP-HPLC

3.2.1 Sample Preparation

1. Dilute the protein standards in ultrapure water.
2. Dilute the protein samples in sodium acetate/acetic acid buffer at pH 4.6 to reach a protein concentration of 0.25% (w/v) (*see* **Note 11**). Separate the isoelectric precipitate by centrifugation at 14,000 × *g* for 30 min at room temperature (*see* **Note 14**).
3. If dilution of the supernatant is necessary, dilute in a mixture of 80% mobile phase A and 20% of mobile phase B (*see* **Note 15**).

4. Discard the pellet and filter the supernatant and the protein standards through 0.45 μm low protein binding and hydrophilic syringe filter (e.g., PES membrane filter type, Sartorius, Göttingen, Germany) directly into the HPLC vials. Fill to the neck of the vial.

3.2.2 HPLC System

The method requires an HPLC separation module with a UV/visible detector and the corresponding software for data analysis.

The results were obtained using a C5 PolymerX RP1 column (*see* **Note 16**) measuring 150 × 4.6 mm from Phenomenex (Torrance, CA, USA). One chromatographic run takes 45 min per sample at a flow rate of 1 mL/min. The injection volume is 20 μL. The column temperature is set at 28 °C.

3.2.3 HPLC Run and Analysis of the Elution Profiles

1. Equilibrate the column with 2–5 column volumes of a mobile phase mixture of 80% mobile phase A and 20% of mobile phase B with a flow rate of 1 mL/min. The absorbance is recorded at 214 and 280 nm (*see* **Note 13**). After equilibration, the absorbance should be constant and changes in absorption close to $\pm 10^{-5}$ AU; extend the equilibration if necessary.
2. Set up the HPLC instrument method to run the gradient detailed in Table 3, at a flow rate of 1 mL/min and at a temperature of 28 °C.
3. Inject 20 μL of a blank (water or mobile phase) at the beginning and end of each set of samples to verify a clean baseline. Inject 20 μL of the samples and the standards. The order for the injection should follow an increasing protein concentration to reduce the risk of cross-contamination.

Table 3
Gradient of elution for the separation of native whey proteins on a C5 PolymerX RP1 column (Phenomenex, Torrance, CA, USA)

Time (min)	% A	% B
0	80	20
3	80	20
13	60	40
33	40	60
35	0	100
40	0	100
40.5	80	20
45	80	20

Solvent A: 0.1% (v/v) TFA; solvent B: 90% (v/v) ACN in 0.1% (v/v) TFA

4. Compare the elution time of the standards to the elution time of the unknown proteins to identify the peak. Use the software functions to integrate the individual peaks and deduce the protein content of each protein from a calibration curve for each protein standard. The amount of denatured protein is calculated as the difference between the initial amount of non-heated protein samples and the residual amount after heating, both determined by this method. Alternatively, the total (native + denatured) amount of protein can be determined by the method described in Subheading 3.1.
5. Anticipated elution profiles of caseinomacropeptide (CMP), α-lactalbumin (α-la), and β-lactoglobulin (β-lg) at 280 nm and 214 nm are shown in Fig. 2.

3.3 Method 3: Determination of the Degree of Protein Aggregation by SEC-HPLC System

Two columns in series, TSK Gel G2000SW$_{XL}$ and TSK Gel G3000SW$_{XL}$ (Tosoh Bioscience GmbH, Griesheim, Germany) are used, preceded by a guard column to prevent potential column blockage. The dimensions of both columns are 7.8 × 300 mm (*see* **Note 17**).

The method requires an HPLC separation module, a UV/visible detector and the corresponding software for the elution analysis. The flow rate is 0.5 mL/min with an isocratic gradient of 20 mM sodium phosphate (pH 7.0). The total duration of the run is 60 min per sample. The injection volume is 20 μL. The column should remain at room temperature without the use of a column oven (*see* **Note 18**).

3.3.1 HPLC Run and Analysis of the Elution Profiles

The samples are standardized to 0.25% (w/v) protein in water (*see* **Note 11**) and filtered through 0.45 μm hydrophilic syringe filters with a low protein binding profile (e.g., PES membrane filter type, Sartorius, Göttingen, Germany). The molecular weight standards were prepared as described in Subheading 2.

1. Equilibrate the column with 2 column volumes of 20 mM of sodium phosphate (pH 7.0; 0.05%, w/v, sodium azide) buffer at a flow rate of 0.5 mL/min. The absorbance is recorded at 214 and 280 nm (*see* **Note 13**). After equilibration, the absorbance at 280 nm should be constant and changes in absorption close to $\pm 10^{-5}$ AU; extend the equilibration if necessary.
2. Set up the HPLC instrument method to run an isocratic gradient of 20 mM sodium phosphate (pH 7.0) at a flow rate of 0.5 mL/min.
3. Inject 20 μL of a blank (water or mobile phase) at the beginning and end of each set of samples to verify a clean baseline. Then, inject 20 μL of the samples and the standards. The order for the injection should follow an increasing protein concentration to reduce the risk of cross-contamination.

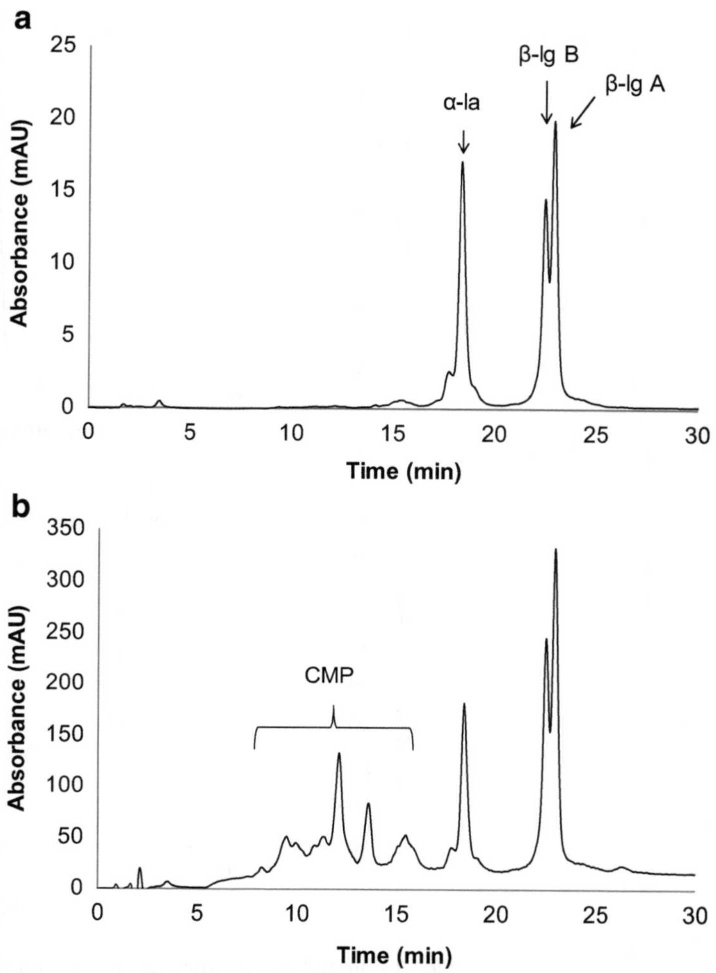

Fig. 2 Anticipated chromatograms (method 2) of native caseinomacropeptide (CMP), α-lactalbumin (α-la), and β-lactoglobulin (β-lg A and B) on a C5 PolymerX RP1 column (Phenomenex, Torrance, CA, USA) at a flow rate of 1 mL/min, detected at 280 nm (**a**) and 214 nm (**b**). The mobile phase A was 0.1% (v/v) TFA in water and the mobile phase B was 90% (v/v) ACN + 0.1% (v/v) TFA in water

4. Calculate the partition coefficient K_{av} of the standards.

   K_{av} is expressed as:

$$K_{av} = (V_e - V_0)/(V_c - V_0),$$

   where V_e is the volume at which the peak was eluted, V_0 is the exclusion volume, and V_c is the volume of the column. V_0 is the elution volume of the blue dextran.

   Plot K_{av} against the logarithm of the molecular weight of the standards.

   Calculate K_{av} of the sample peaks and deduce the molecular weight of the proteins and aggregates using the calibration curve.

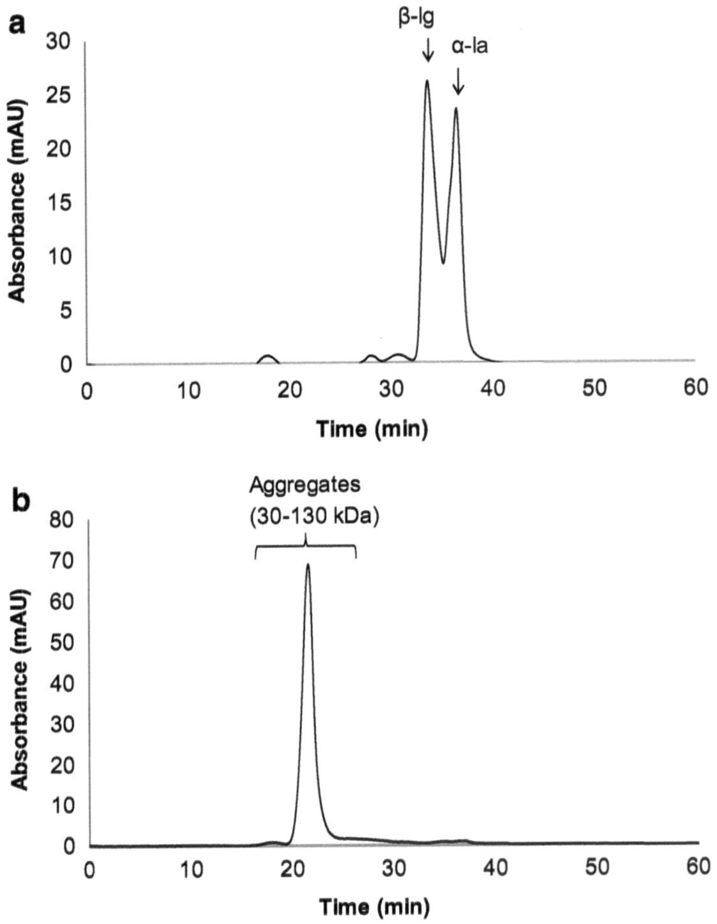

Fig. 3 Chromatograms (method 3) showing the typical profile of (**a**) the whey proteins α-lactalbumin (α-la) and β-lactoglobulin (β-lg), and (**b**) heat-induced aggregates of α-lactalbumin and β-lactoglobulin on TSK Gel G2000SW$_{XL}$ and TSK Gel G3000SW$_{XL}$ in series (Tosoh Bioscience GmbH, Griesheim, Germany) eluted at a flow rate of 0.5 mL/min by SEC-HPLC. The mobile phase was 20 mM sodium phosphate (pH 7.0)

5. An anticipated elution profile of α-lactalbumin (α-la), β-lactoglobulin (β-lg), and heat-induced aggregates of whey proteins at 280 nm is shown in Fig. 3.

4 Notes

1. TFA (trifluoroacetic acid) is an anionic ion-pairing agent interacting with the stationary phase of the column and with the positively charged portions of hydrophilic proteins and peptides, affecting their retention time. TFA is also UV-transparent, which makes it a suitable additive to HPLC

solvents. TFA is very volatile; it is recommended to first add the acetonitrile or water and then the TFA when preparing the mobile phase to avoid loss. Due to its acute toxicity, it must be handled in the fume hood while wearing the appropriate PPE.

2. Buffers containing acetonitrile are very stable and no microbial growth is expected. Thus, the addition of sodium azide is not necessary and the mobile phase can be used for up to 1 year in an air-tight bottle.
3. The addition of 10% (v/v) water reduces the differences in viscosity of the two mobile phases (organic and aqueous) and improves mixing in the HPLC separation module before entering the column.
4. Mixing acetonitrile with water causes an endothermic reaction and a cooling of the mobile phase can be observed. Waiting for the solution to reach room temperature minimizes error in the volume adjustment, thereby improving reproducibility.
5. The addition of TFA reduces the pH and also limits the risk of microbial growth. Thus, the aqueous mobile phase containing TFA can be used for several months after preparation; the addition of sodium azide is not necessary.
6. Sodium phosphate buffer containing 0.05% (w/v) sodium azide can be used for up to one month after preparation.
7. Sodium acetate/acetic acid buffer can be used within a few months due to the low pH of the buffer.
8. Guanidine hydrochloride (6 M) and dithiothreitol (19.5 mM) can be used as denaturing and reducing agent instead of urea and β-mercaptoethanol to improve the separation of some of the proteins [11, 12].
9. The buffer should be freshly prepared and cannot be stored for long due to the high concentration of urea. The prolonged storage of urea leads to the formation of crystals.
10. In case of samples with a low protein concentration, a ratio sample: denaturing buffer of 1:4 can be used.
11. The adjustment of the protein concentration to around 0.25% (w/v) is an indicative figure. We observed a reasonable separation at this protein concentration, but this can be adjusted if necessary.
12. Urea denatures the proteins by disrupting hydrogen bonds. This requires a high concentration of urea. Without heating, β-mercaptoethanol requires more time to reduce the disulfide bonds of the proteins.

13. The choice of the adequate wavelength of detection can be made prior chromatographic separation by measuring an absorption spectrum of the sample with a UV/Vis spectrophotometer. Most proteins and peptides contain aromatic amino acids that absorb at 280 nm. For some polypeptides, such as caseinomacropeptide, and generally shorter peptides, a detection wavelength of 214 nm is recommended, which corresponds to the absorption by the peptide bonds.
14. Centrifugation of the proteins at pH 4.6 allows the isoelectric precipitation of aggregated and denatured whey proteins. The centrifugation speed and duration are chosen to minimize the loss of native proteins in the pellet.
15. Using the original composition of the mobile phase for diluting the sample prevents a potential precipitation of the proteins in the column during the HPLC run. Precipitation of proteins in a chromatography column can cause irreversible blockages and damage to the stationary phase. Measuring the protein content before and after filtration is recommended to verify their solubility. Native proteins can be rehydrated in water unless precipitation is expected in the mobile phase.
16. C4 columns [13] or a PLRP-S column from Latek (Eppelheim, Germany) can also be used as an alternative [14]. The elution gradient and acetonitrile/water ratio of the mobile phases were, in both case, slightly modified.
17. The use of two columns in series increases the number of theoretical plates and thus the quality of the separation. Using two G2000 columns of 300 mm or one G2000 column of 600 mm gives good results to analyze the disappearance of native whey proteins and appearance of soluble aggregated proteins during heat treatment. The exclusion volume of the G3000 column corresponds to a higher molecular weight (5×10^5 Da) and allows the detection of larger aggregates. The OHpak SB-806 HQ-type column from Shodex (Tokyo, Japan), can separate even larger aggregates (exclusion volume corresponding to 2×10^7 Da). It is noteworthy that these SEC columns are very stable if treated with care.
18. The results obtained by SEC-HPLC are less sensitive to temperature variations because hydrophobic interactions with the stationary phase are minimal, contrary to RP-HPLC.

Acknowledgments

The authors would like to thank Bernard Corrigan for providing an example chromatogram of skim milk. This work was supported by the Dairy Levy Research Trust (project MDDT6261 "ProPart"), the Food Institutional Research Measure (FIRM, project no. 08RDTMFRC650) and Enterprise Ireland as part of the Food for Health Ireland Project (Grant Number CC20080001). S.J. Gaspard was funded under the Teagasc Walsh Fellowship Scheme (reference number 2012211).

References

1. Karger BL (1997) HPLC: early and recent perspectives. J Chem Educ 74(1):45–48
2. O'Loughlin IB, Murray BA, Kelly PM, FitzGerald RJ, Brodkorb A (2012) Enzymatic hydrolysis of heat-induced aggregates of whey protein isolate. J Agric Food Chem 60 (19):4895–4904
3. Kehoe JJ, Wang L, Morris ER, Brodkorb A (2011) Formation of non-native β-lactoglobulin during heat-induced denaturation. Food Biophys 6(4):487–496
4. Buggy AK, McManus JJ, Brodkorb A, Carthy NM, Fenelon MA (2016) Stabilising effect of α-lactalbumin on concentrated infant milk formula emulsions heat treated pre- or post-homogenisation. Dair Sci Technol 96:1–15
5. Dupont D, Croguennec T, Brodkorb A, Kouaouci R (2013) Quantitation of proteins in milk and milk products. In: McSweeney PLH, Fox PF (eds) Advanced dairy chemistry: volume 1A: proteins: basic aspects, 4th edn. Springer, Boston, MA, pp 87–134
6. Visser S, Slangen CJ, Rollema HS (1991) Phenotyping of bovine milk proteins by reversed-phase high-performance liquid chromatography. J Chromatogr A 548(Suppl C):361–370
7. Mounsey JS, O'Kennedy BT (2009) Stability of β-lactoglobulin/micellar casein mixtures on heating in simulated milk ultrafiltrate at pH 6.0. Int J Dairy Technol 62(4):493–499
8. Beyer HJ, Kessler HG (1989) Bestimmung des thermischen Denaturierungverhaltens von Molkenproteinen HPLC. GIT Suppl Lebensmittel 2:22–26
9. Tolkach A, Steinle S, Kulozik U (2005) Optimization of thermal pretreatment conditions for the separation of native α-lactalbumin from whey protein concentrates by means of selective denaturation of β-lactoglobulin. J Food Sci 70(9):E557–E566
10. Fox PF, Brodkorb A (2008) The casein micelle: historical aspects, current concepts and significance. Int Dairy J 18(7):677–684
11. Bobe G, Beitz DC, Freeman AE, Lindberg GL (1998) Separation and quantification of bovine milk proteins by reversed-phase high-performance liquid chromatography. J Agric Food Chem 46(2):458–463
12. Bobe G, Beitz DC, Freeman AE, Lindberg GL (1998) Sample preparation affects separation of whey proteins by reversed-phase high-performance liquid chromatography. J Agric Food Chem 46(4):1321–1325
13. Croguennec T, O'Kennedy BT, Mehra R (2004) Heat-induced denaturation/aggregation of β-lactoglobulin A and B: kinetics of the first intermediates formed. Int Dairy J 14 (5):399–409
14. Thomä C, Krause I, Kulozik U (2006) Precipitation behaviour of caseinomacropeptides and their simultaneous determination with whey proteins by RP-HPLC. Int Dairy J 16 (4):285–293

Chapter 9

Differential Scanning Calorimetry to Quantify Heat-Induced Aggregation in Concentrated Protein Solutions

Matthew R. Jacobs, Mark Grace, Alice Blumlein, and Jennifer J. McManus

Abstract

Differential scanning calorimetry (DSC) is an important technique to measure the thermodynamics of protein unfolding (or folding). Information including the temperature for the onset of unfolding, the melt transition temperature (T_m), enthalpy of unfolding (ΔH), and refolding index (RI) are useful for evaluating the heat stability of proteins for a range of biochemical, structural biology, industrial, and pharmaceutical applications. We describe a procedure for careful sample preparation of proteins for DSC measurements and data analysis to determine a range of thermodynamic parameters. In particular, we highlight a measure of protein refolding following complete thermal denaturation (RI), which quantifies the proportion of protein lost to irreversible aggregation after thermal denaturation.

Key words Differential scanning calorimetry, Melt transition temperature, Enthalpy, Refolding index, Protein aggregation, Thermal denaturation

1 Introduction

Protein aggregation is a complex process that occurs by a range of different mechanisms including protein self-association, chemical denaturation, high pressure, interfacial mediated denaturation and high temperature [1–7]. However, aggregation resulting from heat-induced protein denaturation is perhaps the most well studied of these mechanisms [8–13]. The degree of aggregation caused by thermal stress for a given protein concentration can change depending on the solution conditions such as the buffer type and concentration, pH, the presence of inorganic salts or organic modifiers, and total ionic strength [14–16]. The structural and thermal stability of proteins and other biomolecules is often evaluated using DSC, which measures the energy required to increase the temperature of a solution, which provides information on the structural changes that occur during thermally induced protein unfolding [17–19]. DSC provides a range of information including the onset temperature for unfolding (T_{onset}), the melt transition

Jennifer J. McManus (ed.), *Protein Self-Assembly: Methods and Protocols*, Methods in Molecular Biology, vol. 2039, https://doi.org/10.1007/978-1-4939-9678-0_9,

temperature (T_m). Furthermore, DSC provides information on the reversibility of thermal denaturation in the form of the refolding index (RI) [14, 15] and other thermodynamic parameters including the enthalpy associated with protein unfolding (ΔH), the van't Hoff enthalpy (ΔH_{VH}) which provides information on whether unfolding is proceeding via a two-state mechanism and finally the entropy (ΔS) associated with thermal transitions. The thermodynamic parameters measured using DSC can provide useful information about intermolecular and intramolecular interactions between proteins in solution, which impact on the protein thermal stability and the nature of irreversible aggregation that often follows thermal denaturation. These experiments are valuable for optimizing solution conditions for the storage and use of proteins in a variety of fields. In the food industry, the high temperature treatment of dairy products with pasteurization and high temperature processing of food and beverage products which can impact the quality (in terms of taste and nutrition) and shelf life of products [20, 21]. Similarly, DSC is a method used in gaining insight into the heat-induced aggregation of biopharmaceuticals such as monoclonal antibodies, which can cause reduction in the efficacy of these treatments and immunogenicity [22–26]. Optimizing food and pharmaceutical product stability are typically performed by screening a broad range of product conditions where parameters such as salt type, salt concentration, pH, and ionic strength are varied while measuring DSC parameters to determine the environment that maximizes the heat stability of the product. This section details a procedure for measuring a range of DSC parameters using the model protein lysozyme. The thermal behavior of lysozyme has been extensively characterized in a wide variety of solution conditions, and serves as a useful example of the range of the impact that solution conditions can have on DSC parameters [12, 14, 19, 27–29].

2 Materials

1. All solutions should be prepared using ultrapure water (conductivity of 18 MΩ cm^{-1} at 25 °C), which is filtered through a 0.22 μm filter prior to use and all reagents should be analytical grade.
2. Prepare a 10 mM sodium phosphate solution in a 100 ml volumetric flask and adjust to pH 5.
3. Filter the sodium phosphate solution through a 0.22 μm syringe-driven filter into a clean 100 ml glass bottle before use (*see* **Note 1**).
4. A high purity (>99%) source of lysozyme was used as the protein for the present method (*see* **Note 2**), [14, 15, 19, 27, 30].

5. Large volume stainless steel, sealed pans should be used for DSC measurements, since these pans can facilitate up to 60 μl of sample and can be used with the majority of DSC instruments.
6. For the following example a Perkin Elmer Pyris 6 DSC was used, however equivalent DSC instruments from other vendors are also acceptable. Data analysis was performed with Origin 2018b.

3 Methods

3.1 Sample Preparation

1. Weigh 110 mg of lysozyme into a 1.5 ml Eppendorf tube and dissolve the protein in 1 ml of deionized water. Allow the protein to hydrate for 1 h.
2. Filter the sample through a 0.22 μm syringe-driven filter (non-protein binding hydrophilic filter) into a fresh 1.5 ml Eppendorf tube to remove any particulates.
3. It is important to ensure that the pH and ionic strength of the protein solution is accurate. Lyophilized protein will often contain coprecipitated salts, which may affect the pH or ionic strength of solution. This should be minimized by exhaustive dialysis of the protein against the desired buffer. A quick and convenient way to do this is using ultrafiltration.
4. Transfer the sample to a 4 ml centrifugal ultrafiltration device with a molecular weight cutoff of 10 kDa (*see* **Note 3**).
5. Add an additional 3 ml of prefiltered ultrapure water using a micropipette, mixing gently with a pipette (*see* **Note 4**). Centrifuge the sample at 7000 × *g* (for a 23° fixed angle rotor) for 30 min. Repeat this step two additional times for a total of three rinses of lysozyme with deionized water to ensure that all buffers, salts and other modifiers are removed from the sample (*see* **Note 5**).
6. Dilute the protein sample in the ultrafiltration device to a volume of 4 ml with 10 mM sodium phosphate, pH 5. Centrifuge the ultrafiltration tube at 7000 × *g* for approximately 30 min. Repeat this step two additional times for a total of three rinses of protein with buffer (*see* **Note 6**).
7. Concentrate the sample in the ultrafiltration device until the volume is approximately 200 μl.
8. Dilute 1 μl of the concentrated protein in 999 μl of 10 mM sodium phosphate pH 5 and accurately determine the protein concentration using UV-Vis spectroscopy and a mass extinction coefficient for lysozyme of 2.64 ml/mg/cm [31].

The protein concentration can be calculated using the Beer-Lambert law, $A = \varepsilon cl$.

9. Transfer the protein from the ultrafiltration device to a 250 μl Eppendorf tube and adjust the volume with 10 mM sodium phosphate buffer, pH 5 to achieve a protein concentration of 100 mg/ml (*see* **Note 7**).
10. Measure the sample pH and adjust the value to pH 5 if required using sodium hydroxide or hydrochloric acid (*see* **Note 8**).
11. Pipette 55 μl of the protein sample into a large volume stainless steel DSC pan (*see* **Note 9**). For concentrated protein solutions, it is best to transfer to the DSC pan using a positive displacement pipette.
12. Pipette 55 μl of buffer *without* protein into a second DSC pan for use as a reference sample.
13. Hermetically seal the sample and buffer DSC pans.

3.2 DSC

1. Place the sample and buffer DSC pans in the appropriate positions in a Perkin Elmer Pyris 6 DSC instrument and set the initial temperature to 25 °C and allow the sample and buffer blank to equilibrate for 5 min.
2. Set the DSC to heat the sample at a forward scan rate of 1 °C/min from 25 to 95 °C, followed by controlled cooling at a scan at a rate of 1 °C/min from 95 to 25 °C. Repeat this temperature scanning program for a total of two forward and two reverse scans, measuring the power required to maintain the sample and reference cells at the programmed temperature across the measurement range.
3. Export the thermograms generated by the DSC instrument for data processing and import into Origin, or another suitable software package, and perform baseline subtraction using a cubic function [32] as shown in Fig. 1 (*see* **Note 10**).
4. For each thermogram, convert the *y*-axis from units of power or heat flow (typically in mW, note 1 mW = 1 mJ/s) to units of kcal/mol/°C.

   This conversion is achieved by dividing the DSC heat flow (*y*-axis) by calculating the number of moles of protein present in the DSC pan (e.g., 55 μl of 100 mg/ml of lysozyme corresponds to a mass of 5.5 mg protein and dividing this mass by the molecular weight of lysozyme, 14,314 g/mol, yields 3.88×10^{-7} mol). Then divide by the DSC temperature scan rate (note 1 °C min^{-1} = 0.0167 °C/s) and finally convert from mJ to kcal by dividing by 2.39×10^{-7} mJ/kcal.
5. Plot temperature (*x*-axis, °C) versus the normalized heat flow (*y*-axis, kcal/mol/°C). The melt transition temperature (T_m) is defined as the peak maximum of the first heating cycle as shown

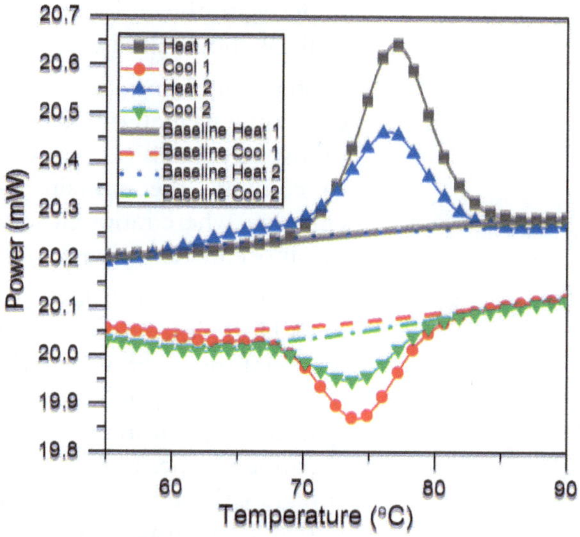

Fig. 1 DSC thermograms of lysozyme, 100 mg/ml in 10 mM sodium phosphate buffer at pH 5 showing two consecutive heating cycles (1 and 2) with both heating and cooling. Baseline correction was performed using a cubic fitting function applied with Origin software

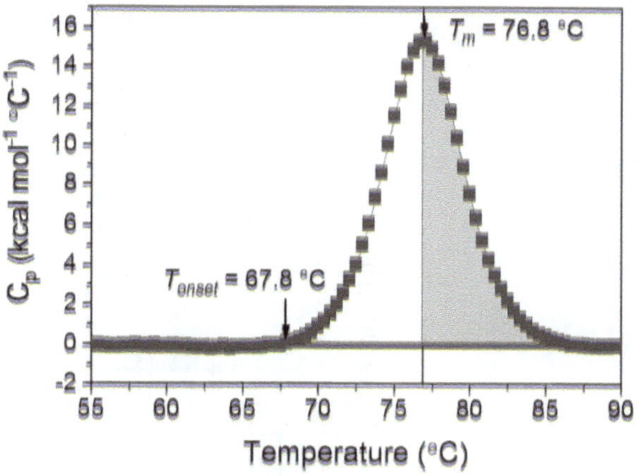

Fig. 2 A baseline subtracted thermogram of the first heating cycle of 100 mg/ml lysozyme in 10 mM sodium phosphate buffer, pH 5. The exothermic peak of the heating cycle is indicative of protein denaturation. The melt transition temperature (T_m = 76.8 °C) and onset of thermal denaturation (T_{onset} = 67.8 °C) are indicated, along with the endothermic portion of the peak that is integrated to determine ΔH

in Fig. 2 (*see* **Note 11**). The onset of protein unfolding (T_{onset}) is determined by measuring the temperature at which protein denaturation begins, which is the inflection point between the baseline of the thermogram and the fitted DSC peak (Fig. 2). For a fully reversible process, the enthalpy (ΔH) required to unfold the protein is obtained by integration of the area underneath the endothermic portion of the DSC thermogram peak, as shown in Fig. 2, and multiplying this quantity by two.

6. Protein unfolding can be a two-stage or multistage process. One way to determine whether the protein unfolds via a two-state process (i.e., folded/native [*N*] or unfolded [*U*]) is to compare the enthalpy of unfolding to the van't Hoff enthalpy, where ratios close to 1 suggest that a two-state transition occurs (*see* **Note 12**).
7. The van't Hoff enthalpy can be determined by constructing a van't Hoff plot. Select five temperatures that cover the range of temperatures over which the thermal unfolding event takes place, as shown in Fig. 3a. It is convenient to select evenly spaced temperatures at equally spaced increments across the thermogram peak (2.5 °C increments are shown in Fig. 3a). Integrate the areas corresponding to unfolded [*U*] and natively folded [*N*] protein at each of the selected temperatures as shown in Fig. 3b, c. Calculate the ratio of unfolded to native protein (K_{eq}) by dividing [*U*] by [*N*] for each of the thermogram temperature divisions as shown in Table 1. Plot the

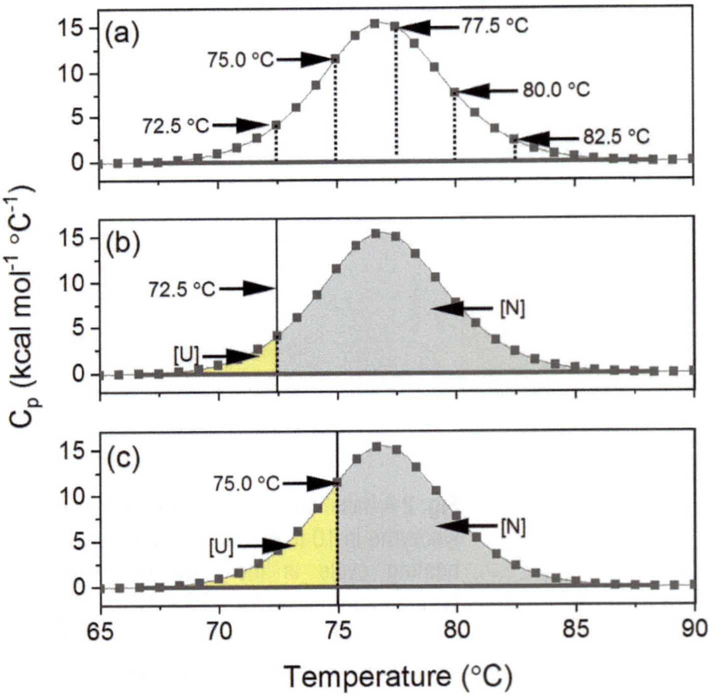

Fig. 3 Diagram showing a DSC peak for 100 mg/ml lysozyme in 10 mM sodium phosphate buffer, pH 5. In (**a**) the peak has been sectioned into six slices at 2.5 °C increments (72.5–82.5 °C). (**b**) shows the integration of the unfolded [*U*] (yellow shaded region) and the native [*N*] (gray-shaded region) areas corresponding to the 72.5 °C peak division. Similarly, (**c**) shows the [*U*] and [*N*] integrated regions for the 75.0 °C peak division

Table 1
Data and calculations for preparation of the van't Hoff plot in Fig. 4

Temperature (°C)	Temperature (K)	1/Temperature (K^{-1})	[*U*] (kcal/mol)	[*M*] (kcal/mol)	ln([*U*]/[*M*])
72.5	345.5	0.00289	6.6	98.4	−2.70
75	348	0.00287	25.1	79.9	−1.16
77.5	350.5	0.00285	60.5	44.5	0.31
80	353	0.00283	89.8	15.3	1.77
82.5	355.5	0.00281	101.4	3.6	3.35

Note that the total area under the thermogram peak was 105 kcal/mol for the 100 mg/ml lysozyme sample in 10 mM sodium phosphate, pH 5

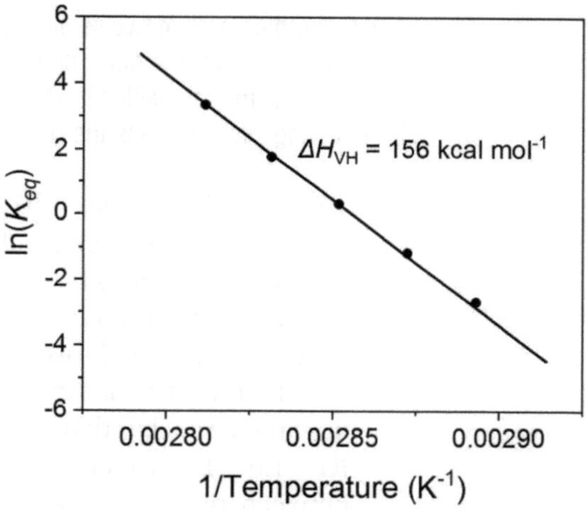

Fig. 4 A van't Hoff plot for lysozyme 100 mg/ml in 10 mM sodium phosphate, pH 5 showing the linear fit of the equilibrium unfolding data ln(K_{eq}) versus reciprocal temperature (1/K), *see* Table 1 for corresponding data

reciprocal of temperature in degrees Kelvin (*x*-axis) for each thermogram division against the natural logarithm of K_{eq} as shown in Fig. 4 and perform linear fitting of the data.

The van't Hoff equation (Eq. 1) is then used to obtain the van't Hoff enthalpy.

$$\ln\left(K_{eq}\right) = \frac{-\Delta H}{RT} + \frac{\Delta S}{R}. \quad (1)$$

The slope of the fitted line, multiplied by −1 and the ideal gas constant, *R*, (1.987×10^{-3} kcal/mol/°C) yields the van't Hoff enthalpy for the protein. A similar van't Hoff enthalpy and integrated enthalpy from the thermogram peak is indicative of a two-state transition.

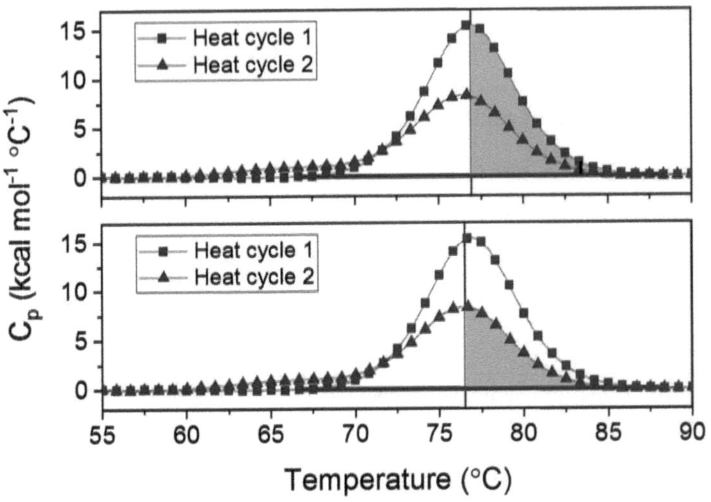

Fig. 5 Thermograms of two sequential heating cycles for a sample of 100 mg/ml lysozyme in 10 mM sodium phosphate pH 5. Integration of the endothermic portion of the melt transition is shown, and the ratio of the areas of heating cycle 2 to heating cycle 1 yields the refolding index (RI)

8. The refolding index (RI) can be calculated if refolding is observed for the protein during consecutive heating cycles, as shown in Fig. 5. This is determined by integrating the area under the thermogram peaks for the first (n) and second ($n + 1$) heating cycles (*see* **Note 13**). Equation 2 is then used to calculate the refolding index by dividing the enthalpy of the second heating cycle by that of the first heating cycle to obtain a RI (Eq. 1), which is the ratio of protein that refolded between those two heating events expressed as a proportion (*see* **Note 14**).

$$\mathrm{RI} = \frac{\Delta H_{\mathrm{me}}(n+1)}{\Delta H_{\mathrm{me}}(n)}. \qquad (2)$$

9. The proportion of protein lost to aggregation during a heating and cooling cycle, can be determined using Eq. 3:

$$\%\mathrm{aggregation} = (1 - \mathrm{RI}) \times 100. \qquad (3)$$

4 Notes

1. If the buffer has been stored for any length of time, check the pH before use.
2. It is essential that the protein being investigated is of high purity, as small amounts of impurities can have a substantial impact on the structural stability (and in particular, the

reversibility of the unfolding process) of the protein evaluated, which directly effects the quality of DSC data and parameters derived from analysis.

3. Select a molecular weight cutoff that is lower than the molecular weight of the protein. Always check the absorbance of the filtrate at 280 nm to ensure that the protein was retained by the membrane of the filter. If protein is detected in the filtrate, select a lower molecular weight cutoff filter.
4. Ultrafiltration should be performed at conditions appropriate to the protein being analyzed. Considerations prior to ultrafiltration include the temperature stability of the protein, propensity to aggregate or precipitate at high concentration, and the impact of centrifugation on solution phase stability. Thermal degradation can be minimized by performing ultrafiltration in a cold room or in a centrifuge equipped with temperature control. Over concentrating some proteins can lead to aggregation or precipitation, which can be controlled by centrifugation at slower rates or for shorter durations. Care should be taken when working with a protein that has not had its stability profile established. Dialysis can be used in place of ultrafiltration for buffer exchange and equilibration at the cost of convenience; however, dialysis does not necessarily provide the capacity for straightforward protein concentration.
5. Proteins in the liquid and solid state are formulated with a range of buffers, salts or organic modifiers to promote stability. The presence of these additives can have a substantial impact on DSC measurements; therefore, it is essential to minimize the impact of undesired additives.
6. Ultrafiltration in should be performed to ensure that the protein has been equilibrated with the desired target buffer. For example, ultrafiltration of 4 ml of sample in *buffer A* using three equilibration steps with *buffer B* where the protein concentrated to 100 μl and then dispersed in 3.9 ml of *buffer B* ensures that is less than 0.01% of *buffer A* remains in the sample.
7. A concentrated sample increases the signal-to-noise ratio of DSC measurements, however not all proteins are soluble at high concentrations. Measurements can be performed at lower concentrations, provided that adequate signal is obtained during a DSC measurement. Ensure that the concentration of protein is accurately known prior to DSC measurement to enable data processing and calculation of ΔH if required.
8. High concentration protein solutions can exceed the buffering capacity of typical buffers, which can lead to a change in pH relative to the initial buffer pH. This can be corrected by adjusting the pH with acid or base as required after performing

the concentration step and then redetermining the protein concentration.

9. DSC sample pans constructed from inert materials such as stainless steel must be used for DSC measurements on proteins. Aluminum pans in particular may react with protein samples to induce aggregation and reduce the sensitivity of DSC, which negatively impacts the quality of results obtained. Use of large volume DSC pans is recommended to increase the signal-to-noise ratio of DSC measurements.

10. Baseline subtraction by fitting and subtracting a cubic function compensates for hysteresis effects that are present in biomolecules like proteins. DSC instruments commonly include software capable of performing cubic baseline subtraction, alternatively many software packages such as Origin, Python, R or MATLAB can be used to perform this step.

11. It is possible for a protein to have multiple DSC peaks that indicate different protein domains unfolding at different temperatures, this behavior is dependent on the protein analyzed, *see* Fig. 6 for an example. Similarly, peak shoulders can indicate the presence of additional unfolding intermediates which preclude a two-state unfolding mechanism.

12. The van't Hoff analysis can only be performed on a DSC thermogram that shows a single thermogram peak, unlike the thermogram shown in Fig. 6, which reveals clear evidence for a multiple stage unfolding pathway.

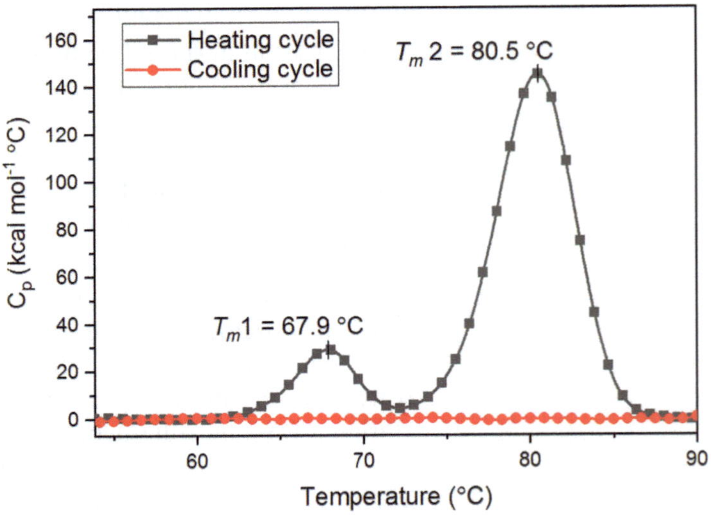

Fig. 6 A thermogram of a monoclonal antibody (100 mg/ml) in 50 mM L-histidine buffer at pH 6, showing the T_m for the unfolding events corresponding to the unfolding of two different domains of this protein at different temperatures (light chains at 67.9 °C and heavy chains at 80.5 °C)

13. If the second heating cycle shows that there is an earlier onset of denaturation (T_{onset}) relative to the first heating cycle, the formation of nonnatively folded protein after the initial heating cycle should be suspected. This contribution to RI calculations is avoided by integrating the second half of the thermogram peak from T_m onward, and then multiplying ΔH obtained by two as shown in Fig. 5.
14. RI can be used to determine the degree of protein refolding between subsequent heating cycles. It is calculated from the ratio of ΔH between two thermal scans (usually the first and second heating events), where an RI of 1 would indicate complete refolding of the protein following a cycle of heating, while an RI of 0 would indicate no protein unfolding was detected during the second thermal cycle; therefore, none of the protein refolded to a native state after the first heating scan.

Acknowledgments

The research work presented in this chapter was made possible through the financial support of Enterprise Ireland under grant number [IP/2015 0358] and the Synthesis and Solid State Pharmaceutical Centre, cofunded under the European Regional Development Fund, and the Department of Agriculture Food and Marine, FIRM grant number [11/F/037], SFI Stokes Lectureship to J.J.M., and Maynooth University Teaching Fellowship to A.B.

References

1. Chi EY, Krishnan S, Randolph TW, Carpenter JF (2003) Physical stability of proteins in aqueous solution: mechanism and driving forces in nonnative protein aggregation. Pharm Res 20:12. https://doi.org/10.1023/A:1025771421906
2. Kamerzell TJ, Esfandiary R, Joshi SB, Middaugh CR, Volkin DB (2011) Protein–excipient interactions: mechanisms and biophysical characterization applied to protein formulation development. Adv Drug Deliv Rev 63:1118–1159. https://doi.org/10.1016/j.addr.2011.07.006
3. Roberts CJ (2014) Protein aggregation and its impact on product quality. Curr Opin Biotechnol 30:211–217. https://doi.org/10.1016/j.copbio.2014.08.001
4. Wu H, Kroe-Barrett R, Singh S, Robinson AS, Roberts CJ (2014) Competing aggregation pathways for monoclonal antibodies. FEBS Lett 588:936–941. https://doi.org/10.1016/j.febslet.2014.01.051
5. Barnett GV, Drenski M, Razinkov V, Reed WF, Roberts CJ (2016) Identifying protein aggregation mechanisms and quantifying aggregation rates from combined monomer depletion and continuous scattering. Anal Biochem 511:80–91. https://doi.org/10.1016/j.ab.2016.08.002
6. Alam P, Siddiqi K, Chturvedi SK, Khan RH (2017) Protein aggregation: from background to inhibition strategies. Int J Biol Macromol 103:208–219. https://doi.org/10.1016/j.ijbiomac.2017.05.048
7. Wang W, Roberts CJ (2018) Protein aggregation – mechanisms, detection, and control. Int J Pharm 550:251–268. https://doi.org/10.1016/j.ijpharm.2018.08.043
8. Privalov PL, Khechinashvili NN (1974) A thermodynamic approach to the problem of stabilization of globular protein structure: a calorimetric study. J Mol Biol 86:665–684. https://doi.org/10.1016/0022-2836(74)90188-0

9. Cooper A (1999) Thermodynamic analysis of biomolecular interactions. Curr Opin Chem Biol 3:557–563. https://doi.org/10.1016/S1367-5931(99)00008-3
10. Johnson CM (2013) Differential scanning calorimetry as a tool for protein folding and stability. Arch Biochem Biophys 531:100–109. https://doi.org/10.1016/j.abb.2012.09.008
11. Bruylants G, Wouters J, Michaux C (2011) Differential scanning calorimetry in life science: thermodynamics, stability, molecular recognition and application in drug design. Curr Med Chem 12:2011–2020. https://doi.org/10.2174/0929867054546564
12. James S, McManus JJ (2012) Thermal and solution stability of lysozyme in the presence of sucrose, glucose, and trehalose. J Phys Chem B 116:10182–10188. https://doi.org/10.1021/jp303898g
13. Rosa M, Roberts CJ, Rodrigues MA (2017) Connecting high-temperature and low-temperature protein stability and aggregation. PLoS One 12:e0176748. https://doi.org/10.1371/journal.pone.0176748
14. Blumlein A, McManus JJ (2013) Reversible and non-reversible thermal denaturation of lysozyme with varying pH at low ionic strength. Biochim Biophys Acta 1834:2064–2070. https://doi.org/10.1016/j.bbapap.2013.06.001
15. Stavropoulos P, Thanassoulas A, Nounesis G (2018) The effect of cations on reversibility and thermodynamic stability during thermal denaturation of lysozyme. J Chem Thermodyn 118:331–337. https://doi.org/10.1016/j.jct.2017.10.006
16. Zbacnik TJ, Holcomb RE, Katayama DS, Murphy BM, Payne RW, Coccaro RC, Evans GJ, Matsuura JE, Henry CS, Manning MC (2017) Role of buffers in protein formulations. J Pharm Sci 106:713–733. https://doi.org/10.1016/j.xphs.2016.11.014
17. Mahler H-C, Friess W, Grauschopf U, Kiese S (2009) Protein aggregation: pathways, induction factors and analysis. J Pharm Sci 98:2909–2934. https://doi.org/10.1002/jps.21566
18. Schön A, Clarkson BR, Siles R, Ross P, Brown RK, Freire E (2015) Denatured state aggregation parameters derived from concentration dependence of protein stability. Anal Biochem 488:45–50. https://doi.org/10.1016/j.ab.2015.07.013
19. Wu S, Ding Y, Zhang G (2015) Mechanic insight into aggregation of lysozyme by ultrasensitive differential scanning calorimetry and sedimentation velocity. J Phys Chem B 119:15789–15795. https://doi.org/10.1021/acs.jpcb.5b08190
20. Peng J, Tang J, Barrett DM, Sablani SS, Anderson N, Powers JR (2017) Thermal pasteurization of ready-to-eat foods and vegetables: critical factors for process design and effects on quality. Crit Rev Food Sci Nutr 57:2970–2995. https://doi.org/10.1080/10408398.2015.1082126
21. Ling B, Tang J, Kong F, Mitcham EJ, Wang S (2015) Kinetics of food quality changes during thermal processing: a review. Food Bioprocess Technol 8:343–358. https://doi.org/10.1007/s11947-014-1398-3
22. Niedziela-Majka A, Kan E, Weissburg P, Mehra U, Sellers S, Sakowicz R (2015) High-throughput screening of formulations to optimize the thermal stability of a therapeutic monoclonal antibody. J Biomol Screen 20:552–559. https://doi.org/10.1177/1087057114557781
23. He F, Woods CE, Trilisky E, Bower KM, Litowski JR, Kerwin BA, Becker GW, Narhi LO, Razinkov VI (2011) Screening of monoclonal antibody formulations based on high-throughput thermostability and viscosity measurements: design of experiment and statistical analysis. J Pharm Sci 100:1330–1340. https://doi.org/10.1002/jps.22384
24. He F, Hogan S, Latypov RF, Narhi LO, Razinkov VI (2010) High throughput thermostability screening of monoclonal antibody formulations. J Pharm Sci 99:1707–1720. https://doi.org/10.1002/jps.21955
25. Harn N, Allan C, Oliver C, Middaugh CR (2007) Highly concentrated monoclonal antibody solutions: direct analysis of physical structure and thermal stability. J Pharm Sci 96:532–546. https://doi.org/10.1002/jps.20753
26. Moussa EM, Panchal JP, Moorthy BS, Blum JS, Joubert MK, Narhi LO, Topp EM (2016) Immunogenicity of therapeutic protein aggregates. J Pharm Sci 105:417–430. https://doi.org/10.1016/j.xphs.2015.11.002
27. Dobson CM, Evans PA, Radford SE (1994) Understanding how proteins fold: the lysozyme story so far. Trends Biochem Sci 19:31–37. https://doi.org/10.1016/0968-0004(94)90171-6
28. Esposito A, Comez L, Cinelli S, Scarponi F, Onori G (2009) Influence of glycerol on the structure and thermal stability of lysozyme: a dynamic light scattering and circular dichroism study. J Phys Chem B 113:16420–16424. https://doi.org/10.1021/jp906739v

29. Meng-Lund H, Friis N, van de Weert M, Rantanen J, Poso A, Grohganz H, Jorgensen L (2017) Correlation between calculated molecular descriptors of excipient amino acids and experimentally observed thermal stability of lysozyme. Int J Pharm 523:238–245. https://doi.org/10.1016/j.ijpharm.2017.03.043
30. Alam MT, Rizvi A, Rauf MA, Owais M, Naeem A (2018) Thermal unfolding of human lysozyme induces aggregation: recognition of the aggregates by antisera against the native protein. Int J Biol Macromol 113:976–982. https://doi.org/10.1016/j.ijbiomac.2018.02.095
31. Sophianopoulos AJ, Rhodes CK, Holcomb DN, van Holde KE (1962) Physical studies of lysozyme: I. Characerization. J Biol Chem 237:1107–1112
32. Harding SE, Chowdhry B (2000) Protein-ligand interactions: hydrodynamics and calorimetry: a practical approach. In: Oxford University Press. Oxford, New York, NY

Chapter 10

Nanoparticle Tracking Analysis to Examine the Temperature-Induced Aggregation of Proteins

Svenja Sladek, Kate McComiskey, Anne Marie Healy, and Lidia Tajber

Abstract

In recent years, nanoparticle tracking analysis (NTA) has emerged as an alternative tool for particle size characterization. Especially when examining polydisperse systems, individual particle to particle tracking allows for higher peak resolution than dynamic light scattering techniques. However, NTA requires an experienced user with a good insight into how the different settings can affect the determination of particle size and size distributions. This chapter provides a guideline for protein aggregation studies using the example of temperature-induced aggregation of IgG at low concentration.

Key words Nanoparticle tracking analysis, Protein aggregation, Size distribution, Polydisperse samples, Particle visualization

1 Introduction

Over the last decade, nanoparticle tracking analysis (NTA) has emerged as an alternative tool for the characterization of submicron-sized particles in a liquid dispersion, by overcoming certain limitations of other particle sizing techniques. NTA combines laser light scattering with a high-resolution camera attached to a microscope. Particles, which are illuminated by the laser, scatter light, enabling their Brownian motion to be recorded in real time. In the processing step, the software then identifies and individually tracks each individual particle in two dimensions. This allows the determination of the particle diffusion coefficient (Dt) and subsequently the sphere-equivalent hydrodynamic diameter (d) of each particle is calculated using the Stokes–Einstein equation, Eq. 1 [1]:

$$\mathrm{Dt} = \frac{Tk_{\mathrm{B}}}{3\pi\eta d} \quad (1)$$

where T is temperature, k_B is Boltzmann's constant, and η is solvent viscosity.

Jennifer J. McManus (ed.), *Protein Self-Assembly: Methods and Protocols*, Methods in Molecular Biology, vol. 2039,
https://doi.org/10.1007/978-1-4939-9678-0_10,

This approach of particle-by-particle tracking is highly advantageous when working with polydisperse samples such as protein dispersions, as it allows a higher peak resolution of particle size and intensity distributions than dynamic light scattering (DLS) [2]. Additionally, the known sample volume allows for the calculation of particle number per ml in the sample. So far, NTA has been successfully employed for characterization of a number of submicron-sized particles such as liposomes [3], extracellular vesicles [4, 5], and protein/enzyme aggregation [2, 6, 7]. Most of the abovementioned NTA studies were conducted in combination with a complementary technique such as DLS or flow field-flow fractionation.

Downsides of the NTA technique include data acquisition and analysis being very dependent on the user's settings selection and experience [2, 4, 5, 8]. Furthermore, particle concentration and size detection limits need to be taken into consideration. Overall, the predominant opinion is that NTA is an information-rich technique requiring an experienced operator, as the chosen settings for video capture and analysis can lead to underrepresentation or overemphasis of certain particle populations.

Here we describe the procedure of measuring the temperature-induced aggregation of IgG, as a model protein, using NTA. It should be noted that capture and analysis settings are extremely equipment, sample, and user dependent. For this reason, information may not be relevant or suitable for other instrument types and all samples under analysis.

2 Materials

The IgG used for this experiment was delivered in TRIS buffer pH 7 (*see* **Note 1**). In order to achieve a higher concentration of IgG in this buffer, the sample was subject to centrifugation-filtration (Amicon filter devices, MWCO 10 kDa). Two concentration cycles were carried out by centrifuging 4 ml of sample at 3220 × *g* (swinging bucket rotor) and 20 °C. At the end of the concentration process the volume was made up with TRIS buffer pH 7 to achieve the protein concentration of interest (5–10 mg/ml). The same method can be employed for buffer exchange. Centrifugation steps need to be repeated approximately 2–3 times until the pH of the protein solution matches the buffer pH (*see* **Note 2**). Due to sample preparation and concentration steps protein aggregates/assemblies can be present in the sample prior to performing the NTA studies. Therefore, the sample needs to be filtered through syringe filters (0.22 μm PES or 0.02 μm Anotop).

1. Prepare all solutions using ultrapure Milli-Q water and analytical grade reagents.

2. Filter appropriately all solutions/liquids used in NTA experiments or for cleaning (*see* **Note 3**).
3. Filters, low protein binding (0.2 μm polyethersulfone (PES) filters or 0.02 μm Anotop filters).
4. Amicon filter devices, 4 ml (MWCO: 10 kDa).
5. 50 mM Tris buffer pH 7 (use HCl for pH adjustment).
6. Centrifuge with a swinging bucket rotor.
7. IgG.
8. pH meter.
9. NanoSight NS300.
10. 10% v/v ethanol in Milli-Q water.
11. 1 ml tubes (Protein LoBind Tubes, Eppendorf).
12. Compressed air duster cans.
13. Lint-free wipes.
14. Cotton buds.
15. Block heater.

3 Method

This method is NanoSight NS300 specific; requiring a certain level of understanding and knowledge regarding the use of the hardware as well as software tools (NTA 3.2). Therefore, first-time users are advised to make themselves familiar with the software (NanoSight NS300 software manual). For this experiment IgG in 50 mM Tris buffer, pH 7 was used as a model protein. Stress conditions consisted of temperature-induced aggregation at 70 °C. However, this method can be applied to various other proteins and stress conditions (such as temperature, pH or stirring) over time (*see* **Note 4**).

3.1 Sample Preparation

1. Place 0.5 ml of the filtered protein solution in a low protein binding Eppendorf tube. Allow one tube per time point that you want to measure.
2. The sample aliquots are then placed in a block heater and heated up to 70 °C prior to the experiment. After the predetermined length of incubation (5, 10, 20, 30, 60, and 120 min), the samples are taken for NTA measurements.
3. Prior to NTA measurements it might be necessary to dilute the sample to achieve a particle concentration of 10^7–10^9 particles/ml (20–100 particles per frame). Depending on sample protein concentration, this may require some initial dilution studies for method optimization (*see* **Note 5**).

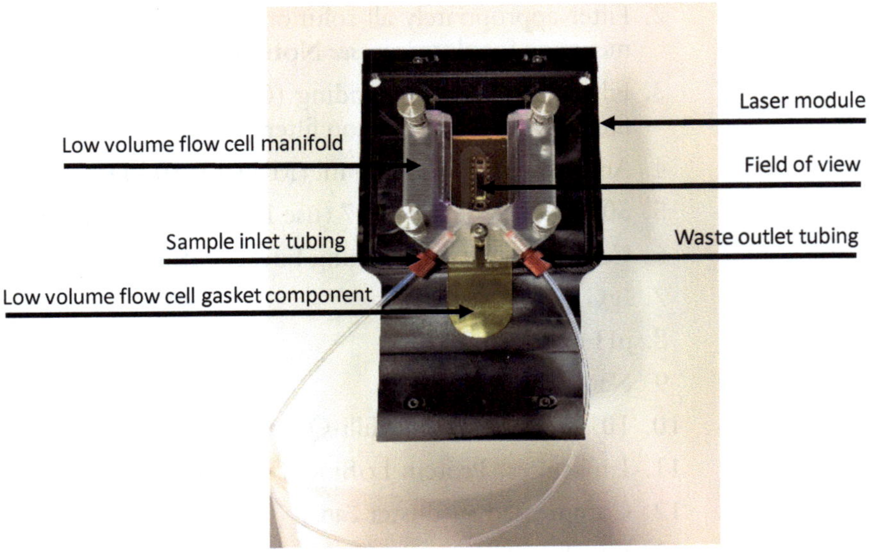

Fig. 1 Low volume flow cell top-plate mounted onto laser module

3.2 Preparation of the NTA Instrument (NanoSight NS300)

1. Switch on the NS300 module before running the NanoSight NTA software. Make sure that the temperature sensor is detected prior to using the equipment, otherwise the temperature and subsequently the viscosity of the solvent is not factored into the size calculations correctly.
2. Assemble the module (Fig. 1) using the flow-cell top plate and connect the tubing as directed in the NanoSight user manual.

3.3 Sample Loading

1. The sample is loaded into the chamber with a 1 ml syringe using the inlet tubing. Before injecting the first sample, the tubing should be primed with buffer. The first sample is best introduced with the module held outside of the instrument, to see and avoid air pockets in the viewing field.
2. Prior to loading sample, ensure all air bubbles are removed from the syringe to avoid interference with measurements. Attach the syringe (not entirely filled) to the Luer outlet and pull slightly on the plunger joining the menisci of sample and buffer. This will be noticeable by an air bubble in the syringe. Then, the sample is slowly introduced into the chamber holding the syringe upright to avoid air bubbles going through the tubing (*see* **Note 6**).
3. Place the chamber (laser module) into the slide and gently push forward until it connects with the power connector inside and turn the red lever until it is in a vertical position. Close the access door.
4. Advance the sample until it is visible in the viewing field (*see* **Note 7**).

3.4 Video Capture

Before capturing the first video of each sample, capture settings need to be optimized. Therefore, after loading the sample and directing the software to start the measurement, an initial live image is displayed. The optimization of the capture settings is an iterative process and may require repeated adjustments (*see* **Note 8**).

1. Under "SOP" (SOP—standard operating procedure; bottom middle of the screen) select "Standard measurement." For temperature-induced aggregation of IgG measurements three captures per sample are chosen with a capture duration of 60 s each. Samples are measured at room temperature; therefore, temperature control and target temperature are not needed (*see* **Note 9**).
2. Optimize the beam position by moving the image up and down using the cursor. Make sure the beam is central in the field of view, so the illuminated particles fill the capture screen.
3. The focus can be adjusted using the focus dial on the NS300 instrument (for coarse adjustments), but also in the software under "Hardware"—"Focus" using the slider control. For the accuracy of the measurement it is important that the particles look as sharp as possible or appear with a spherical halo around them. Figure 2 shows the optimum focus for IgG in buffer (*see* **Note 10**).
4. The "Camera Level" under "Capture" needs to be adjusted so all particles are visible, but the majority of the particles are not saturated (signified by red pixels). There are 16 set camera levels (combination of camera gain and camera shutter speed) ranging from the least sensitive (level 1) to most sensitive (level 16). Proteins have a low refractive index which makes them appear very dim in relation to the image background. In this case there is an option to individually set the camera gain and shutter speed under "Hardware"—"Advanced Camera Settings" in order to optimize contrast between the particles and background. For IgG samples in TRIS pH 7 buffer camera level settings between 10 and 13 were chosen, depending on the incubation time (*see* **Note 11**).
5. The user also needs to make sure that the particle concentration of the sample is within the limits discussed earlier (10^7–10^9 particles/ml). If not, the sample should be diluted with 50 mM Tris, pH 7 or another buffer of choice. This was not necessary for concentrations of around 10 mg/ml IgG in 50 mM Tris, pH 7.
6. Once all settings are optimized, click "ok" and advance the sample for video captures. Avoid touching the syringe or the NanoSight instrument during captures to minimize interference caused by vibration.

7. In between captures advance the sample when prompted by the software. This allows for a good representation of the whole sample. Avoid the application of too much pressure when advancing the sample as this could lead to the destruction of aggregates, but may also break the seal between the top plate and the laser module.
8. Start the next capture once the drift of particles induced by the pressure has stopped.

3.5 Video Processing

By default, the software starts processing the video immediately after finishing all captures.

1. Before processing the video, the detection threshold needs to be set so only distinct particles are detected by the software (as shown in Fig. 3). Blue crosses signify noise and a detection threshold that is too low. Particles without a red cross in the center indicate that the detection threshold is too high. Do not change the detection threshold between captures of the same samples (*see* **Note 12**).
2. Processed data is displayed in three different plots: size versus concentration plot, intensity versus size scatter plot and a 3D plot of size versus concentration versus intensity. Data can be viewed for each single capture, but also as an overlay of all captures and averaged. A right click on each graph gives more display and export options. The scale of the *x*-axis of the 2D plots can be changed by a left click followed by a left or right motion while holding.
3. At the end of the capture processing, data is exported to the base file as a PDF document containing size and intensity distribution as well as all measurement settings. CSV files of data points can be saved to plot data in other graphing software. Captures can be saved as compressed WMV files. Additionally, pictures of each graph can be exported (*see* **Note 13**).

Fig. 2 Image focus. Image **a** and **b** show poorly focused particles due to the stage being too low or too high, respectively. Particles as shown in image **c** are acceptable. Ideally particles should appear as clearly defined spherical shapes

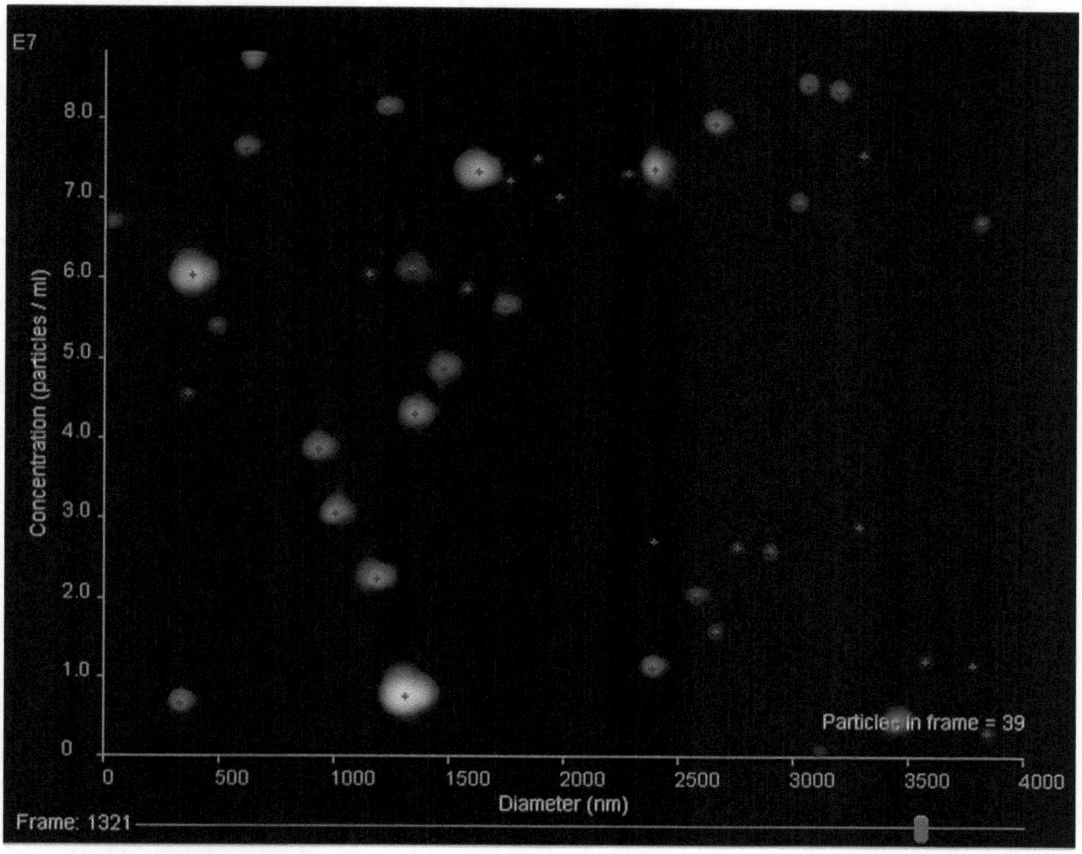

Fig. 3 Video still of 10 mg/ml IgG in 50 mM Tris buffer, pH 7 after 10 min incubation at 70 °C. Detection threshold was set at 23

3.6 Cleaning Procedure

1. When using the flow cell top plate, the module does not have to be disassembled in between samples.
2. Flush the system 3–4 times with a 1 ml syringe using 50 mM Tris, pH 7 (or the buffer of choice) as described in Subheading 3.3, **step 2**.
3. For the final flush, after finishing the study, use water and a 10% v/v ethanol/water mixture (3–4 times each).
4. Disassemble the module and tubing. Clean and dry using lint-free wipes, cotton buds and compressed air cans.

4 Notes

1. When using protein in a solid form, it can simply be dissolved in the buffer of choice at the concentration of interest.

2. Protein concentration in buffer can easily be checked using Bradford or BCA assays. Ensure that the buffer is not interfering with the protein assay.
3. Prior to measurements, it must be verified that all solvents, diluents, or other media are particle free.
4. NTA is able to accurately size particles between 30 and 1000 nm. However, the lower detection limit depends on the refractive index of the particles. This is especially important when examining protein assemblies, as proteins have very low refractive indices. Therefore, NTA is not able to size protein monomers or small oligomers [2].
5. It is important to do this after the incubation, so that the dilution step does not interfere with the aggregation behavior of the protein [2]. Under advanced settings, viscosity and dilution of the sample can be edited for consideration in calculations.

   The presence of fewer particles than the lower concentration limit requires longer capture times to obtain statistically reproducible results. With concentrations above the higher limit, the likelihood of the motion of neighboring particles interfering with each other becomes very high.
6. Air bubbles in the sample chamber can cause high background scatter and sample drift. They can be removed by rinsing the chamber or taking the chamber apart and cleaning it.
7. There should be a noticeable difference between buffer/cleaning solution and sample in the viewing field. At least 0.3 ml should be introduced to avoid dilution of the sample by buffer/cleaning solution.
8. Key parameters for video capture and subsequent analysis are the capture number and duration, gain and shutter speed of the camera (standard settings under camera level) as well as detection threshold [9]. Potential sources of error or incorrect measurements include inaccurate temperature measurement, vibration and cleanliness of glass optical surface [5].
9. The duration and amount of captures need adjustment depending on sample quality. Lower particle concentrations require longer captures times. Also, polydisperse samples require longer capture times and repeated captures to gain a better overview over the whole sample.

   If choosing automatic processing of data after capture, ensure there is enough time between incubation time points.

   The use of temperature control at room temperature can lead to overheating of the laser module and subsequently lead to software crashes.

10. Fuzzy-looking particles are out of focus and cannot be sized accurately.
11. Screen gain under "Capture" or "Process" does not change the data processing. The screen gain is used to make the image appear brighter or dimmer to the user. However, when taking video stills the same screen gain settings should be used for all images to make them comparable.
12. Detection threshold determines the minimum intensity value of an image necessary to qualify as a particle to be tracked for analysis. This setting is going to change from sample to sample. Try to avoid red crosses in nondistinct particles and too many blue crosses. In each frame 10–100 particle centers should be detected (number at bottom right of analysis screen).
13. Images of each graph and video stills can be taken before, during or after data processing by right clicking on the image.

Acknowledgments

This chapter has emanated from research conducted with the financial support of Enterprise Ireland under Grant Number [IP/2015 0358] and the Synthesis and Solid State Pharmaceutical Centre, funded by Science Foundation Ireland (SFI) under Grant Number [12/RC/2275], cofunded under the European Regional Development Fund.

References

1. Malloy A, Carr B (2006) Nanoparticle tracking analysis – the halo™ system. Part Part Syst Charact 23:197–204
2. Filipe V, Hawe A, Jiskoot W (2010) Critical evaluation of nanoparticle tracking analysis (NTA) by NanoSight for the measurement of nanoparticles and protein aggregates. Pharm Res 27:796–810
3. Ribeiro LN de M, Couto VM, Fraceto LF et al (2018) Use of nanoparticle concentration as a tool to understand the structural properties of colloids. Sci Rep 8:982
4. Maas SLN, De Vrij, Van Der Vlist EJ et al (2015) Possibilities and limitations of current technologies for quantification of biological extracellular vesicles and synthetic mimics. J Control Release 200:87–96
5. Gardiner C, Ferreira YJ, Dragovic RA et al (2013) Extracellular vesicle sezing and enumeration by nanoparticle tracking analysis. J Extracell Vesicles 2:19671
6. Rather GM, Mukherjee J, Halling PJ et al (2012) Activation of alpha chymotrypsin by three phase partitioning is accompanied by aggregation. PLoS One 7:e49241
7. Sediq AS, Nejadnik MR, El Bialy I et al (2015) Protein–polyelectrolyte interactions: monitoring particle formation and growth by nanoparticle tracking analysis and flow imaging microscopy. Eur J Pharm Biopharm 93:339–345
8. Krueger AB, Carnell P, Carpenter JF (2016) Characterization of factors affecting nanoparticle tracking analysis results with synthetic and protein nanoparticles. J Pharm Sci 105:1434–1443
9. Patois E, Capelle MAH, Palais C et al (2012) Evaluation of nanoparticle tracking analysis (NTA) in the characterization of therapeutic antibodies and seasonal influenza vaccines: pros and cons. J Drug Deliv Sci Technol 22:427–433

Chapter 11

Evaluation of Temporal Aggregation Processes Using Spatial Intensity Distribution Analysis

Zahra Rattray, Egor Zindy, Kara M. Buzza, and Alain Pluen

Abstract

Small proteinaceous oligomeric species contribute to the formation of larger aggregates, a phenomenon that is of direct relevance to the characterization of protein aggregation in biopharmaceuticals and understanding the underlying processes contributing to neurodegenerative diseases.

The ability to monitor in situ oligomerization and aggregation processes renders imaging and image analysis an attractive approach for gaining a mechanistic insight into early processes contributing to the formation of larger aggregates in disease models and biologics. The combination of image analysis tools enables the detection of both oligomeric and larger aggregate subtype in contrast to conventional kinetic-based approaches that lack the ability to resolve dimers from monomeric moieties in samples containing mixed populations.

In this chapter, we describe the process for confocal time series image acquisition for monitoring the in situ loss of monomers, and the subsequent analysis pipeline using spatial intensity distribution analysis (SpIDA) to evaluate oligomer content.

Key words Monomer loss, Protein aggregation, Light scattering, Microscopy, Image analysis, SpIDA

1 Introduction

Biopharmaceutical proteins constitute a growing family of medicines for many therapeutic areas. However, there are a number of associated challenges in their formulation and manufacture that include the prediction and control of reversible and irreversible aggregate formation. Identifying aggregation-prone biopharmaceutical proteins during early stages of product development is important for the biopharmaceutical industry. Protein aggregation mechanisms are also important in disease especially neurodegenerative diseases such as Alzheimer, Parkinson's, or Huntington's disease [1]. For these amyloid forming proteins and peptides, intermolecular interactions resulting in self-association lead to the formation of early oligomeric species, often rich in β-sheet structures that rapidly convert into protofibrils. Eventually, protofibrils

Jennifer J. McManus (ed.), *Protein Self-Assembly: Methods and Protocols*, Methods in Molecular Biology, vol. 2039,
https://doi.org/10.1007/978-1-4939-9678-0_11,

assemble into elongated structures of mature amyloids. While visible and subvisible aggregate detection has made considerable progress in the recent years, a remaining significant challenge is understanding the molecular pathways implicated in protein aggregation, and associated determinants. The formation of small transient oligomeric intermediates has been identified as the initial step contributing to the formation of both amorphous and fibrillar structures following manufacture and purification in the production of protein-based biopharmaceutical preparations, or in processes underlying the formation of neurodegenerative amyloid plaques [2]. Spatiotemporal changes in receptor oligomerization are also central to many endogenous signal transduction processes that occur in vivo [3].

Oligomeric species are often challenging to characterize owing to their reversibility as a consequence of thermodynamic unfavorability that limits in situ biophysical detection and characterization of such species. Furthermore, heterogeneities in oligomer formation contribute to uncertainties in the assessment of their structure. This has stimulated recent interest in the development of appropriate mathematical models, synthesis of stable irreversible oligomers and novel technologies for profiling their formation, characteristics, and role in aggregation processes to study oligomer formation [2, 4].

Non-native aggregation of proteins is a multistage process generally understood to be initiated by native monomer partial unfolding following an intermediate conformational transition that may render the monomer reactive to association with other unfolded monomers.

To support an understanding of the underpinning stages implicated in the formation of larger aggregates, mathematical models have been developed and applied to the determination of oligomerization and polymerization processes [5–8].

Current analytical approaches utilized in the formation of oligomer species have centered on light scattering detection (i.e., static light scattering, multiangle light scattering and dynamic light scattering) and size exclusion chromatography (SEC), or a combination of both. However, disadvantages associated with the dilution of aggregates have been deemed responsible for reversing oligomers when analyzing monomer loss using SEC coupled to light scattering-based detectors, and dynamic light scattering does not possess the ability to resolve dimer populations from monomers [9].

Limitations in technologies enabling the real-time monitoring of kinetics of monomer loss and subsequent larger aggregate formation have posed an obstacle to real-time studies of monomer loss in protein-based samples to date. Hence, in this study SpIDA was utilized to study oligomerization of bovine serum albumin (BSA) in confocal image time series obtained from BSA samples subjected to thermal stress. Results obtained using these approaches were subsequently compared against dynamic light scattering and fluorescence correlation spectroscopy data to facilitate the comparison of image analysis tools with previously-utilized approaches.

Spatial intensity distribution analysis, SpIDA, is an approach based on super-Poissonian fitting of fluorescence intensity histograms calculated from confocal images, and yields information on the number of fluorescent particles and their quantal brightness. For a defined region of interest (ROI) within an image, the intensity histogram is determined from counting the frequency of pixels for each integrated fluorescence intensity value that are collected from fluorescence emitted from fluorophores following excitation by the laser beam within a certain region (i.e., per confocal volume). Subsequently, the intensity histogram of all potential configurations is plotted as a function of their weighted probability assuming a Poissonian spatial distribution. The underlying mathematical basis of SpIDA and associated equations are described elsewhere [3]. This method has previously been used for characterization of receptor tyrosine kinase oligomerization [10], and the quantification of fluorophore accumulation for a model compound [11] and transporter expression in immunofluorescent specimens [12].

2 Materials

2.1 Sample Preparation for Analysis

All samples prepared and analyzed in the present chapter were a combination of 1 μM labeled protein (i.e., BSA-Alexa Fluor 488) and unlabeled protein (i.e., unlabeled BSA) to the target concentration (in the present example, Hamrang et al. [13] studied a final BSA concentration of 0.4 and 1 mg/mL)

1. Formulation or system buffer (10 mM phosphate-buffered saline and a citrate buffer were used in the present protocol).
2. Sodium chloride for ionic strength adjustment (or any other salt/denaturant to initiate unfolding/conformational changes, optional).
3. Fluorescently-labeled monomeric protein (BSA-Alexa Fluor 488® is used in this example) was reconstituted in PBS. The sample was purified to the monomeric form by size exclusion chromatography (SEC), using an appropriate high resolution column at 0.5 mg/mL flow rate. The concentration was determined spectrophotometrically at 280 nm assuming a molar extinction coefficient of 48,824 M^{-1} cm^{-1}. The labeling ratio and quantal brightness for the analyte must be characterized for this approach.

4. Unlabeled monomeric BSA. Purification of unlabeled BSA is performed using the aforementioned approach for fluorescently-labeled protein.
5. Perform any additional pH and Ionic strength adjustment immediately prior to sample transfer in the flow chamber at time zero, and image acquisition.
6. Similar to labeled protein purification, ensure that the excess stock unlabeled solutions are also subjected to preparative chromatography and the monomeric/oligomeric fractions collected.
7. Store all solutions reported at 4 °C at all time prior to experimentation and avoid freeze–thaw stress. All buffers and solutions utilized in the present protocol were prefiltered using a 0.2 μm pore sized filter in order to remove any potential particulates contributing to seeding.

2.2 Image Acquisition and Analysis

1. In the present protocol, a Zeiss 510 ConfoCor 2 confocal microscope was used and the diffusion of Rhodamine Green (or alternative dye of interest depending on excitation laser source) was used to measure the laser beam waist radius for a 488 nm argon laser excitation source.
2. A pre-bleached slide for measurement of photomultiplier tube (PMT) shot noise.
3. Microscope heated stage with Peltier control (for temperature control or increasing temperature in thermal stress experiments).
4. Custom chamber for sample placement.
5. Confocal microscope (for this study, a Zeiss 510 ConfoCor 2 was utilized for all image acquisition).
6. Spatial Intensity Distribution Analysis, SpIDA, software (can be downloaded freely from https://neurophotonics.ca/software) which contains a user guide document.
7. A data analysis software such as Origin or GraphPad Prism.

3 Methods

3.1 Characterization of System Performance for Image Analysis

Image acquisition parameters utilized during confocal imaging are critical to the reliability of output measurements obtained from subsequent analysis using SpIDA. Hence, prior to performance of imaging experiments, it is imperative to characterize system attributes (laser beam waist size, white noise, laser power intensity, etc.) prior to performing kinetic studies. Some of these system tests and a stepwise guide to their performance are included below.

3.1.1 Measurement of Laser Power Intensity

Throughout the laser lifetime variations in laser output may occur when using the same intensity. Hence, it is recommended that the laser output is periodically characterized (i.e., bi-monthly), and the laser power adjusted accordingly to account for such variations. A laser power meter should be used to confirm laser power (*see* **Note 1**).

3.1.2 Measurement of Laser Beam Waist Size

To measure the beam waist radius size for a 488 nm argon laser, using the ConfoCor 2 LSM510 setup (*see* **Note 2**), the diffusion time of a dye excited by the respective laser excitation source (in this example, Rhodamine Green) may be measured. The following protocol was applied to an experiment in which a 40×/1.2 NA water-immersion objective lens was used.

1. Using the ConfoCor 2 setup, acquire 30 runs each of 10 s duration to measure Rhodamine Green diffusion in solution (*see* **Notes 3** and **4**).
2. Using A single-component fit on the ConfoCor2 software, derive the measured diffusion time and use the following equation to determine the laser waist beam size;

$$\tau_D = \frac{\omega_0^2}{4D}$$

   where ω_0, is the laser beam waist size for the Argon excitation laser derived from the autocorrelation function (ACF) obtained from Rhodamine Green™ diffusion through the confocal volume.
3. Using previously-reported diffusion coefficients of Rhodamine 6G (2.8×10^{-6} cm^2/s) or any other dye of interest, determine the laser waist beam radius [14, 15].

   For the setup described in the present protocol, the diffusion time obtained from FCS analysis of Rhodamine Green using the ConfoCor 2 LSM510 setup was 18.9 ± 2.6 μs (ω_0: 0.139 ± 0.001 μm).

3.1.3 Measurement of Photomultiplier Tube Shot Noise (Detector Calibration)

Variation of PMT voltage and laser power intensity may be used to assess the conditions under which direct linearity exists between the photoelectric current and measured fluorescence intensity (*see* **Note 5**). This is significant since the derived brightness and number of particles parameters following SpIDA analysis of confocal images are strongly influenced by shot noise. It is well-known that PMTs do not respond to light in a constant manner [16].

1. Using the laser of interest (in the case of Alexa Fluor® 488, the Argon excitation laser), image immobilized pre-bleached beads (Zeiss, Jena, Germany). In this case a c-Apochromat 40×/NA 1.2 water-immersion objective was utilized to capture images.

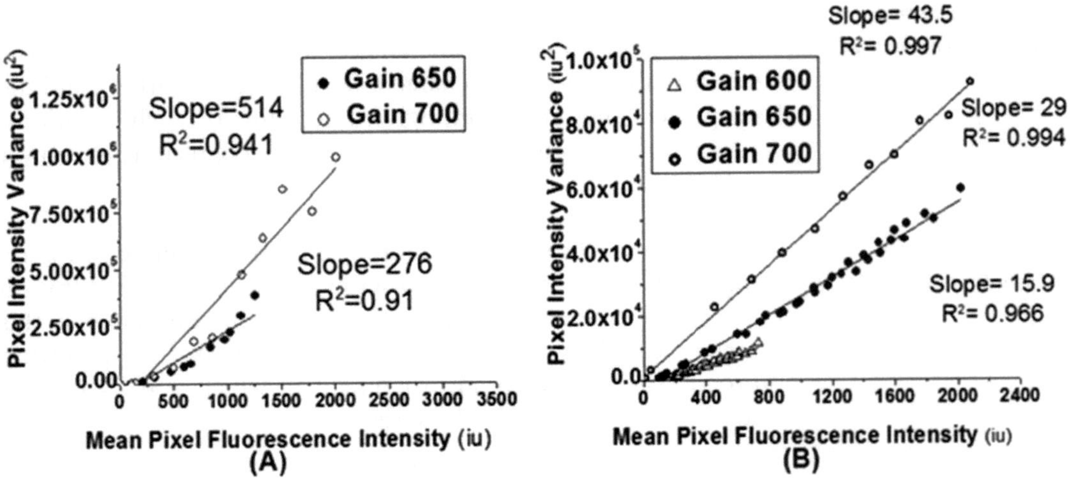

Fig. 1 Example plots of pixel intensity variance versus mean pixel intensity for (**a**) a 543 nm helium–neon and (**b**) a 488 nm argon laser for various detector gain settings ($n = 1024$) using the setup described in the present protocol

2. Excite the beads over a range of laser powers, at different PMT gains (i.e., 600, 650, 700, and 750 for a Zeiss system), pixel dwell times (to be determined from pixel dwell times used for future image acquisition experiments) and average the results for 1024 points (i.e., pixels). Example plots are presented in Fig. 1 to demonstrate values obtained using the protocol example setup.

3.1.4 Measurement of White Noise Contribution

The background contribution of buffer or extracellular space within the sample can be assessed by measuring the fluorescence intensity of these regions of interest within a confocal image.

In this experiment, the contribution of buffer was measured through the acquisition of images using representative parameters (e.g., pixel size and scan speed) at various gains.

1. Acquire images of blank sample buffer or extracellular regions using representative image acquisition parameters (objective, pixel size, resolution, and dwell time) at various PMT gains or the PMT gain setting that will be used to acquire all experimental image time series (e.g., 500, 550, 600, 650, 700, and 750), Fig. 2.
2. Measure the resultant mean fluorescence intensity from these images in Image J by selecting the corresponding ROIs and use of the Analyze → Measure function.
3. The measured fluorescence intensity from these ROIs will be used as the white noise for input into the SpIDA user interface.

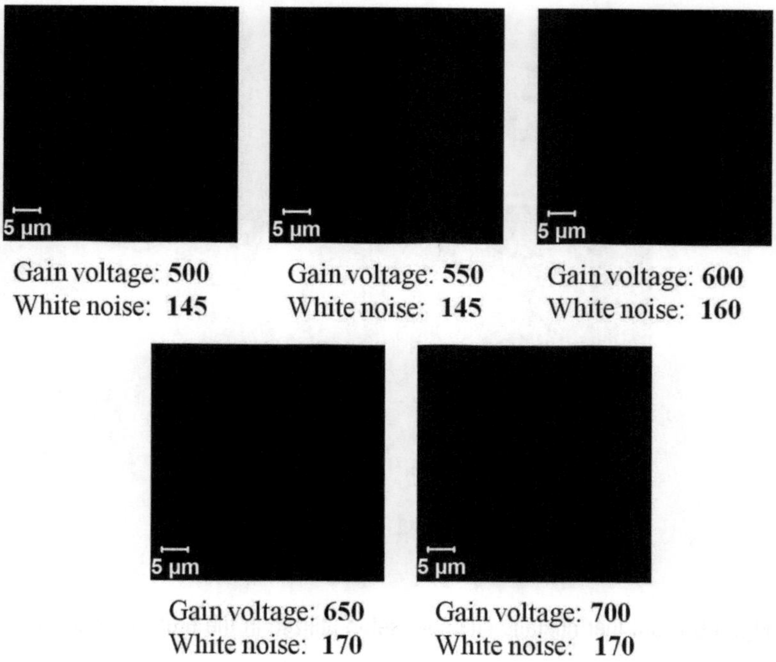

Fig. 2 Example images of 1024 × 1024 pixel resolution images captured from citrate-phosphate buffer, and resultant fluorescence intensity as measured in Image J

3.1.5 Quantification of Monomer Quantal Brightness

SpIDA exploits the direct relationship between the quantal brightness of a monomeric entity and a dimer ($\varepsilon = 2\varepsilon_0$) to determine the spatiotemporal evolution of subpopulations in samples (i.e., image time series) consistent with aggregation. When two fluorescent populations with different molecular quantal brightness are present within a sample (i.e., a confocal image) and not spatially segregated, or in the presence of autofluorescence, the total histogram becomes a convolution of the two distributions obtained from each species (*see* **Note 6**).

Application of a one-population model in a mixed sample will yield a resultant quantal brightness intermediate between the species present in the sample, while following performance of a monomeric ε control test it is possible to extract information about the populations using a two-population (i.e., monomer–oligomer) model. Thereby, through appropriate knowledge of monomeric ε it is possible to determine spatiotemporal aggregation profiles from confocal image time series [3].

1. Immediately following purification of the monomeric fraction, prepare samples for imaging.
2. Acquire images from monomeric sample, and analyse using SpIDA over 100 frames.

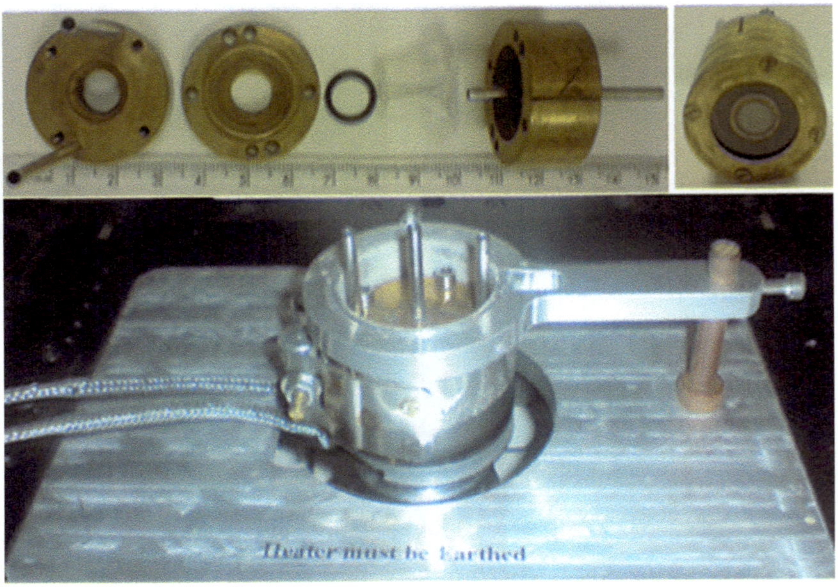

Fig. 3 Custom thermostatic chamber setup utilized in the present protocol

3. The determined quantal brightness will be used for all future measurements.

3.2 Confocal Image Acquisition

Following the performance of preliminary image acquisition experiments and characterization of system attributes under these conditions, it is important to maintain all image acquisition parameters between experiments that are to be compared (*see* **Note 1**).

1. Transfer monomeric samples (i.e., 500 μL) to a thermostatic chamber positioned in a (custom) stage heater preset to 50 °C (*see* Fig. 3); this temperature was intentionally selected below the BSA melting temperature (T_m) (*see* **Note 7**).
2. Use a pinhole diameter of one Airy Unit, and a raster scan to capture images (images of 1024 × 1024 pixel resolution are recommended).
3. Set the optical zoom so that a pixel size (x, y-sampling size) of approximately 40–90 nm is achieved (in the present protocol, a zoom factor resulting in a pixel size of 40 nm was utilized for acquiring all images). Acquiring images with such pixel sizes will achieve oversampling and optimize the detection of fluorescent species moving in and out of the confocal volume.
4. Select an appropriate laser excitation intensity that will not result in photobleaching of the fluorophore of interest, and avoid oversaturation of the detector. In the case of the present setup, a 30 mW argon excitation laser power of 5% was utilized and the absence of photobleaching verified through

monitoring the intensity histogram during image time series acquisition (*see* **Notes 5** and **6**).

5. Verify that the amplifier gain is 1 and the offset is set to zero.
6. Temporal changes in monomeric samples of BSA-AF488 concentrations (i.e., 0.4 and 1 mg/mL) at different NaCl concentrations (i.e., 50, 150, and 500 mM) were recorded at 50 °C over a period of 240 min as a confocal image time series. Determine the duration of image acquisition based upon expected changes in the system under examination (*see* **Note 8–10**).
7. Analyze resultant image time series using SpIDA to evaluate the temporal evolution of (in this example, BSA) monomer loss and the formation of dimers and higher order aggregates.

3.3 SpIDA Analysis

Following acquisition of confocal image time series, the images may be opened using the SpIDA graphical user interface (GUI) in MATLAB. A discussion of the underpinning algorithms and principles behind SpIDA analysis is outside the scope of the present protocol, and the reader is referred to [3].

1. In order to load the image onto the GUI, the user is prompted to enter the pixel size and laser beam waist size (radius) (*see* **Notes 3** and **4**).
2. When the image time series loads on the GUI, input the predetermined system parameters (white noise, PMT shot noise, etc.).
3. Following input of the system parameters, select the ROI for analysis. At this stage it is important to confirm ROI attributes within a single image frame.
4. Fit the histogram to determine the number of monomeric or oligomeric species per beam area.
5. Save data using "save all" button. Data are saved with a .dat extension (*see* **Notes 11** and **12**).

3.4 Data Analysis

SpIDA data files can be analyzed on any computer as the files a .dat extension opened with Microsoft Excel.

1. Open file using Excel (*see* Fig. 4).
2. Select the columns (*see* Fig. 4) of interest, that is, density population for monomer, dimer, etc. For instance, the density population is given in column 2 for the first population, columns 4 for the second population, etc. Data are provided as population density per beam area.
3. Depending on the focus of the study, different populations can be considered. For example, during the study of the

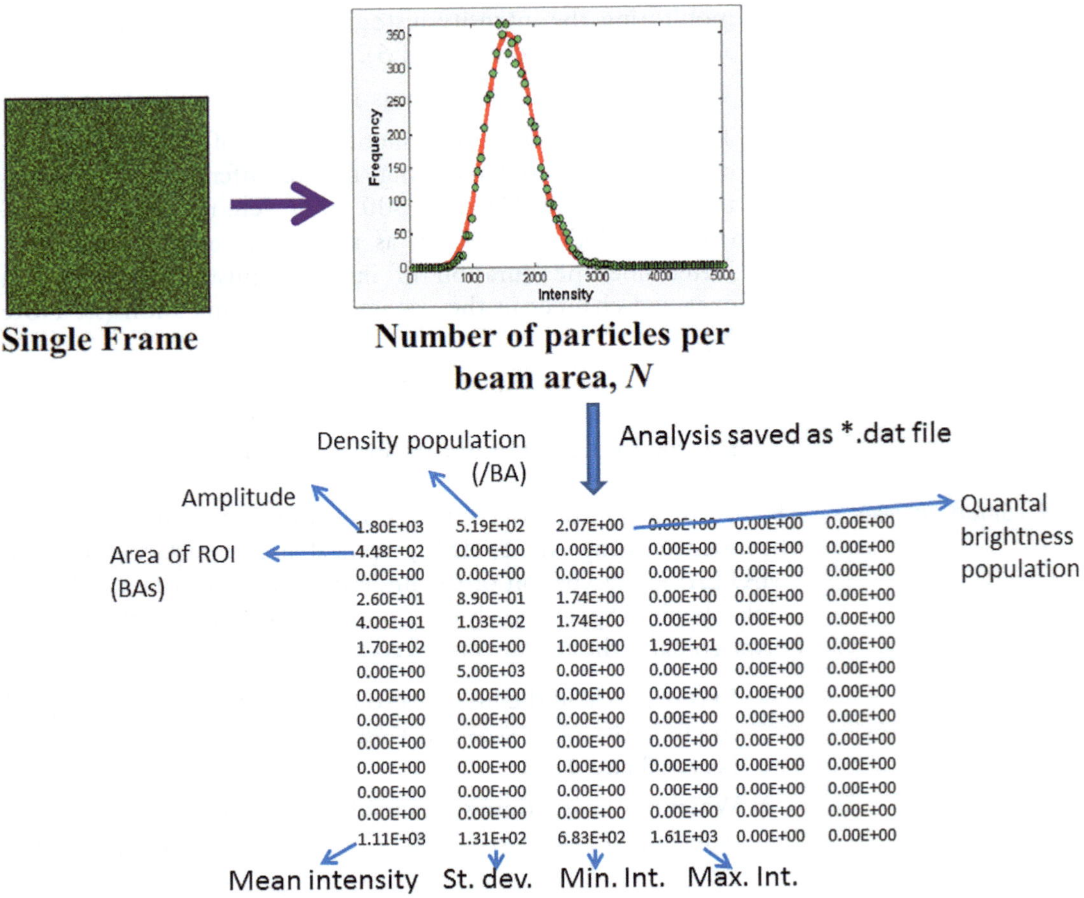

1.80E+03	5.19E+02	2.07E+00	0.00E+00	0.00E+00	0.00E+00
4.48E+02	0.00E+00	0.00E+00	0.00E+00	0.00E+00	0.00E+00
0.00E+00	0.00E+00	0.00E+00	0.00E+00	0.00E+00	0.00E+00
2.60E+01	8.90E+01	1.74E+00	0.00E+00	0.00E+00	0.00E+00
4.00E+01	1.03E+02	1.74E+00	0.00E+00	0.00E+00	0.00E+00
1.70E+02	0.00E+00	1.00E+00	1.90E+01	0.00E+00	0.00E+00
0.00E+00	5.00E+03	0.00E+00	0.00E+00	0.00E+00	0.00E+00
0.00E+00	0.00E+00	0.00E+00	0.00E+00	0.00E+00	0.00E+00
0.00E+00	0.00E+00	0.00E+00	0.00E+00	0.00E+00	0.00E+00
0.00E+00	0.00E+00	0.00E+00	0.00E+00	0.00E+00	0.00E+00
0.00E+00	0.00E+00	0.00E+00	0.00E+00	0.00E+00	0.00E+00
0.00E+00	0.00E+00	0.00E+00	0.00E+00	0.00E+00	0.00E+00
0.00E+00	0.00E+00	0.00E+00	0.00E+00	0.00E+00	0.00E+00
1.11E+03	1.31E+02	6.83E+02	1.61E+03	0.00E+00	0.00E+00

Fig. 4 Example analysis of a BSA solution confocal micrograph and information obtained from the ".dat" file (adapted from GUI SpIDA user guide)

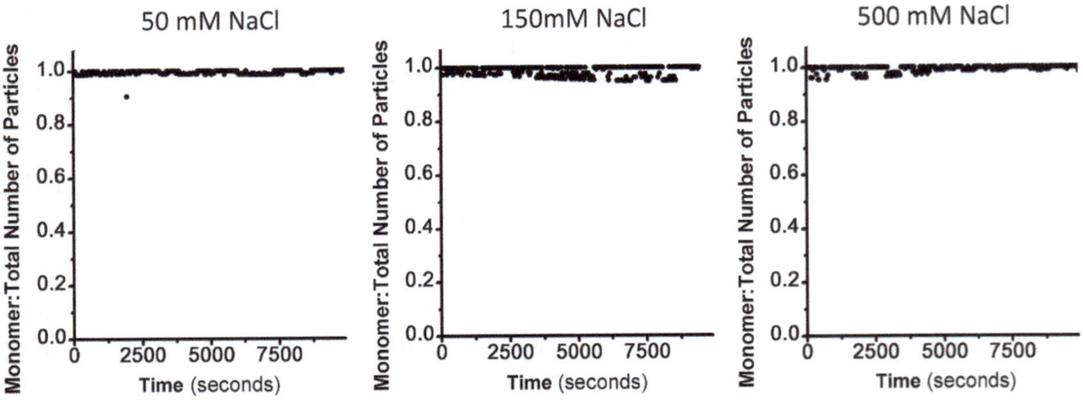

Fig. 5 Temporal evolution of monomer loss (relative to the total number of particles) in a 1 mg/mL sample of monomeric BSA-AF488 (pH 7.0 ± 0.2) maintained at 50 °C in the absence of agitation. Monomer ratios were determined using SpIDA from a confocal image time series acquired with a pixel dwell time of 6.4 μs, resolution of 1024 pixels, and pixel size of 44 nm for 50, 150, and 500 mM NaCl samples (adapted from Hamrang et al. [13])

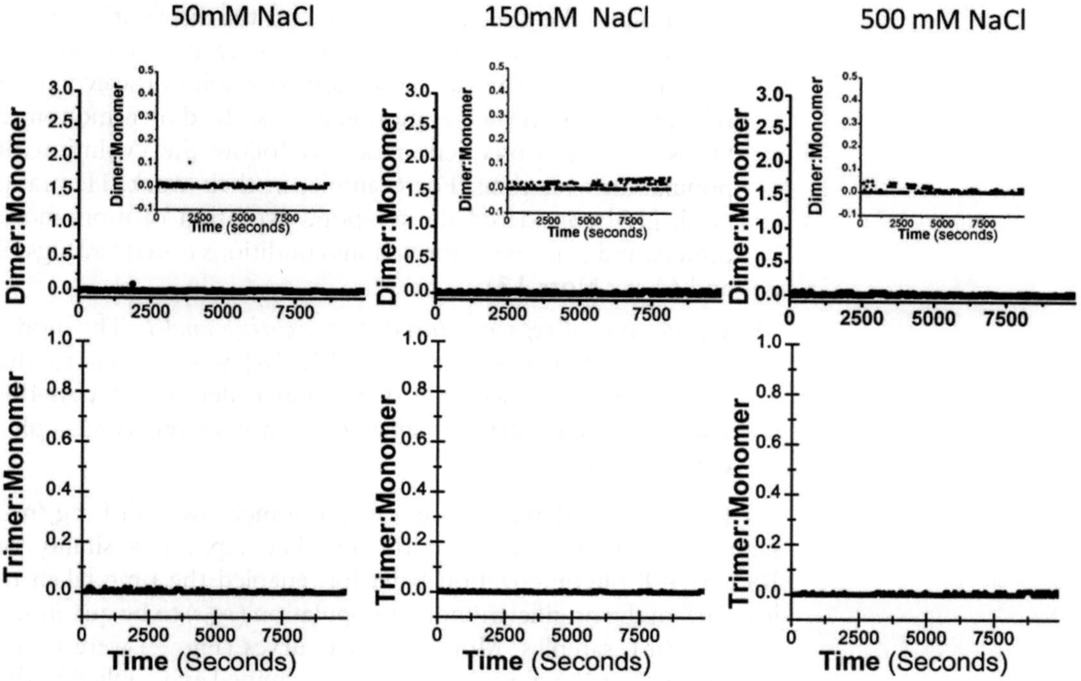

Fig. 6 Real-time evolution of dimer formation expressed as dimer-to-monomer ratio (top row) and trimer formation expressed as trimer-to-monomer ratio (bottom row) of 1 mg/mL BSA-AF488 subjected to thermal stress at 50 °C and subsequent analysis with SpIDA at indicated NaCl concentrations (adapted from Hamrang et al. [13])

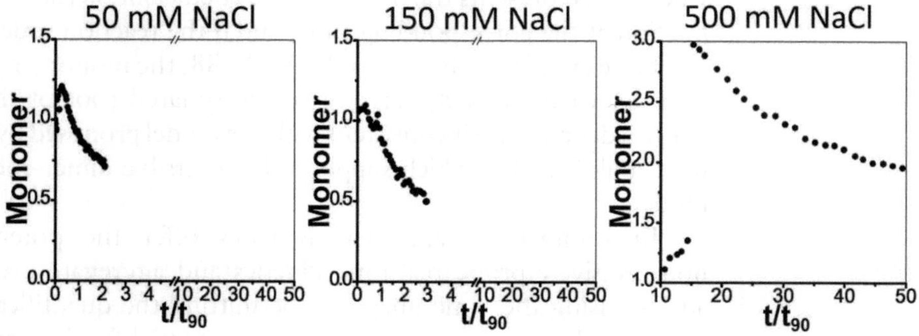

Fig. 7 Temporal evolution of monomer loss (relative to the monomer concentration at the start of the experiment) in a 1 mg/mL sample of monomeric BSA-AF488 (pH 7.0 ± 0.2) maintained at 50 °C in the absence of agitation. Monomer concentration was determined using SpIDA applied to a confocal image time series acquired with a pixel dwell time of 6.4 μs, resolution of 1024 pixels, pixel size of 44 nm for 50, 150, and 500 mM NaCl samples (adapted from Hamrang et al. [13])

oligomerization of BSA-AF488, the population of monomer, dimers, and trimers was determined for each time value.

4. Using different ROIs on the pictograms and experimental repeats, both mean density and its corresponding standard

deviation can be directly determined or, the distribution of the population densities can be plotted for each condition as a function of time. This can be performed to follow the evolution of subpopulations such as monomer loss, the dimer–monomer ratios, or any subsequent ratios to follow the evolution of population over time. For example, in their work, Hamrang et al. [13] considered the temporal evolution of monomers, dimers, and trimers for the various conditions tested (*see* Figs. 5 and 6) (*see* **Note 13**).

5. *Comparison of experimental data to existing models.* The model described by Brummitt et al. [17, 18] was applied to the assessment of monomer loss reaction orders in BSA-AF488 samples through determination of monomer versus t/t_{90} plot curve slopes.

Data presented in Fig. 7 indicate monomer loss in all 1 mg/mL samples throughout the experiment. The rapid reversibility of BSA-AF488 oligomerization behavior, enabled the time taken to lose 10% of the original monomer population ($t_{90}s$) to be quantified in 1 mg/mL samples. Monomer loss curves (Fig. 7) were transformed in a log-log plot; slopes were determined and applied to the analysis of monomer loss kinetic reaction order determination using the following equation from Brummitt et al. [18]:

$$\frac{dm}{dt} = -k_{obs} m^{\alpha}$$

where k_{obs} represents the observed rate coefficient, m the monomer fraction at the corresponding time, and α the reaction order.

For example, in the case of BSA-AF488, the monomer population decreased with the inverse of the squared root of the ratio t/t_{90}, a dependence inconsistent with the model proposed by Brummitt et al. [17, 18] which supports the reversible dimer–monomer model.

In summary, image analysis tools offer the potential to non-invasively probe real-time kinetics and aggregate profiles of analyte using the same image set permitting the quantification of oligomer distributions. This offers the potential for further exploration in unstable systems with a higher propensity to form reversible soluble oligomers. Furthermore, cross-comparison of data between SpIDA and other techniques such as DLS and RICS has demonstrated complementarity between all methods, each providing unique information on temporal sample evolution following exposure to thermal stress. Furthermore, rapid data acquisition through confocal imaging permits the direct real-time monitoring of changes in comparison to traditionally-utilized methods such as size exclusion chromatography that may distort the equilibrium of soluble aggregates through their reversal due to dilution effects or the lengthy duration of their separation. Since oligomer formation is recognized as a principal contributory component to aggregation in biopharmaceutical preparations and neurodegenerative disorders, analysis of real-time oligomerization with confocal microscopy may prove useful in the interpretation of the dynamics and equilibria of oligomer formation and subsequent higher order aggregate formation over a broad concentration and particle size range as demonstrated here.

4 Notes

1. It is important to optimize image acquisition parameters prior to performing any experiments, so that the setup can be characterized for relevant setup parameters (laser, laser power, gain, resolution, scan speed, etc.).
2. Please note that any confocal setup can be used to generate the image time series required for analysis of spatiotemporal oligomerization status.
3. For measurement of the laser beam waist size using the diffusion coefficient of Rhodamine Green, it is assumed that given the molecular weight of Rhodamine Green (507 g/mol) is similar to that of Rhodamine 6G (479 g/mol), the diffusion coefficient would be the same.
4. Alternative approaches to that described in the present manuscript may be used to determine the laser beam waist size. For example, acquisition of *z*-stacks from sub-diffraction sized 100 nm fluorescently labeled beads may be used which is described elsewhere.
5. In contrast to image correlation spectroscopies, SpIDA is not susceptible to photobleaching from the laser excitation source.
6. The quantal brightness of a quality control sample must be periodically verified to confirm the absence of drift in sample quantal brightness.
7. The use of additives that are autofluorescent within the laser source excitation/emission range should be avoided where possible, as these may reduce the signal-to-noise ratio for such experiments.
8. During optimization of experimental conditions and subsequent image analysis pipeline, it may be useful to validate preliminary data against orthogonal approaches (i.e., fluorescence correlation spectroscopy or size-exclusion chromatography).
9. Condensation of chamber windows at temperatures in excess of 50 °C, and consequent loss of sample due to evaporation may adversely impact image acquisition. It is advised that

sample volume prior to addition to the system, and following imaging is recorded.

10. Following image acquisition, samples can be retained for end-point analysis using orthogonal sizing approaches (e.g., dynamic light scattering and size exclusion chromatography).
11. Although the software allows for saving data and information following analysis, keeping notes of ROI coordinates and settings (e.g., beam size, pixel size, slope, white noise, bin) separately is important as it can be very cumbersome to derive these from individual data files.
12. Please note that SpIDA is not a commercial software.
13. A limitation of this approach is that larger oligomers (larger than trimers) cannot be characterized using SpIDA. However, correlation-based approaches such as Raster image correlation spectroscopy [19] that are capable of differentiating changes greater than 3–4 times in molecular weight can be used to assess the formation of larger intermediates.

Acknowledgments

The authors would like to thank the University of Manchester Alumni Fund for supporting this project.

References

1. Theillet FX et al (2014) Physicochemical properties of cells and their effects on intrinsically disordered proteins (IDPs). Chem Rev 114:6661–6714. https://doi.org/10.1021/cr400695p
2. Ahmad B, Winkelmann J, Tiribilli B, Chiti F (2010) Searching for conditions to form stable protein oligomers with amyloid-like characteristics: the unexplored basic pH. Biochim Biophys Acta 1804:223–234. https://doi.org/10.1016/j.bbapap.2009.10.005
3. Barbeau A, Godin AG, Swift JL, De Koninck Y, Wiseman PW, Beaulieu JM (2013) Quantification of receptor tyrosine kinase activation and transactivation by G-protein-coupled receptors using spatial intensity distribution analysis (SpIDA). In: Conn PM (ed) G protein coupled receptors: modeling, activation, interactions and virtual screening, Methods in enzymology, vol 522. Elsevier, London, pp 109–131. https://doi.org/10.1016/b978-0-12-407865-9.00007-8
4. Andrews JM, Roberts CJ (2007) A Lumry-Eyring nucleated polymerization model of protein aggregation kinetics: 1. Aggregation with pre-equilibrated unfolding. J Phys Chem B 111:7897–7913. https://doi.org/10.1021/jp070212j
5. Lee J et al (2013) Functional consequences of complementarity-determining region deactivation in a multifunctional anti-nucleic acid antibody. J Biol Chem 288:35877–35885. https://doi.org/10.1074/jbc.M113.508499
6. Li Y, Mach H, Blue JT (2011) High throughput formulation screening for global aggregation behaviors of three monoclonal antibodies. J Pharm Sci 100:2120–2135. https://doi.org/10.1002/jps.22450
7. Li Y, Ogunnaike BA, Roberts CI (2010) Multivariate approach to global protein aggregation behavior and kinetics: effects of pH, NaCl, and temperature for alpha-chymotrypsinogen A. J Pharm Sci 99:645–662. https://doi.org/10.1002/jps.21869
8. Roberts CJ (2003) Kinetics of irreversible protein aggregation: analysis of extended Lumry-Eyring models and implications for predicting protein shelf life. J Phys Chem B 107:1194–1207. https://doi.org/10.1021/jp026827s

9. Carpenter JF, Randolph TW, Jiskoot W, Crommelin DJA, Middaugh CR, Winter G (2010) Potential inaccurate quantitation and sizing of protein aggregates by size exclusion chromatography: essential need to use orthogonal methods to assure the quality of therapeutic protein products. J Pharm Sci 99:2200–2208. https://doi.org/10.1002/jps.21989
10. Swift JL et al (2011) Quantification of receptor tyrosine kinase transactivation through direct dimerization and surface density measurements in single cells. Proc Natl Acad Sci U S A 108:7016–7021. https://doi.org/10.1073/pnas.1018280108
11. Hamrang Z, McGlynn HJ, Clarke D, Penny J, Pluen A (2014) Monitoring the kinetics of CellTrace (TM) calcein red-orange AM intracellular accumulation with spatial intensity distribution analysis. BBA-Gen Subjects 1840:2914–2923. https://doi.org/10.1016/j.bbagen.2014.05.014
12. Hamrang Z, Arthanari Y, Clarke D, Pluen A (2014) Quantitative assessment of P-glycoprotein expression and function using confocal image analysis. Microsc Microanal 20:1329–1339. https://doi.org/10.1017/s1431927614013014
13. Hamrang Z, Zindy E, Clarke D, Pluen A (2014) Real-time evaluation of aggregation using confocal imaging and image analysis tools. Analyst 139:564–568. https://doi.org/10.1039/c3an01693e
14. Petrasek Z, Schwille P (2008) Precise measurement of diffusion coefficients using scanning fluorescence correlation spectroscopy. Biophys J 94:1437–1448. https://doi.org/10.1529/biophysj.107.108811
15. Schwille P, Haupts U, Maiti S, Webb WW (1999) Molecular dynamics in living cells observed by fluorescence correlation spectroscopy with one- and two-photon excitation. Biophys J 77:2251–2265. https://doi.org/10.1016/s0006-3495(99)77065-7
16. Ward RJ, Marsango S, Pediani JD, Milligan G (2017) The use of spatial intensity distribution analysis to examine G protein-coupled receptor oligomerization. In: Herrick-Davis K, Milligan G, DiGiovanni G (eds) G-Protein-coupled receptor dimers, Receptors series, vol 33. Springer, New York, NY, pp 15–38. https://doi.org/10.1007/978-3-319-60174-8_2
17. Brummitt RK, Nesta DP, Chang LQ, Kroetsch AM, Roberts CJ (2011) Nonnative aggregation of an IgG1 antibody in acidic conditions, part 2: nucleation and growth kinetics with competing growth mechanisms. J Pharm Sci 100:2104–2119. https://doi.org/10.1002/jps.22447
18. Brummitt RK, Nesta DP, Roberts CJ (2011) Predicting accelerated aggregation rates for monoclonal antibody formulations, and challenges for low-temperature predictions. J Pharm Sci 100:4234–4243. https://doi.org/10.1002/jps.22633
19. Hamrang Z, Pluen A, Zindy E, Clarke D (2012) Raster image correlation spectroscopy as a novel tool for the quantitative assessment of protein diffusional behaviour in solution. J Pharm Sci 101:2082–2093. https://doi.org/10.1002/jps.23105

Chapter 12

Fluorescence Correlation Spectroscopy for Particle Sizing in Highly Concentrated Protein Solutions

Judith J. Mittag, Matthew R. Jacobs, and Jennifer J. McManus

Abstract

Highly concentrated solutions of biomolecules play an increasingly important role in biopharmaceutical drug development. In these systems, the formation of reversible aggregates by self-association creates a significant analytical challenge, since dilution is often required for techniques such as HPLC/liquid chromatography and analytical ultracentrifugation. There is a growing demand for methods capable of analyzing these assemblies, ideally under formulation conditions (i.e., in the presence of excipients). One approach that addresses this need is based on fluorescence correlation spectroscopy (FCS), which is a flexible and powerful technique to measure the diffusion of fluorescently labeled particles. It is particularly suited to measuring the size distribution of reversible aggregates of proteins or peptides in highly concentrated formulations, since it overcomes some of the challenges associated with other methods. In this protocol, we describe state-of-the-art measurement and analysis of protein self-assembly by determination of particle size distributions in highly concentrated protein solutions using FCS.

Key words Fluorescence correlation spectroscopy, Protein self-assembly, Size distribution, High concentration, Polydispersity, Gaussian distribution model, Formulation

1 Introduction

The development of highly concentrated protein liquid formulations leads to both formulation and analytical challenges during the physicochemical characterization required in development and by regulatory authorities to ensure the safety of the product [1–5]. While challenging in all protein solutions, at high protein concentrations, characterization of assemblies and aggregates is nontrivial since fewer analytical techniques are appropriate for high concentration solutions. Irreversible aggregates can be characterized/quantified by standard analysis methods such as size exclusion chromatography (SEC) and gel electrophoresis, since they persist after dilution [6]. Reversible aggregates formed by self-association are more challenging to characterize. Techniques such as SEC and SDS-PAGE, sometimes require the dilution of

Jennifer J. McManus (ed.), *Protein Self-Assembly: Methods and Protocols*, Methods in Molecular Biology, vol. 2039, https://doi.org/10.1007/978-1-4939-9678-0_12, © Springer Science+Business Media, LLC, part of Springer Nature 2019

samples or a change of buffer, often different to the initial formulation conditions [1, 4]. Fluorescence correlation spectroscopy (FCS) is a promising method to overcome the limitations imposed by the requirement to dilute a sample or change the solution conditions. FCS has been used to measure protein aggregate sizes and size distributions of disease relevant proteins in dilute aqueous solutions [7–9]. Its low sample consumption makes it especially attractive for efficient preformulation and formulation studies.

Here we describe a protocol to measure the size ranges of protein assemblies in highly concentrated protein solutions containing IgG1 using FCS and advanced analysis methods. FCS has primarily been optimized for measurements of the diffusion of species in low concentrations solutions (down to picomolar concentrations) [10] or in cells [11, 12]. However, in concentrated solutions, an increase in the refractive index of the sample and varying solution viscosities make these measurements a little more challenging. We describe in some detail options for the passivation of the surfaces of the measurement chambers, which is important in obtaining reliable results. Briefly, we present data analysis approaches that can be performed on these heterogeneous solutions, including maximum entropy fitting (MEMFCS) and multi-Gaussian fitting (GDM), which recognize that protein formulations often contain a distribution of different sized species and that resolution of these different components is possible using advanced analysis methods.

2 Materials

Prepare all solutions using ultrapure Milli-Q water and using analytical grade reagents.

1. Unlabeled protein—IgG1 is used here (*see* **Note 1**).
2. Fluorescently labeled IgG1 (or a labeling kit) (*see* **Note 2**). Here we use Dylight 488.
3. Fluorescent dye for reference/calibration (e.g., Alexa 488 or the dye that is used for labeling) (*see* **Note 3** and Table 1).
4. Measurement chambers (e.g., LabTek, Nunc 8-well-plates, borosilicate bottom, 200–400 μl sample volume per well (Thermo Scientific), Sensoplate plus, 20–100 μl sample volume per well (384-well, black, 175 μm glass bottom, Greiner Bio-One) or 3D-printed wells (*see* **Note 4**).
5. Coating agents such as BSA, poly-L-lysine, or UHT low-fat milk for surface passivation (*see* **Note 5** and Table 2).
6. Milli-Q water or immersion oil (*see* **Note 6**).

Table 1
Properties for a number of fluorescent small molecules that can be used for protein labeling (NHS-esters) and excitation with $\lambda = 488$ nm

Dye	$A_{max, dye}$ (nm)	Correction factor CF	Extinction coefficient ε_{max} (M^{-1} cm^{-1})	Supplier
Alexa 488	495	0.11	73,000	Thermo Fisher Scientific
Atto 488	500	0.09	90,000	Atto-Tec GmbH
Dylight 488	493	0.147	70,000	Thermo Fisher Scientific
Fluorescein	494	0.3	70,000	Sigma-Aldrich

Table 2
IgG1 labeled with Atto 488, dye–protein ratio = 0.25

Surface treatment	Labeled IgG1
Poly-L-lysine (15 μg/ml)	Protein sticks
Pluronic (10 mg/ml)	Protein sticks
Ficoll (1%)	Protein sticks
PVA (1%)	Protein sticks
Sucrose (2%)	Protein sticks
Tween 20 (1%)	Stable for 10 min, then protein sticks
Casein (1%)	Stable for 10 min
Milk (pure)	Stable overnight

Wells were incubated with each solution for 1 h at RT, afterward rinsed extensively with MilliQ and dried at RT, Sticking = loss of particles of 1/3 or more within the first 10 min of measuring

7. 50 mM sodium acetate buffer, 0.02% sodium azide, pH 5 (*see* **Note 7**).
8. UV Quartz cuvette.
9. UV-Vis spectrophotometer.
10. Benchtop centrifuge.
11. Ultrafiltration unit with a 10 kDa molecular weight cutoff (e.g., Amicon ultrafiltration devices), or syringe-driven filters with a 0.02 μm pore size (Whatman Anotop).
12. Eppendorf tubes.
13. Aluminum foil.
14. Kimwipes.

15. Data Analysis Software (e.g., Origin, Matlab, Igor, QuickFit [13]).
16. Refractometer (*see* **Note 8**).
17. Glycerol (*see* **Note 8**).

2.1 Sample Preparation

1. Prepare unlabeled IgG1 by exhaustive dialysis against 50 mM sodium acetate buffer.
2. Determine the protein concentration by measuring the UV absorbance at 280 nm using the formula $c = A/\varepsilon l$ where A is absorbance, ε is the molar extinction coefficient of the protein (equal 210,000 M^{-1} cm^{-1} for a typical IgG1) and l is the path length in cm (*see* **Note 9**).
3. Dissolve the fluorescently labeled IgG1 as per the supplier's instructions (*see* **Note 2** and Table 1).
4. For labeled protein, calculate the molar concentration of protein using the formula:

$$c = \left(\frac{A_{280} - (A_{\text{max,dye}} - \text{CF})}{\varepsilon}\right) \cdot \text{Dilution factor}$$

where $A_{\text{max, dye}}$ = absorbance maximum of dye and CF is the correction factor for the dye (supplied by the manufacturer; *see* Table 1).

5. Determine the unlabeled to labeled protein ratio using the formula:

$$\frac{\text{Dyes [moles]}}{\text{Protein [moles]}} = \left(\frac{A_{\text{max, labeled}}}{\varepsilon' \cdot c \cdot \text{Dilution factor}}\right)$$

where ε' is the extinction coefficient of the labeled protein (*see* **Notes 10** and **11** and Table 1).

6. To begin concentrating the protein, wash two ultrafiltration units with 10 kDa molecular weight cutoff filters (4 ml) with Milli-Q water.
7. Divide the unlabeled IgG1 between the two ultrafiltration units (adding approximately 1.5 ml to each unit).
8. To one of these ultrafiltration units, add enough labeled IgG1 so that the concentration of labeled protein in the final (concentrated) solution is 10–20 nM. The other unlabeled protein solution will be used for reference measurements and should be concentrated at the same time as the labeled protein.
9. Concentrate the protein in each unit by centrifugation at 6000 × g for 20 min (*see* **Note 12**).
10. Discard the buffer in the collection tube. Carefully mix the protein solution in the filter tube using a pipette to obtain a homogeneous solution, trying to avoid bubble formation.

11. Determine the concentration of the protein in the retentate by UV absorbance. If a higher concentration is required, continue concentrating until the desired concentration is achieved (*see* **Note 13**).
12. Once the desired protein concentration has been reached (approx. 150 mg/ml for the experiments described here), mix the protein solution in the filter tube with a pipette to homogenize the solution and transfer to an Eppendorf tube for storage. Wrap the tubes with aluminum foil to protect the samples from light.

3 Methods

3.1 Preparation of Measurement Chambers (See Note 5)

Proteins interact strongly with surfaces. Since FCS is capable of measuring changes in protein concentration during measurements, as proteins adsorb to the surface, this will result in a reduction in the apparent concentration (since the number of diffusing proteins will decrease as they stick to the surface of the chamber) or even a complete loss of the measurement signal (Fig. 1). There is also a possibility that proteins stuck to the chamber surface may seed surface nucleation and lead to aggregation. To minimize these effects, using treated surfaces for FCS measurements is important. Standard measurement chambers can be treated to minimize protein interactions with the surface (passivation). A broad range of coating agents for surface passivation can be used, including BSA, casein, and Ficoll. The effectiveness will depend on the protein under consideration. As an example, we describe the procedure using UHT low-fat milk, but other coating agents can be used with the same procedure.

1. Open a fresh bottle of UHT low-fat milk and fill each chamber to the top with milk.
2. Incubate at room temperature for 1 h in the dark.
3. Aspirate the low-fat milk completely from the chambers.
4. Rinse the chambers carefully with Milli-Q water 20 times to remove the residual milk.
5. Dry the underside of the chamber slide carefully with a Kimwipe to avoid water marks.
6. Allow the chamber to dry in air at room temperature (or use a gentle stream of filtered nitrogen gas).

3.2 FCS Setup and Optimization

The steps required for instrument setup will depend on the instrument manufacturer and model (*see* the manufacturer's instructions for assistance in performing these steps).

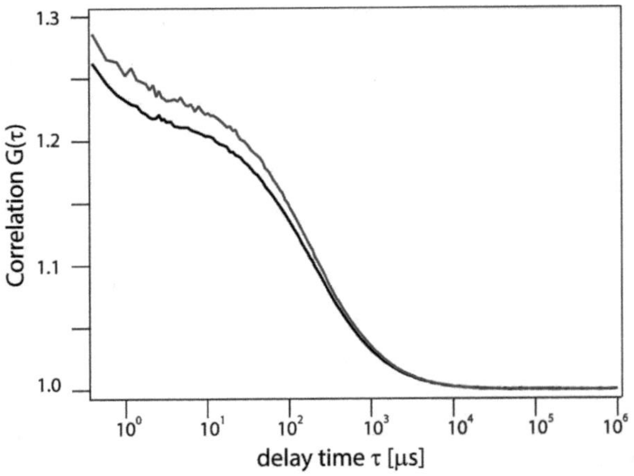

Fig. 1 Correlation function of 35 nM IgG1 labeled with Atto 488 in 100 mM phosphate buffer, pH 7 measured in a well coated with milk (black) and 1% PVA (gray). The increase in the height of the amplitude indicates adsorption of the protein to the surface of the measurement chamber coated with 1% PVA

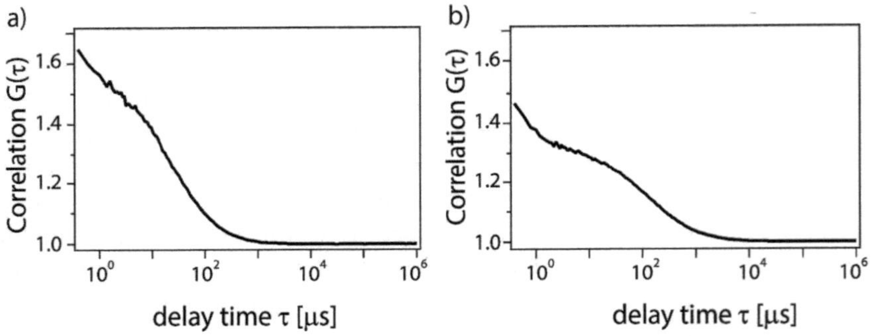

Fig. 2 Dylight 488 in (**a**) Milli-Q water and (**b**) in IgG1 in 50 mM sodium acetate (pH 5) at 150 mg/ml. The FCS setup was adjusted individually for each sample

1. Switch on the laser for 1 h before instrument setup (if required).
2. Place a droplet of water or oil on the immersion objective of the microscope and place the treated chamber on the stage.
3. The calibration of the confocal volume is usually carried out using a fluorescent dye dissolved in Milli-Q water and further diluted in the sample buffer for measurements performed at low concentrations. At higher protein concentrations, the refractive index of the solution is considerably different to that of the buffer and the calibration should be performed in a protein solution at the same concentration that you will use for measurements (or a solution with the same refractive index) (Fig. 2).

4. To perform the calibration, freshly prepare a reference solution containing unlabeled protein (at the same concentration that you will measure at later) and a reference dye (e.g., Alexa 488) to a final concentration of 10–50 nM (of free dye). Add 200 μl of this solution to one of the chambers. It is important that the final protein concentration of the reference solution is the same as for the sample to be measured (*see* **Note 14**).
5. Find the second reflection, optimize the quality of the signal by positioning the confocal volume at an appropriate height above the bottom of the well and with the objective collar ring and adjust the laser power to a suitable level (*see* **Note 15**).
6. Perform the pinhole alignment.
7. Take 10 × 60 s measurements of the dye–protein reference solution.
8. Perform a one-component fit of the averaged autocorrelation function using the instrumentation software to determine the diffusion time (τ_D), and the structure parameter (S). Note these values and fix the structure parameter in the FCS software for the measurement of all other samples with the same protein concentration and buffer.
9. The instrument setup is now fixed and any changes will require the calibration to be performed again.

3.3 Measuring an Autocorrelation Function in a Concentrated Protein Solution

1. Measurements are performed using a mixture of unlabeled and labeled protein. The proportion of labeled protein should be kept to the absolute minimum to ensure that what is measured by FCS reflects that would happen in an unlabeled protein solution. However, the proportion of labeled protein used also needs to be high enough to ensure that the signal to noise ratio is sufficient to obtain good quality data.
2. Add a freshly prepared labeled protein sample (a solution that is spiked with labeled protein—often less than 1% labeled protein is more than enough) to a sample chamber and seal with Parafilm or other adhesive film to reduce evaporation, if the will run over several hours.
3. Take a measurement. 10 × 60 s measurements per time point in the experiment are usually enough (*see* **Notes 16–18**).

3.4 Data Analysis

Usually, the software supplied with the instrument is sufficient for performing data analysis on monodisperse solutions. Fitting procedures for multicomponent, polydisperse solutions (e.g., MEMFCS and multi-Gaussian models) require adequate computer power and data analysis software (e.g., Igor, Matlab, or Quickfit [13]).

3.4.1 A Single Component Fit

To extract physically relevant information from the autocorrelation curve, the data is analyzed by fitting an appropriate model function to the experimental data. The simplest fitting formula is for a single component which is a freely diffusing species in three dimensions:

$$G(\tau) = \frac{1}{N}\left(\frac{1}{1+\frac{\tau}{\tau_D}}\right)\left(\frac{1}{1+\frac{\tau}{S^2\tau_D}}\right)^{\frac{1}{2}}$$

where N is the average number of particles inside the confocal volume, τ is the correlation time, S is the structure parameter, and τ_D is the translational diffusion time of the molecule. An example of a one-component fit for the dye–protein reference is shown Fig. 2. For this specific example, we expect that the dye molecule does not interact with the protein and that the diffusion time relates to a molecule freely diffusing as a monomer in the solution. The single component fit here results in a diffusion time of 130 µs. Use this single component fit to extract the diffusion time for the dye in the reference sample.

3.4.2 Higher Order Fitting: MEMFCS and GDM (Multicomponent Fitting) (See **Note 19**)

For a protein solution which contains oligomers or small aggregates, the solution will contain a number of components with different diffusion times. A single component fit is not appropriate and higher order fits are required. There are several options and approaches that can be used for data analysis in solutions that contain species of different sizes.

Three main approaches for multicomponent fits are used: CONTIN [14], MEMFCS [15, 16], and GDM [7, 17]. Here we will only describe MEMFCS and GDM approaches. The basis of each analysis method is the assumption of a quasi-continuous distribution of a large number of diffusing components. The major advantage of MEMFCS is that it does not make a priori assumptions and thereby reduces the risk of over-interpreting the data for polydisperse systems [18]. GDM works in a similar way, but requires an assumption of the form of the amplitude distribution (i.e., the size ranges of the distributions used for the fit) [2, 13]. While care must be exercised when using GDM, it allows for more refined estimates of particle sizes than MEMFCS (*see* **Note 20**). In our experiments, we tend to use both methods to ensure consistency.

For MEMFCS, the correlation function is fitting is performed using:

$$G(\tau) = \sum_{i=1}^{n} a_i\left(\frac{1}{1+\frac{\tau}{\tau_{Di}}}\right)\left(\frac{1}{1+\frac{\tau}{S^2\tau_{Di}}}\right)^{1/2}$$

where n is the number of freely diffusing species and a_i and τ_{Di} are the relative amplitude and diffusion time of the ith component, respectively.

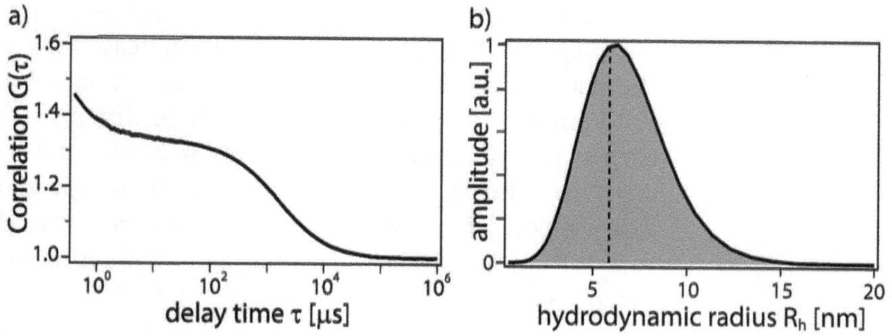

Fig. 3 IgG1 in 50 mM sodium acetate (pH 5) at 150 mg/ml, sample spiked with IgG1 labeled with Dylight 488 (**a**) Correlation function and (**b**) the corresponding normalized size distribution determined using MEMFCS. The dashed lines represent the result of one-component fitting

MEMFCS looks for a distribution of diffusion times $a_i\ (\tau_{Di})$ that maximizes the entropy

$$H = \sum_i p_i \ln p_i$$

with $p_i = a_i\ (\tau_{Di})/\Sigma\ a_i\ (\tau_{Di})$ being the probability of finding a certain component i inside the confocal volume.

In Fig. 3a, we show an autocorrelation function for an IgG1 solution at 150 mg/ml. Figure 3b shows the size distribution obtained using MEMFCS. The dashed line indicates the size determined by a single component fit.

GDM analysis is performed using the following equation:

$$G(\tau) = \sum_{i=1}^{n} a_i(\tau_{Di}) \left(\frac{1}{1 + \frac{\tau}{\tau_{Di}}} \right) \left(\frac{1}{1 + \frac{\tau}{S2\ \tau_{Di}}} \right)^{1/2}$$

with the amplitude distribution

$$a_i(\tau_{Di}) = \sum_{n=1}^{k} A_n \exp\left[-\left(\frac{\ln(\tau_{Di}) - \ln(\tau_{Pn})}{b_n} \right)^2 \right]$$

where A_n is the relative amplitude of the components, τ_{Pn} is the peak diffusion time of the nth component, and b_n is related to the width of the distribution.

This is an iterative process. The first step is to define the number of size ranges to be used in the fit. It is helpful to perform an MEMFCS fit before trying GDM, since this will provide an estimate of the maximum size range of particles present in your sample. Then follow the following procedure:

1. Start with a single size distribution with a relatively narrow size range (e.g., 4–10 nm for IgG1).
2. Increase/decrease the range of the size distribution to improve the quality of the fit.

3. If a single component fit with a broad distribution gives rise to an acceptable fit, it may be possible to increase the resolution of the particle size estimates by introducing a larger number of different size ranges (e.g., two or three different size ranges, each with a Gaussian profile). This is performed by trial and error to achieve the most acceptable fit to the data. In our experience, if larger particles sizes are not present in the sample, two or more GDM distributions will not produce an acceptable fit.
4. The size ranges for the required number of Gaussian distributions can be refined at this point.

3.4.3 Triplet Decay

Triplet decay gives rise to fluctuating fluorescence intensities at very short timescales as labeled molecules passing through the confocal volume decay to the dark state. These fluctuations can contribute an additional component to the autocorrelation curve. This can be accounted for by including the following equation in the fit to the autocorrelation function:

$$G_{\text{triplet}}(\tau) = \left(1 + \frac{T}{1-T}\exp\left(-\frac{\tau}{\tau_{\text{T}}}\right)\right)$$

where τ_{T} is the triplet state relaxation time and T is the fraction of fluorophores in the dark state. The autocorrelation curve then becomes a product of the triplet function and the model $G(\tau)$ as follows:

$$G_{\text{total}}(\tau) = G_{\text{triplet}}(\tau)\cdot G(\tau)$$

3.4.4 Determination of the Hydrodynamic Radius

The determination of the hydrodynamic radius for proteins in a concentrated solution, as measured by FCS is not straightforward. At higher protein concentrations, one cannot assume that the solution viscosity is the same as for water and the standard form of the Stokes–Einstein equation is not useful (unless the solution viscosity and the diffusion coefficient of the dye in this medium are known). To overcome this, we determine the particle size by comparing the diffusion time for the reference dye (here Alexa 488), which has a known hydrodynamic radius to the diffusion time of the labeled protein in solution. The hydrodynamic radius of the labeled protein can then be determined using the following equation:

$$R_{\text{h, protein}} = \frac{\tau_{\text{D, protein}}}{\tau_{\text{D, reference dye}}}\cdot R_{\text{h, reference dye}}.$$

This relation is only valid, if the reference dye has not bound to the protein under formulation conditions and the reference and sample are measured under the same conditions (temperature, buffer, protein concentration, viscosity, etc.).

4 Notes

1. Other proteins can also be prepared in the same way. It is important to consider the selection of pH and the ionic strength of the buffer for each specific protein.
2. FCS measurements at high protein concentrations are usually performed using only a very small fraction of labeled protein (and hence a mixture of labeled and unlabeled protein is used). The lowest possible amount of labeled protein should be used. If a labeling kit is used to tag the protein of interest, the standard option is covalent attachment of a fluorescent molecule, usually by conjugation to either a primary amine or to a free cysteine. The choice of the specific dye will depend on the amino acid sequence of the protein being examined. For methods that require covalent attachment of a dye, it is important to remove all excess (nonconjugated) dye. Exhaustive washing is required. The washings should be tested by measuring fluorescence intensity at the emission maximum of the dye in a fluorometer to ensure that all excess dye has been removed. For both prelabeled protein and protein labeled using a kit, the characteristics of the fluorescent label chosen should also be carefully evaluated for brightness, photostability (to ensure little or no bleaching), quantum efficiency, the size of dye (~1 nm) vs. the size of protein (often a few nm), the ability to determine the concentration of the labeled protein, and dye–-protein ratio.
3. To calibrate the FCS instrument, the confocal volume is determined by measuring the diffusion of a fluorophore that does not interact directly with the protein. Select a dye best suited to the excitation lasers available and the filter set installed for emission. The excitation maximum for the calibration dye should match the excitation maximum for the fluorescent label used with the protein.
4. A number of different sample environments and chambers are commercially available for FCS measurements, even using very low volume if there is limited sample. At a minimum, the chamber should preferably have a glass base (but not too thick, since this will prohibit the adjustment of the collar ring of the objective lens). If no suitable product is commercially available, chambers can be 3D-printed and glued to a suitable microscope slide.
5. FCS measurements are often performed at very low concentrations of fluorescently labeled protein (even if the total protein concentration is high). If the protein binds to the sample

chamber walls, this can reduce the bulk concentration significantly or indeed be responsible for the assembly of the protein (*via* surface nucleation). The concentration of protein (in a nonaggregating solution at nanomolar concentration of the fluorescently labeled protein) can be monitored over time from the particle concentration data gathered in the autocorrelation function (as $1/N$). If the intercept of the autocorrelation function increases over time (without a concurrent increase in the diffusion time), protein binding to the chamber walls may be occurring. This surface binding can be minimized by coating the wells with a variety of other reagents. A procedure using UTH low-fat milk has been described here, but poly-L-lysine, PEG, lipids, or BSA can also be used (using the same procedure).

6. The Milli-Q water used for the objective should be dust free. If present, the measurements may be affected. Filter the water through 0.22 μm filters prior to use.
7. Prepare a suitable buffer using analytical grade reagents. Often buffers can contain stabilizers such as salts, amino acids or sugars. Filter through 0.22 μm filters prior to use. It is important that complex buffers that contain excipients are free of large particles (i.e., flocculation of amino acids or salts), as this will lead to light scattering.
8. To determine the refractive index of a protein formulation you can use a refractometer and prepare a refractive index-matching solution with glycerol and Milli-Q water [19]. A solution with a refractive index matched to the protein solution refractive index can reduce the amount of protein material required for measurements. Note that the refractive index matching solution does not necessarily match the viscosity of the protein sample.
9. To obtain reliable concentration values, use a positive replacement pipet for viscous samples. Furthermore, wipe the outside of the tip with the aspirated protein solution before you release the protein into the buffer for dilution. The amount of protein sticking to the outside of the tip can lead to an increased concentration of the diluted sample and an overestimation of the final concentration.
10. Be aware that this is only an estimate of the number of dye molecules per protein. The equation is only valid if the extinction coefficient of the free dye ε is the same as the one of the protein-bound dye ε'. This is not automatically the case, but most often these changes are very small and therefore negligible [20].
11. The dye–protein ratio includes unlabeled, monolabeled, and multilabeled proteins and assumes each are present in a Poisson

distribution [21, 22]. Poisson statistics are appropriate here since a protein can only be labeled with an integer number of dyes *k*. Assuming that the dye–protein ratio corresponds to the mean of the Poisson distribution μ (dye–protein ratio $= \mu$) the following description is obtained:

$$\text{Percentage } (k) = \left[\frac{(\mu)^k}{k!} e^{-(\mu)}\right] \cdot 100\%$$

For example, the amount of unlabeled protein ($k = 0$) is:

$$\text{Percentage of unlabelled protein } (k = 0) = \left[e^{-(\mu)}\right] \cdot 100\%$$

12. The spinning time depends on the viscosity of the solution, which is dependent on the starting and final protein concentrations, if aggregates are present and the viscosity of the formulation buffer. A very high final protein concentration in a viscous buffer can lead to prolonged spinning times.
13. For a sample containing a tiny amount of labeled protein, we can assume that the contribution of the label to the absorbance at 280 nm is negligible. Most instruments are not sensitive enough to detect a deviation from the baseline at the excitation maximum of the dye; therefore, the correction described in **step 4** in Subheading 2.1 is not applicable.
14. Ensure that the fluorescent dye and the buffer are compatible. The dye brightness can depend on pH or the presence of specific ions (e.g., calcium).
15. Depending on the concentration of the protein solution, it may be better to position the focus closer to the glass base to avoid distortion of the confocal volume due to the change of the refractive index between the water droplet and the highly concentrated protein solution. Stay at least 20 μm above the glass base. More details can be found in Müller et al. and Banachowicz et al. [19, 23].
16. If measurements will take several hours (typical for kinetic studies of protein assembly), sealing the sample chamber is important since evaporation will lead to an increase in concentration of the solutes and a change in viscosity. Starting with a larger sample volume will help to ensure that evaporation does not dramatically alter the sample concentration.
17. For longer measurements, check the water droplet on the lens regularly to ensure it does not evaporate. Using oil instead of water immersion can help.
18. If dust or large particles are present in the sample, this will produce significant spikes in the intensity. "Dust filters" integrated into most FCS software platforms can be used to

exclude count rates above a certain value (which can be defined by the user), ensuring that these counts arising from these spikes are not used in the autocorrelation function.

19. Distinguishing between monomers and small oligomers is very challenging. The autocorrelation function is an average of the time correlated intensity fluctuations of the diffusing species, where the intensity is proportional to the sixth power of the hydrodynamic radius. In general, to distinguish one species from another, one component should have twice the hydrodynamic radius of the other. However, as a broad rule of thumb, a doubling in the hydrodynamic radius is equivalent to eightfold increase in molecular weight.
20. In general, polydispersity in the system is indicated when there are significant deviations between the data obtained experimentally and the one-component fit. Even if the initial fit is satisfactory (as indicated by the residuals), a multicomponent fit may still be warranted if the sizes produced by the one-component fits are physically unrealistic (e.g., indicate a hydrodynamic radius smaller than a protein monomer).

Acknowledgments

This work has emanated from research supported by the Synthesis and Solid State Pharmaceutical Centre, the National University of Ireland Maynooth and funded by a research grant from Enterprise Ireland (EI) under Grant Number IP 2015 0358. We thank J.O. Rädler for use of his laboratory and the FCS setup at LMU Munich (Germany).

References

1. Shire SJ, Shahrokh Z, Liu J (2014) Challenges in the development of high protein concentration formulations. J Pharm Sci 93 (6):1390–1402
2. Harris RJ, Shire SJ, Winter C (2004) Commercial manufacturing scale formulation and analytical characterization of therapeutic recombinant antibodies. Drug Dev Res 61 (3):137–154
3. Daugherty AL, Mrsny RJ (2006) Formulation and delivery issues for monoclonal antibody therapeutics. Adv Drug Deliv Rev 58 (5–6):686–706
4. Lowe D, Dudgeon K, Rouet R, Schofield P, Jermutus L, Christ D (2011) Aggregation, stability, and formulation of human antibody therapeutics. Adv Protein Chem Struct Biol 84:41–61
5. Staub A, Guillarme D, Schappler J, Veuthey J-L, Rudaz S (2011) Intact protein analysis in the biopharmaceutical field. J Pharm Biomed Anal 55(4):810–822
6. Muneeruddin K, Thomas JJ, Salinas PA, Kaltashov IA (2014) Characterization of small protein aggregates and oligomers using size exclusion chromatography with online detection by native electrospray ionization mass spectrometry. Anal Chem 86 (21):10962–10999
7. Mittag JJ, Milani S, Walsh DM, Rädler JO, McManus JJ (2014) Simultaneous measurement of a range of particle sizes during $A\beta_{1-42}$

fibrillogenesis quantified using fluorescence correlation spectroscopy. Biochem Biophys Res Comm 448(2):195–199

8. Mittag JJ, Rädler JO, McManus JJ (2018) Peptide self-assembly measured using fluorescence correlation spectroscopy. Methods Mol Biol 1777:159–171
9. Wolff M, Mittag JJ, Herling T, de Genst E, Dobson CM, Knowles TPJ, Braun D, Buell AK (2016) Quantitative thermophoretic study of disease-related protein aggregates. Sci Rep 6:22829
10. Elson EL (2011) Fluorescence correlation spectroscopy: past, present and future. Biophys J 101:2855–2870
11. Kim SA, Heinze KG, Schwille P (2007) Fluorescence correlation spectroscopy in living cells. Nat Methods 4:963–973
12. Engelke H, Heinrich D, Rädler JO (2010) Probing GFP-actin diffusion in living cells using fluorescence correlation spectroscopy. Phys Biol 7(4):046014
13. Krieger JW, Langowski J (2015) QuickFit 3.0 (status: beta, compiled: 2015-03-18, SVN: 3891): a data evaluation application for biophysics. http://www.dkfz.de/Macromol/quickfit/. Accessed 2 Jan 2018
14. Provencher SW (1982) Contin: a general purpose constrained regularization program for inverting noisy linear algebraic and integral equations. Comput Phys Commun 27:229–242
15. Nyeo SL, Chu B (1989) Maximum-entropy analysis of photon correlation spectroscopy data. Macromolecules 22(10):3998–4009
16. Garai K, Sahoo B, Sengupta P, Maiti S (2008) Quasihomogeneous nucleation of amyloid beta yields numerical bounds for the critical radius, the surface tension, and the free energy barrier for nucleus formation. J Chem Phys 128 (4):045102-1-7
17. Pal N, Verma SD, Singh MK, Singh MK, Sobhan S (2011) Fluorescence correlation spectroscopy: an efficient tool for measuring size, size-distribution and polydispersity of microemulsion droplets in solution. Anal Chem 83 (20):7736–7744
18. Sengupta P, Garai K, Balaji J, Periasamy N, Maiti S (2003) Measuring size distribution in highly heterogeneous systems with fluorescence correlation spectroscopy. Biophys J 84 (3):1977–1984
19. Banachowicz E, Patkowski A, Meier G, Klamecka K, Gapiński J (2014) Successful FCS experiment in nonstandard conditions. Langmuir 30(29):8945–8955
20. Vira S, Mekhedov E, Humphrey G, Blank PS (2010) Fluorescent labeled antibodies – balancing functionality and degree of labeling. Anal Biochem 402(2):146–150
21. Cilliers C, Nessler I, Christodolu N, Thurber GM (2017) Tracking antibody distribution with near-infrared fluorescent dyes: impact of dye structure and degree of labeling on plasma clearance. Mol Pharm 14:1623–1633
22. Brinkley (1992) A brief survey of methods for preparing protein conjugate with dyes, haptens, and cross linking reagents. Bioconjug Chem 3:2
23. Müller C, Eckert T, Loman A, Enderlein J, Richtering W (2008) Dual-focus fluorescence correlation spectroscopy: a robust tool for studying molecular crowding. Soft Matter 5:1358–1366

Chapter 13

Size Determination of Protein Oligomers/Aggregates Using Diffusion NMR Spectroscopy

Pancham S. Kandiyal, Ji Yoon Kim, Daniel L. Fortunati, and K. H. Mok

Abstract

Diffusion-ordered spectroscopy (DOSY) is a widely used NMR technique for the identification of different chemical moieties/compounds contained in mixtures and has been successfully employed for the separation of small molecules based on hydrodynamic radii. Herein we show that DOSY can also be applied for the size determination of larger biomolecules such as proteins and protein oligomers/aggregates. Proof-of-principle is first shown with a cross-linked oligomeric protein mixture where the hydrodynamic volumes of each component are estimated and subsequently verified with size-exclusion HPLC and SDS polyacrylamide gel electrophoresis. We then determine the sizes of protein oligomers contained in a protein solution subjected under amyloid fibrillogenesis conditions. These studies aim to provide insight into the kinetics behind protein aggregation involved in amyloidosis as well as to determine the hydrodynamic radii of proteins within the mixture.

Key words DOSY, Hydrodynamic radii, Protein, Oligomers, Aggregation, Diffusion coefficient, Pulsed-field gradient (PFG), Stokes–Einstein equation

1 Introduction

Nuclear magnetic resonance (NMR) spectroscopy is used for characterization of conformational and atomic level details of small molecules, peptides and proteins in solution. Availability of various multidimensional, multinuclear experiments and isotopically labeled techniques have made life easier to study the 3D structure of macromolecules using NMR spectroscopy [1]. Diffusion is an intrinsic property of all molecules and varies on the basis of a particle's size, shape, solubility, charge distribution, etc. [2]. The theoretical and experimental demonstration for self-diffusion property of molecules was first given by Stejskal and Tanner in 1965 [3]. Advancement in hardware and software technology in NMR have made the performing of diffusion experiments straightforward for small molecules, leading to (1) the determination of the

Jennifer J. McManus (ed.), *Protein Self-Assembly: Methods and Protocols*, Methods in Molecular Biology, vol. 2039, https://doi.org/10.1007/978-1-4939-9678-0_13, © Springer Science+Business Media, LLC, part of Springer Nature 2019

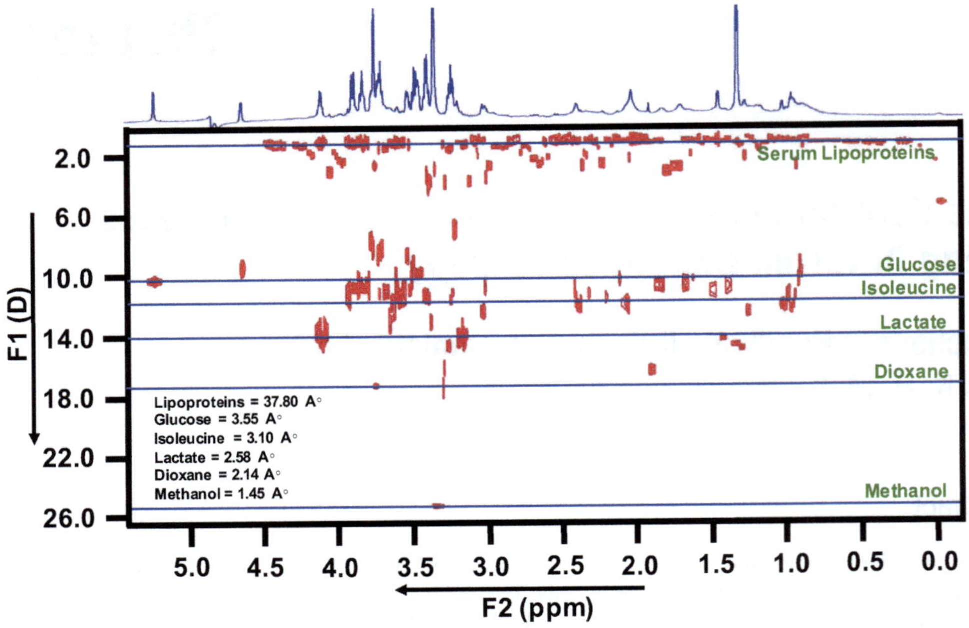

Fig. 1 Separation of metabolites and lipoproteins in a serum sample using DOSY NMR

mobility of a compound, (2) binding of small molecules to protein, (3) aggregation of molecules, and (4) identification of individual compounds in a mixture (Fig. 1) [4].

One of the key technical features in a diffusion NMR experiment pulse sequence is the utilization of pulsed-field gradients (PFGs) that spatially encode the solutes, the magnetizations of which are then refocused through either a spin echo or stimulated-echo of the spins (PFGSE or PFGSTE) [5]. For the past two decades, DOSY has been implemented for the characterization of mixtures containing small molecules [6, 7]. In contrast, its application for macromolecules has been less due to the shorter relaxation times and the overlapping of chemical shifts, despite the fact that in principle, DOSY NMR can be very helpful to understand the rate of protein aggregation and oligomer formation. While previous work with DOSY NMR has been able to show the separation of proteins and lipids, samples containing a mixture of different size proteins have not been shown with any frequency [8]. Here we have implemented 2D DOSY NMR to determine the size of an intentionally prepared, chemically cross-linked oligomeric mixture of a protein by addition of a cross-linker [9]. For our experiment, we used DOSY bipolar pulse pair stimulated echo (Dbppste) pulse sequence [10] and optimized the experiment by varying different parameters such as the diffusion gradient length, gradient diffusion delay time, relaxation delay and gradient recovery time.

When samples contain a mixture of different proteins, it is hard to readily distinguish all of the different sized-proteins according as a function of their diffusion coefficient value (D value) due to the large overlapping of their chemical shifts. For our cross-linked sample contained a mixture of overlapping signals originating from monomers, dimers and trimers of a protein, extracting the diffusion coefficients from requires deconvolution of the data. By plotting the intensities/peak areas obtained against the square of the gradient strength (G^2) according to the Stejskal–Tanner equation and fitting the data points with multiple Gaussian curves, it was possible to capture more subtle differences in diffusion coefficients (D) which would not be accessible using a less involved analysis. The D value used further for calculating the hydrodynamic radii using the Stokes–Einstein equation [11].

From these data and the exponential/Gaussian fitting of them, we were able to extract three different sizes of component protein, that is, monomer, dimer, and trimer for lysozyme, despite the fact that the overlapping NMR signals of the oligomers made it impossible to consider these proteins as separate components from the 2D DOSY experiment alone, and as a result making it difficult to calculate their different diffusion coefficient values.

2 Materials

All reagents should be prepared using ultrapure water.

1. Proteins; Hen egg-white lysozyme (HEWL; lyophilized powder, protein ≥90%, ≥40,000 units/mg protein), bovine thyroglobulin, bovine γ-globulins, human serum albumin, β-lactoglobulin, horse cytochrome *c*, L-tryptophan.
2. Deuterium oxide (D_2O).
3. 1,4-Dioxane.
4. The water-soluble, homobifunctional cross-linker bis(sulfosuccinimidyl) substrate (BS^3). BS^3 contains sulfo-NHS (*N*-hydroxysuccinimide) esters on both carboxylate ends of the molecule, providing greater solubility due to the negatively charged sulfonate groups, making the molecule very hydrophilic.
5. Chromatography was performed on a Dionex Ultimate HPLC 3000 Standard System running Chromeleon 6 software (Dionex, Thermo Scientific), and the size exclusion-HPLC (SE-HPLC) was a Superdex 200 10/300 GL column (10 mm × 30 cm, Particle size 13 μm, GE Healthcare).
6. SDS electrophoresis gels, tank, and power supply.
7. NMR experiments were carried out on an Agilent Technologies 18.8T (800 MHz) DD2 Premium Compact spectrometer

with a triple-resonance, 5 mm enhanced cold probe; however, any high-field magnet with pulsed-field gradient capabilities will be sufficient. (The interest is more in monitoring the decay of signal as a function of applied gradient than the resolution of the spectra.)

8. HPLC elution buffer; 0.05 M sodium phosphate at pH 7.0 containing 0.1 M Na_2SO_4, filtered through a 0.45 μm (or 0.2 μm) pore-size membrane and degassed.
9. Conjugation buffer and; 0.1 M sodium phosphate buffer at pH 7.0 containing 0.15 M NaCl, filtered through a 0.45 μm filter.
10. Quenching buffer for cross-linking with BS^3; 1 M Tris–HCl, pH 7.5 filtered through a 0.45 μm (or 0.2 μm) pore-size membrane.
11. Solutions for NMR; 0.05 M sodium phosphate buffer at pH 7.0 filtered through a 0.45 μm (or 0.2 μm) pore-size membrane. Other buffers are also acceptable provided that the buffer does not contain components (usually of aliphatic nature) that have strong, unexchangeable ^1H spectral signals. All solutions for NMR should contain at least 10% of D_2O that will be used as a deuterium lock signal.

3 Methods

3.1 Initial Diffusion Coefficient Value Estimation from SE-HPLC Data

An initial, rough estimate of the expected diffusion coefficients of the cross-linked HEWL species can be obtained from analysis of size exclusion-HPLC (SE-HPLC) data.

1. Dissolve the calibration standards at a concentration of 0.1 mM with the elution buffer and inject onto the SE-HPLC column eluting with 0.05 M sodium phosphate buffer pH 7.0 containing 0.1 M Na_2SO_4 at a flow rate of 0.50 mL/min with detection at 280 nm.
2. Generate a standard calibration curve using a suitable range of protein molecular weight standards (for example, bovine thyroglobulin, bovine γ-globulins, human serum albumin, β-lactoglobulin, horse cytochrome *c*, L-tryptophan) including human serum albumin (HSA), which has a hydrodynamic radius (R_H) of 40 Å (Figs. 2 and 3). The calibration curve is a plot of R_H values of these proteins vs. the elution volume obtained from the SE-HPLC.
3. Prepare cross-linked forms of HEWL using BS^3. For a HEWL concentration of 5 mg/mL, use a tenfold molar excess of BS^3 cross-linker and react at room temperature for 3 h. Quench the reaction using the quenching buffer to a final concentration of

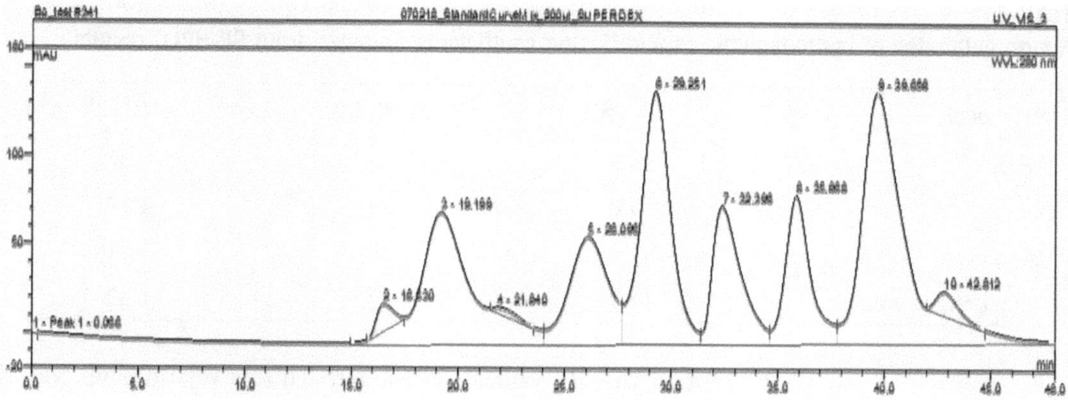

Fig. 2 Protein calibration standard chromatogram using size-exclusion HPLC

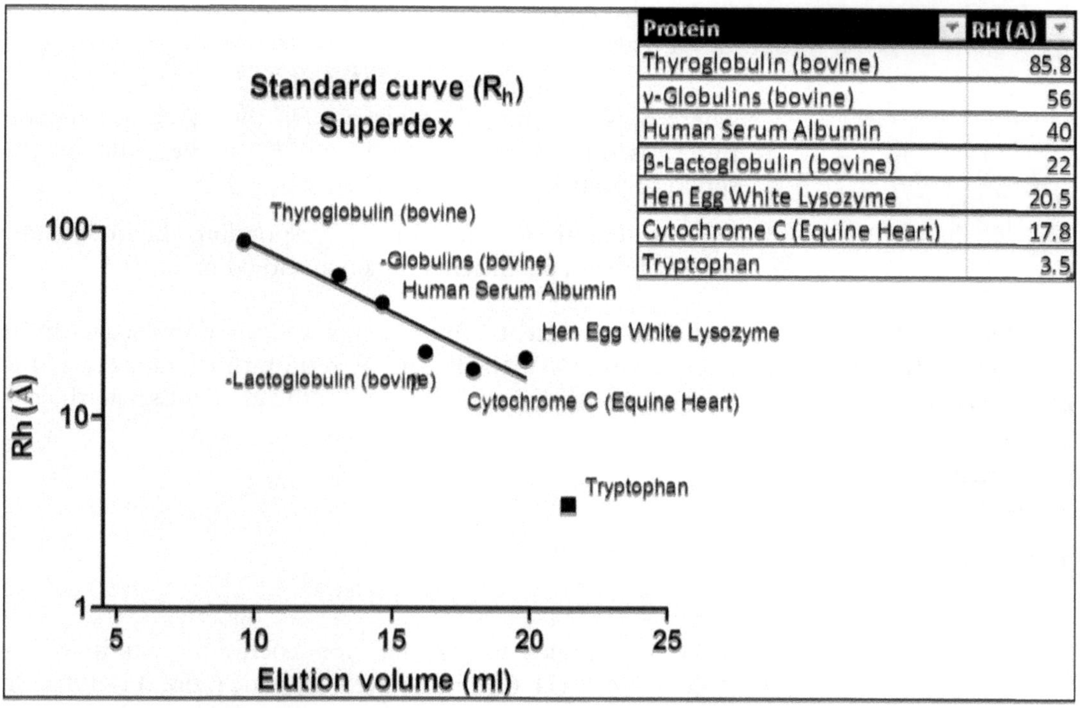

Fig. 3 SE-HPLC standard curve

20–50 mM Tris. Dialyze the protein solution with 0.05 M sodium phosphate buffer pH 7.0 (or any buffer that will keep the protein well solubilized).

4. Run a SDS-PAGE to observe the extent of cross-linking.
5. Inject an approximate 0.1 mM concentration sample of cross-linked HEWL onto the SE-HPLC and measure the retention times (*see* **Note 1**).

Table 1
Rough estimates of hydrodynamic radii/diffusion coefficients obtained from SE-HPLC results

HEWL peak	R_t (min)	R_H (A)	D (10^{10} m^2/s)
Trimer	36.5	28	0.87
Dimer	39.0	23	1.10
Monomer	41.0	20	1.21
Compacted monomer	42.9	17	1.42

6. Obtain the R_h values by using the linear relationship found on the standard calibration curve (Fig. 3). From the Stokes–Einstein equation, a rough estimate of the diffusion coefficients (D) for each species is as follows:

$$D = \frac{k_b T}{6\pi\eta R_H}$$

where k_b is Boltzmann's constant, T is the absolute temperature, η is the dynamic viscosity, and R_H is the radius of the spherical particle.

Typical retention times and corresponding hydrodynamic radii/diffusion coefficients obtained shown in Table 1.

3.2 DOSY-NMR and the Identification of Protein Oligomer/Aggregate Components in a Mixture

3.2.1 Background

The diffusion coefficient (D) of a single species of molecule can be extracted from DOSY-NMR data by plotting the Intensity of the peaks (I) obtained against the square of the gradient strength (g^2), according to the Stejskal–Tanner Eq. 3:

$$I = I_0 e^{-D\gamma^2 g^2 \delta^2 (\Delta - \delta/3)} \quad (1)$$

which takes the general form

$$y = Ae^{-Qx^2} \quad (2)$$

where Q corresponds to the diffusion coefficient multiplied by $(\gamma^2\,\delta^2\,(\Delta - \delta)/3)$. (Here, γ is the gyromagnetic ratio, Δ is diffusion time given, and δ is the time for the gradient pulse.)

For a mixture of different components, the graph will represent a summation of different Gaussian curves (poly-Gaussian), so the equation will become:

$$\sum i = A_1 e^{-Q_1 x^2} + A_2 e^{-Q_2 x^2} + \ldots + A_i e^{-Q_i x^2} \quad (3)$$

3.2.2 Practical

To fit the DOSY-NMR data of a mixture of components with an equation of this type, a sequential approach is taken.

1. Dissolve the HEWL samples in 50 mM sodium phosphate buffer (pH 7.4, 90% H_2O, 10% D_2O), but the pH and buffer of the sample should be optimized to the conditions necessary for the oligomers/aggregates (*see* **Note 2**).
2. Perform diffusion-ordered spectroscopy (DOSY) measurements at 293 K. The DgscteSL_dpfgsc DOSY pulse program (Agilent VNMR) is used, which consists of gradient compensated stimulated echo with spin lock using the excitation sculpting solvent suppression method [12]. A spectral window of 13,020 Hz was used, with an acquisition time of 2.46 s with a relaxation delay of 3 s. The FIDs were collected with 32,000 complex data points with 64 scans. Logarithmically the gradient pulse strength was increased from 3% to 86% of the maximum strength of 32,767 G/cm in 60 steps. A diffusion time (Δ) of 100 ms and bipolar half-sine-shaped gradient pulses (δ) of 5 ms were applied. 1,4-Dioxane, which is known to behave independently of protein concentration and the folded state of the protein, was used as an internal chemical shift reference and hydrodynamic radius calibration reference (3.75 ppm; $R_H = 2.12$ Å) [11, 13]. Three replicate acquisitions were given for each sample, and the resulting diffusion coefficient (D) values calculated.
3. Upon acquisition of the NMR spectra, fit first a sample of neat dioxane with a single Gaussian to obtain the diffusion value for dioxane (Fig. 4). By referencing to dioxane it is then possible to obtain relative diffusion values/hydrodynamic radii for other species without having to correct for all the other factors contained in Q (which is instead necessary to obtain absolute measurements) (*see* **Note 3**).
4. Running the identical experiment and observing dioxane in a protein mixture sample clearly results in a sum of different poly-

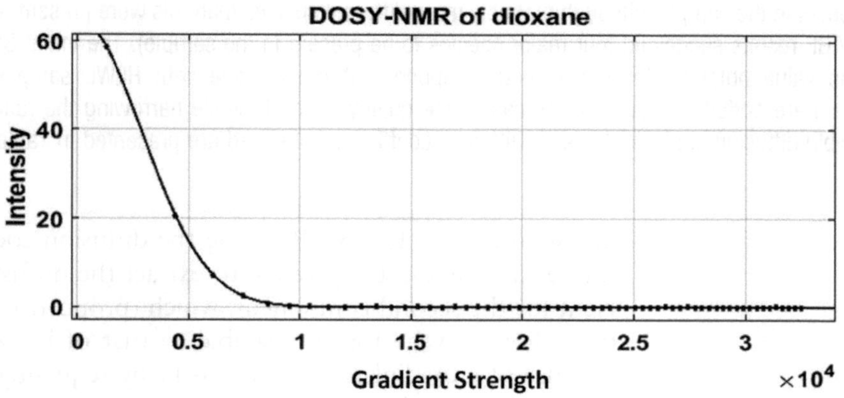

Fig. 4 DOSY-NMR fitting of pure 1.4-dioxane. Pure dioxane fit with a single curve (Eq. 2). The diffusion coefficient (un-adjusted) obtained from this fit was 5.868×10^{-8}

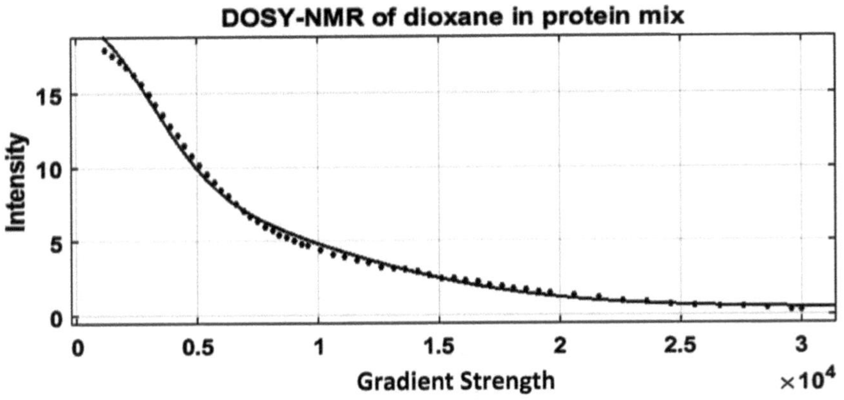

Fig. 5 DOSY-NMR fitting of 1,4-dioxane in a protein solution. Dioxane from the protein mix, fit with a poly-Gaussian (Eq. 3). In this case, it is clear that the fit could not be a single Gaussian like in the case of the pure dioxane sample, probably due to protein peaks overlapping in the dioxane region. By fixing the diffusion value of one of the components of the poly-Gaussian, it was possible to obtain a diffusion values for a second component, equal to 6.137×10^{-9}

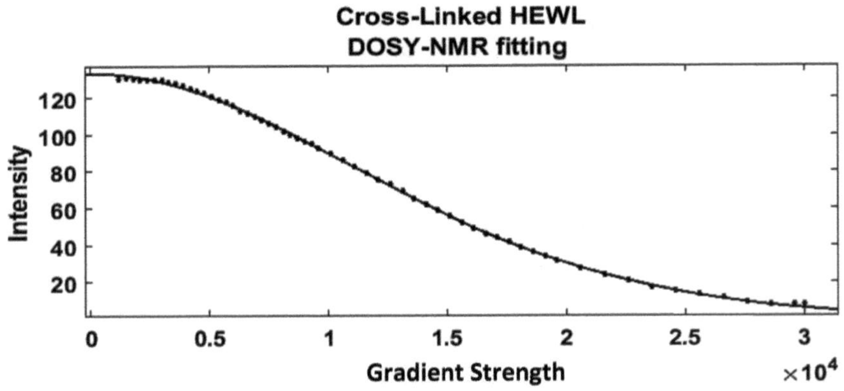

Fig. 6 DOSY-NMR fitting of cross-linked hen egg white lysozyme. Poly-Gaussian fitting of cross-linked HEWL DOSY-NMR data. A four component poly-Gaussian was chosen to approximate the population of cross-linked species present in the sample (although probably more different species than this were present, the SE-HPLC and SDS PAGE results suggested four major species to be present in the sample). The initial Gaussian was fixed at the value obtained for the second component of the dioxane with HEWL sample, and other components were added sequentially, increasing the quality of the fit while narrowing the range of values of the possible diffusion coefficients used. Diffusion coefficients obtained are presented in Table 2

Gaussian curves (Fig. 5). By fixing the diffusion coefficient of one of the values, it is possible to extract the diffusion coefficient for the second component, which (proportionally to the pure dioxane value) is comparable to that of lysozyme. The mixture of cross-linked HEWL was fit by sequentially adding curves to an initial (imperfect) fit, fixing values as fits got better and narrowing the range toward the values obtained through SE-HPLC (Fig. 6).

Table 2
Diffusion coefficients/hydrodynamic radii obtained from DOSY-NMR

Species	$D \times 10^9$ (a.u.)	R_h (A)	I_0 (a.u.)
Dioxane	58.7	2.12	64.8
HEWL1 (trimer)	3.0	41.4	82.5
HEWL2 (dimer)	4.8	25.9	16.19
HEWL3 (mono)	6.1	20.3	34.97
HEWL4 (compact monomer)	6.5	19.1	0.4

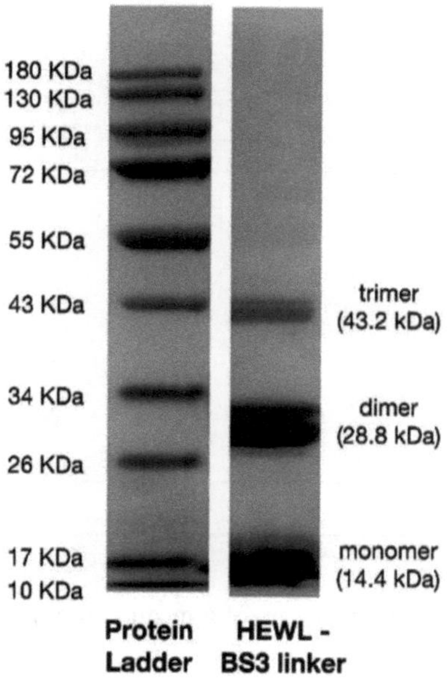

Fig. 7 SDS-PAGE gel of the selected HEWL-BS^3 cross-linked mixture sample

4 Notes

1. Compared to the SDS-PAGE results, which reflect disulfide bridge-reduced and fully denatured proteins (Fig. 7), the separation of monomer, dimer, and trimer of HEWL showed that the hydrodynamic radii of these species (in their nondenatured/reduced state) were not as widely distributed (Figs. 8 and 9). Nevertheless, not only was it possible to distinguish different species in the shoulders of the peak obtained by deconvolution (Figs. 9 and 10), but one is able to identify a further-compacted monomer which appeared to have resulted from intramolecular covalent cross-linkages.

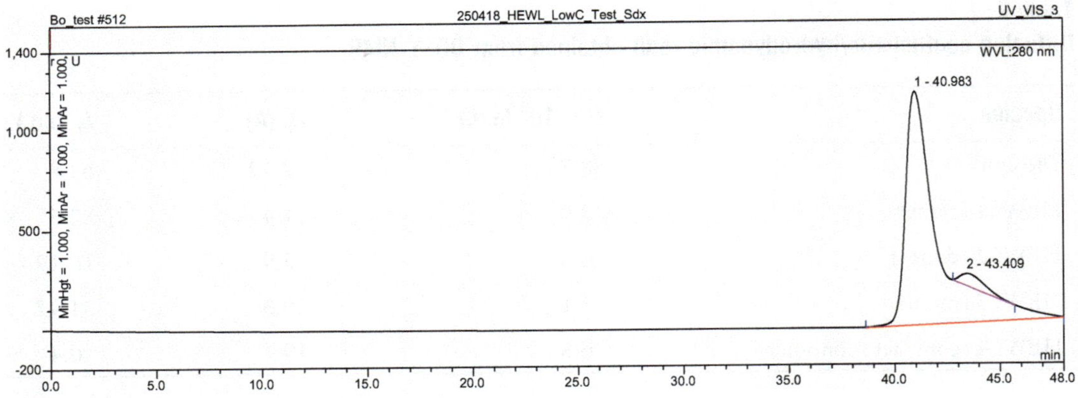

Fig. 8 Size-exclusion chromatogram of monomeric HEWL

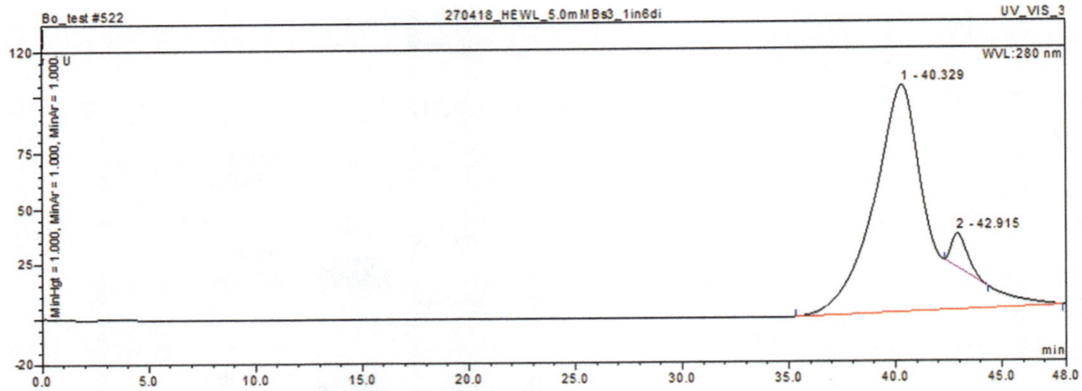

Fig. 9 Size-exclusion chromatogram of a mixture of cross-linked HEWL (trimer, dimer, and monomer)

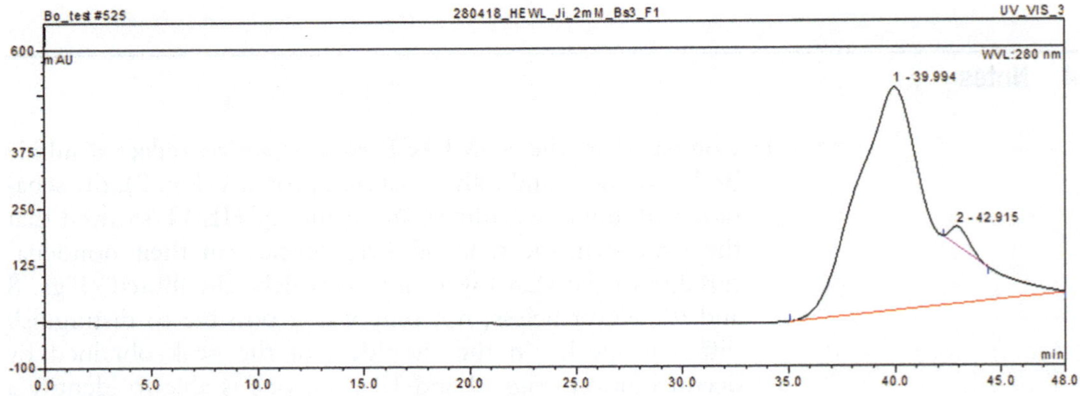

Fig. 10 Size-exclusion chromatogram of mixture of cross-linked HEWL (trimer and dimer enriched via centrifugal concentration)

2. When performing NMR, it would be advantageous to use a buffer which does not have interfering signals with the sample of interest.
3. The D (diffusion coefficient) values for 1,4-dioxane are slightly variable dependent upon the cosolute, which rightly reflects the different solution microenvironment conditions that both solutes are mutually experiencing for each sample.

References

1. Cavanagh J, Fairbrother WJ, Palmer AG III, Rance M, Skelton NJ (2006) Protein NMR spectroscopy (2nd Ed): principles and practice. Academic, New York, NY
2. Li D, Keresztes I, Hopson R, Williard PG (2009) Characterization of reactive intermediates by multinuclear diffusion-ordered NMR spectroscopy (DOSY). Acc Chem Res 42 (2):270–280
3. Stejskal EO, Tanner JE (1965) Spin diffusion measurements: spin echoes in the presence of a time-dependent field gradient. J Chem Phys 42:288–292
4. Dehner A, Kessler H (2005) Diffusion NMR spectroscopy: folding and aggregation of domains in p53. Chembiochem 6:1550–1565
5. Pagès G, Gilard V, Martinob R, Malet-Martino M (2017) Pulsed-field gradient nuclear magnetic resonance measurements (PFG NMR) for diffusion ordered spectroscopy (DOSY) mapping. Analyst 142:3771
6. Morris KF, Johnson CS Jr (1992) Diffusion-ordered two-dimensional nuclear magnetic resonance spectroscopy. JACS 114:3139–3141
7. Barjat H, Morris GA, Smart SC, Swanson AG, Williams SCR (1995) High resolution diffusion ordered 2D spectroscopy (HR-DOSY)—a new tool for the analysis of complex mixtures. J Magn Reson B 108:170–171
8. Balayssac S, Delsuc M-A, Gilard V, Prigent Y, Malet-Martino M (2009) Two-dimensional DOSY experiment with excitation sculpting water suppression for the analysis of natural and biological media. J Magn Reson 196:78–83
9. Arora B, Tandon R, Attri P, Bhatia R (2017) Chemical crosslinking: role in protein and peptide science. Curr Protein Pept Sci 18 (9):946–955
10. Wu D, Chen A, Johnson CS Jr (1995) An improved diffusion-ordered spectroscopy experiment incorporating bipolar-gradient pulses. J Magn Reson A 115(2):260–226
11. Jones JA, Wilkins DK, Smith LJ (1997) Characterization of protein unfolding by NMR diffusion measurement. J Biomol NMR 10:199–203
12. Pelta MD, Barjat H, Morris GA, Davis AL, Hammond SJ (1998) Pulse sequences for high-resolution diffusion-ordered spectroscopy (HR-DOSY). Magn Reson Chem 36:706–714
13. Shimizu A, Ikeguchi M, Sugai S (1994) Appropriateness of DSS and TSP as internal references for (1)H NMR studies of molten globule proteins in aqueous media. J Biomol NMR 4:859–862

Part III

Computational Approaches to Measure Protein Self-Assembly

Chapter 14

Patchy Particle Models to Understand Protein Phase Behavior

Nicoletta Gnan, Francesco Sciortino, and Emanuela Zaccarelli

Abstract

In this chapter we describe numerical procedures to evaluate the phase behavior of coarse-grained models for globular proteins. Specifically we focus on models based on hard spheres complemented with "patchy-like" anisotropic interactions that mimic the attractive regions on the surface of the proteins. We introduce the basic elements of grand canonical Monte Carlo simulations for these types of models in which rotational and translational moves need to be accounted for. We describe the techniques for the estimation of the fluid–fluid critical point, coexistence curve, and fluid–crystal boundaries. We also discuss an efficient method for the evaluation of the fluid–fluid phase diagram: the successive umbrella sampling technique. Finally we briefly describe how to exploit the same tools for the calculation of the phase behavior of protein binary mixtures.

Key words Anisotropic interactions, Patchy particles, Globular proteins, Phase behavior, Critical point, Coexistence curve, Monte Carlo simulations

1 Introduction

Soft matter physics aims at studying the physical properties and the phase behavior of systems made by mesoscopic constituents that can be synthesized in the laboratory or can be found in nature. In particular, the advances in chemical synthesis nowadays provide the possibility to generate a wide range of colloidal particles with different shapes and interactions [1], which can form a variety of states such as crystals, gels, liquid-crystals, and glasses. In addition, the size of colloidal particles allows the dynamics in the system to be measured with several experimental techniques, even at the single particle level as in the case of confocal microscopy [2]. For all these reasons colloids represent the favorite model systems of physicists for investigating new states and phases and to provide useful insights into the behavior of more complex systems such as biological ones.

Jennifer J. McManus (ed.), *Protein Self-Assembly: Methods and Protocols*, Methods in Molecular Biology, vol. 2039, https://doi.org/10.1007/978-1-4939-9678-0_14,

For instance attractive colloids, i.e., colloids that interact with an attractive potential, display a phase behavior that strongly depends on the range of the attraction. It has been shown [3] that if the attraction is sufficiently short-ranged, the fluid–fluid coexistence line becomes metastable with respect to the fluid–solid coexistence. A metastable fluid–fluid coexistence curve is also typical of several globular proteins [4–7] and, for this reason, models of colloids with short-range attraction have been employed for the investigation of globular proteins phase behavior [8–10]. These models generally provide a good qualitative agreement with the behavior of globular proteins, but still present some major problems: (1) a quantitative comparison is difficult to achieve when the attraction is purely isotropic, often resulting in a too narrow fluid–fluid coexistence curve; (2) such models neglect the presence of specific crystal contacts in real proteins that have a fundamental role in the phase behavior and in the crystallization process; (3) experimental observations show that globular proteins have non-homogeneous surface patterns made by a distribution of charge and hydrophobic residues, the latter mainly buried in the core of the proteins [7]. Such heterogeneous distribution translates into highly selective interactions between proteins that are clearly orientation-dependent, which need to be incorporated in the models as they are responsible for significant variations with respect to the phase behavior of isotropic models, namely the presence of low-density critical points [11] and crystals [12]. In addition, single point mutations [13] or the addition of a fluorescent dye that bind to specific sites of the protein [14] have been shown to deeply modify the phase behavior, again providing indications toward the need to go beyond isotropic models. Recently, the behavior of eye lens proteins of squids [15], pidan protein gels [16], and monoclonal antibody suspensions [17, 18] has been interpreted in terms of patchy models, clearly showing the importance of directional interactions for properly describing protein phase behavior. For these reasons recent coarse-grained numerical simulations of proteins all incorporate the anisotropic aspect of the interactions [19–22] by mainly relying on models of patchy particles [11], i.e., hard-sphere particles with anisotropic attractive sites. The advantage in exploiting such models is twofold: on one hand, they provide a better description of the phase behavior of globular proteins and are able to capture the variations in the phase behavior when mutations are induced in real proteins [22]. On the other hand, the investigation of new patchy models can also help to develop new strategies to manipulate proteins, for instance, for producing high-quality crystals or for the mutagenesis of native proteins to control their interaction.

This chapter is conceived for providing a walkthrough description of the numerical methods for performing computer simulations of patchy-particle models for proteins and for analyzing the results to build their phase diagram. The chapter is organized as follows: in the first part we will describe the interaction potential of patchy particles and we will introduce the framework of the Monte Carlo method for simulating particles with anisotropic interactions. Then we will describe, step-by-step, the main passages to locate the fluid–fluid critical point and the corresponding coexistence line from grand canonical Monte Carlo simulations, both using "standard" simulations and with a more advanced technique, named successive umbrella sampling. Finally we will also discuss how to calculate the fluid–crystal phase boundaries as well as how to generalize these concepts for the investigation of the phase behavior of binary mixtures.

2 Materials

In the following (*see* Subheading 3.5.1) we will show that the most efficient way to calculate numerically the phase behavior of the models proposed is to run a set of parallel simulations on different cores. The amount of cores needed depends on the size of the system considered and the density at which the critical point is located. For the data shown in this chapter we have employed 150 cores (for the largest system size) on 12-cores Intel(R) Xeon (R) CPU X5680 @3.33 GHz machines.

3 Methods

3.1 Patchy Models

In a patchy colloidal approach, a globular protein is modeled as a hard sphere (HS) of diameter σ decorated with attractive sites (patches). The HS potential [23] takes into account the excluded volume of the protein and forbids two proteins i and j with relative distance r_{ij} to interpenetrate as

$$V_{\mathrm{HS}}(r_{ij}) = \begin{cases} \infty & \text{if} \quad r_{ij} \leq \sigma \\ 0 & \text{if} \quad r_{ij} > \sigma. \end{cases} \tag{1}$$

The directional attraction can be modeled in several ways (*see* **Note 1**). Here we focus on the so-called Kern–Frenkel (KF) potential [24, 25] in which a patch is represented by a cone with the tip placed at the center of the particle as shown in Fig. 1. Given a patch α on particle i, the normal vector identifying the patch orientation is indicated as $\hat{n}_i^{\alpha}$. Additional parameters that are important for modulating the patch–patch interaction are the angular width 2θ and the range δ.

An attractive interaction between two patches α and β on particles i and j only occurs when the vector $\mathbf{r}_{ij}$ connecting the centers of i and j lies inside the cones of both patches. This is mathematically described by the angular function

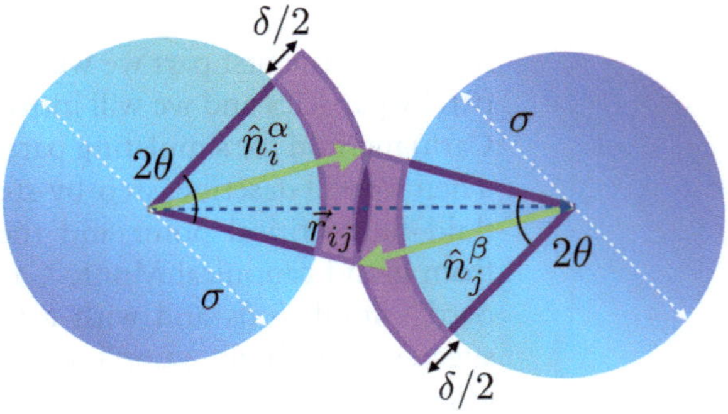

Fig. 1 Cartoon of patchy particles where patches (purple areas) are modeled via the Kern–Frenkel model

$$f(\hat{n}_i^\alpha, \hat{n}_j^\beta) = \begin{cases} 1 & \text{if} \quad \begin{cases} \hat{r}_{ij} \cdot \hat{n}_i^\alpha > \cos(\theta) \\ \hat{r}_{ij} \cdot \hat{n}_j^\beta > \cos(\theta) \end{cases} \\ 0 & \text{otherwise.} \end{cases} \tag{2}$$

In addition, the two patches must overlap and hence the center-to-center distance between particles i and j has to be smaller than $\sigma + \delta$ where δ is the interaction range. When also this condition is satisfied, patches form a bond of energy ε, which has the form of a square-well (SW) potential:

$$V_{\mathrm{SW}}(r_{ij}) = \begin{cases} \infty & \text{if} \quad r_{ij} \leq \sigma \\ -\varepsilon & \text{if} \quad \sigma < r_{ij} \leq \sigma + \delta \\ 0 & \text{if} \quad r_{ij} > \sigma + \delta. \end{cases} \tag{3}$$

Given Eq. 2 and Eq. 3 the Kern–Frenkel potential can be written as

$$V_{\mathrm{KF}}(r_{ij}, \hat{n}_i^\alpha, \hat{n}_j^\beta) = V_{\mathrm{SW}}(r_{ij}) \cdot f(\hat{n}_i^\alpha, \hat{n}_j^\beta) \tag{4}$$

i.e., as a SW potential modulated by an angular function. It follows that the total interaction potential between two patchy particles is

$$V(r_{ij}, \hat{n}_i^\alpha, \hat{n}_j^\beta) = V_{\mathrm{HS}}(r_{ij}) + \sum_{\alpha=1}^{M_i} \sum_{\beta=1}^{M_j} V_{\mathrm{KF}}(r_{ij}, \hat{n}_i^\alpha, \hat{n}_j^\beta) \tag{5}$$

where the two sums run over all patches M_i and M_j of particles i and j.

To ensure the single-bond per patch condition (*see* **Note 2**), it is necessary to control the bonding volume of the patch (i.e., the purple area of Fig. 1) by imposing that δ and θ satisfy the relation

$$\sin(\theta) \leq \frac{1}{2(1 + \delta/\sigma)} \tag{6}$$

It has been found that the combination of a patchy attraction plus an isotropic square-well potential is able to reproduce experimental data for the phase coexistence of lysozyme and of bovine and human γD (HGD) crystalline protein [14, 20]. In the following we will focus on such a case as a representative example to illustrate the numerical methods to evaluate the fluid phase diagram, also in comparison with HGD experimental results.

3.2 Monte Carlo Simulations of Patchy Particles

Suppose that we want to estimate, for a given temperature and volume, the bonding probability $p_b = -\langle U \rangle / (N \cdot f)$ of the system which is defined as the average number of bonds formed by particles, where f is the valence and $\langle U \rangle$ is the average potential energy. To sample p_b for the relevant configurations of particles in the phase space one needs to calculate the ensemble average of the potential energy. As discussed in **Note 3**, the ensemble average of a given observable can be calculated by performing an importance-weighted random walk in phase space, which is the strategy adopted in Monte Carlo (MC) simulations, described below.

3.2.1 Displacement Moves

In order to explore the phase space in MC simulations, the Metropolis algorithm [26] is employed to control the transition rule from a configuration to another. Specifically, given the n-tuple of particle positions $\{\mathbf{r}^N\}$, the probability p to go from one configuration n ($\{\mathbf{r}^N\}$) associated with its Boltzmann factor $e^{-\beta U(\{\mathbf{r}^N\})}$ to a new one $n'(\{\mathbf{r}'^N\})$ with $e^{-\beta U(\{\mathbf{r}'^N\})}$ is given by:

$$\begin{aligned} p(n \to n') &= \alpha_{n,n'}\ acc(n \to n') \qquad &\text{if} \quad \rho_n/\rho_{n'} \geq 1 \\ &= \alpha_{n,n'} &\text{if} \quad \rho_n/\rho_{n'} < 1 \end{aligned} \tag{7}$$

where $\rho_n = \exp[-\beta U(\{\mathbf{r}^N)\}] / \int_V d\mathbf{r} \exp[-\beta U(\{\mathbf{r}^\mathbf{N}\})]$ is the probability density of the configuration n, $\alpha_{n,n'}$ is the probability that a trial move from n to n' is attempted, and $acc(n \to n')$ is the probability that the move is accepted. Since the denominator of ρ_n and $\rho_{n'}$ is the same, the condition in Eq. 7 reduces to the calculation of $U(\{\mathbf{r}'^N\}) - U(\{\mathbf{r}^N\})$. The Metropolis scheme can thus be described as follows:

1. Starting from an initial configuration n with energy $U(\{\mathbf{r}^N\})$, generate a trial configuration n' with energy $U(\{\mathbf{r}'^N\})$ by picking randomly a particle i and attempting to displace it of $r'_i = r_i + \Delta_{\text{MAX}}\mathbf{v}$, being Δ_{MAX} the maximum displacement and $\mathbf{v}$ a random vector whose components are uniformly distributed between $[-1, 1]$.
2. Calculate the energy difference $\delta U_{n,n'} = [U(\{\mathbf{r}'^N\}) - U(\{\mathbf{r}^N\})]$. If $\delta U_{n,n'} < 0$, then $\exp(-\beta\delta U_{n,n'}) > 1$ and the move is accepted. On the other hand, if $\delta U_{n,n'} > 0$, the move is accepted with a probability $\exp(-\beta\delta U_{n,n'})$. In the latter case a random number $R \in [0, 1]$ is extracted and the move is accepted if $\exp(-\beta\delta U_{n,n'}) \geq R$.

The maximum interval of displacement is directly connected to the acceptance rate, defined as the ratio between the attempted and the accepted moves. For example, for a dense system in which particles can make only small steps to move, if Δ_{MAX} is too large, then the probability of accepting the trial move will be very small. On the other hand, if the step is too small, then the time required for sampling all the relevant configurations in the phase space would be too large. A reasonable choice is to perform exploratory runs and set the acceptance rate to 0.5. Some works [27, 28] suggest that the efficiency of the MC simulation is improved if the acceptance rate is smaller than 30%.

3.2.2 Rotation Moves

For particles with anisotropic interactions, rotations must also be implemented in the MC moves. Hence the Metropolis scheme has to be adapted to include the rotation of a particle around a vector. The procedure can be summarized as follows:COMP: Please set the below equation within the page width.

1. Randomly extract a particle i and attempt to displace it by virtually changing its position $r_i' = r_i + \Delta_{\text{MAX}}\mathbf{v}$.
2. Before evaluating the new energy of the configuration $U(\{\mathbf{r}'^N\})$, randomly generate a versor $\hat{u}$ on a sphere [29] and rotate the normal vectors identifying each patch of the particle by a random angle $\theta \in [-\theta_{\text{MAX}}, \theta_{\text{MAX}}]$. In particular given a patch α on particle i and its normal vector $\hat{n}_i^\alpha$, the rotational matrix

$$R(\theta,\hat{u}) = \begin{bmatrix} \cos\theta + u_x^2(1-\cos\theta) & u_x u_y(1-\cos\theta) - u_z\sin\theta & u_x u_z(1-\cos\theta) + u_y\sin\theta \\ u_y u_x(1-\cos\theta) + u_z\sin\theta & \cos\theta + u_y^2(1-\cos\theta) & u_y u_z(1-\cos\theta) - u_x\sin\theta \\ u_z u_x(1-\cos\theta) - u_y\sin\theta & u_z u_y(1-\cos\theta) + u_x\sin\theta & \cos\theta + u_z^2(1-\cos\theta) \end{bmatrix}$$

   is calculated and then the patch vector is rotated through the transformation

$$\hat{n}_i'^\alpha = \mathcal{R}(\theta,\hat{u}) \cdot \hat{n}_i^\alpha. \tag{8}$$

3. Calculate the energy difference $\delta U_{n,n'} = [U(\{\mathbf{r}'^N\}) - U(\{\mathbf{r}^N\})]$ and accept or reject the rototranslation according to the Metropolis scheme described in Subheading 3.2.1.

Another way to handle rotations is through quaternions [30] which we do not discuss here. Also for rotation moves it is common to set the angular displacement in order to have an acceptance ratio 0.3. For standard patchy particles at low and moderate densities, a convenient choice is $\theta_{\text{MAX}} \sim 0.1$ rad and $\Delta_{\text{MAX}} \sim 0.05\sigma$.

3.3 Grand Canonical Simulations

Once the main framework of MC simulations has been discussed we need to account for the case in which the system is in equilibrium with a reservoir of particles. In this case the referring ensemble is the grand canonical (GC) ensemble and the thermodynamic parameters are μ, V, and T where μ is the chemical potential which describes the energy cost to insert a particle into the volume V at a given temperature T. In this ensemble, the number of particles fluctuates and extra MC moves have to be introduced to account for it, namely insertion and removal moves. In the first case a particle is inserted with a random position and orientation; in the second case instead a randomly picked particle is removed from the simulation box. We can summarize the new Monte Carlo grand-canonical (MC-GC) scheme as follows:

Extract a random number $R_1 \in [0, 1]$.

- if $R_1 < (\mathcal{N} - 1)/\mathcal{N}$ (where N is and integer, describing how frequently rototranslations are performed with respect to insertion/removal moves) perform a rototranslations as for the canonical ensemble
- else extract a random number $R \in [0, 1]$; if $R > 0.5$ try to insert a new particle with random position and orientation and with a probability:

$$acc_{\text{insertion}}(n \to n') = \min\left\{1, \frac{\exp[\beta\mu]\, V}{(N+1)\sigma^3} \exp\left[-\beta\left(U(\{\mathbf{r}'^{N+1}\}) - U(\{\mathbf{r}^N\})\right)\right]\right\} \quad (9)$$

if $R \leq 0.5$ try to remove a randomly chosen particle from the simulation box with an acceptance probability

$$acc_{\text{removal}}(n \to n') = \min\left\{1, \frac{N\sigma^3}{\exp[\beta\mu]\, V} \exp\left[-\beta\left(U(\{\mathbf{r}'^{N-1}\}) - U(\{\mathbf{r}^N\})\right)\right]\right\}. \quad (10)$$

To the best of our knowledge there are no rules to a priori decide the value of N. A commonly employed choice is $N = 500$ (*see* also **Note 4**).

3.4 Critical Point Estimation

A first hint of the presence of a second order critical point detected by MC-GC simulations is provided by strong fluctuations of the number of particles (and hence of the density). As a consequence the density distribution becomes bimodal, with two peaks corresponding to a low-density and a high-density phase, respectively, as shown representatively in Fig. 2.

However, a precise estimation of the critical point location is obtained by performing a finite size scaling analysis of near-critical fluids [31]. Indeed, at the fluid-fluid critical point the distribution of the ordering operator $\mathcal{M}$ (discussed below) follows a universal curve, shown in Fig. 2 (right panel), which is characteristic of the Ising universality class:

$$P(\mathcal{M}) = a_{\mathcal{M}}^{-1} L^{\beta/\nu} P_{\text{universal}}(a_{\mathcal{M}}^{-1} L^{\beta/\nu}[\mathcal{M} - \mathcal{M}_c]). \quad (11)$$

In Eq. 11 $\mathcal{M}_c$ is the ordering operator evaluated at the critical point, $a_{\mathcal{M}}^{-1} L^{\beta/\nu}$ is the scaling factor which depends on the size of the system

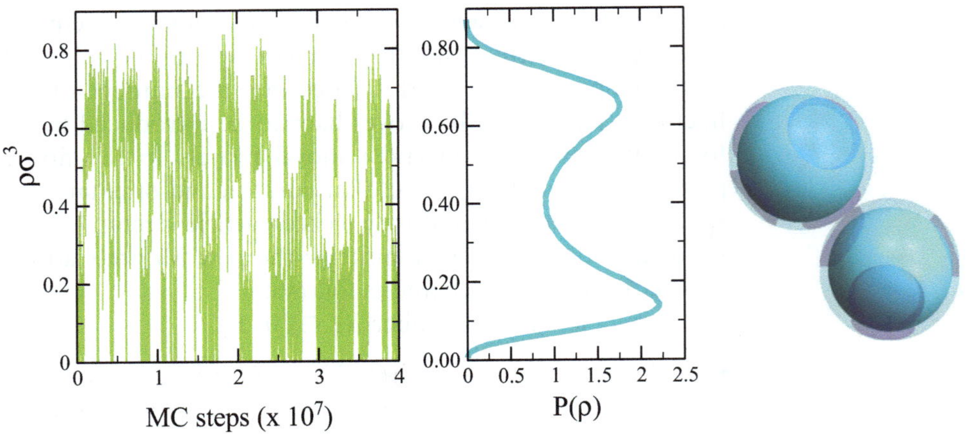

Fig. 2 Grand canonical simulations of a patchy-particle model of globular proteins at its critical point [20]. In the considered case, particles are modeled with an isotropic SW attraction complemented by very short-range attractive patches placed randomly on the surface (see snapshot on the right) [14, 20]. The critical parameters are ($T_c = 0.8185$, $\mu_c = -2.4667$) and the edge of the simulation box is $L = 5\sigma$. (Left panel) critical density fluctuations within a GMC_GC simulation run; (right panel) distribution of density fluctuations. The model describes a protein as a hard sphere of diameter σ with four Kern–Frenkel patches randomly distributed on the surface complemented by an isotropic square well of depth u_0 and width 0.5σ. The KF parameters are $\cos\theta = 0.95$ and $\delta = 0.05\sigma$ and $\epsilon = 5u_0$. The same model is used in the following figures

L through the critical exponents β and ν, and gives unit variance through $a_{\mathcal{M}}^{-1}$. For the $3d$ Ising universality class $\nu = 0.629$ and $\beta = 0.326$. The universal probability distribution can thus be numerically obtained and a good approximation to it is given by the formula,

$$P_{\text{universal}}(\mathcal{M}) \propto \exp[-(\gamma\mathcal{M}^2 - 1)^2(a\gamma\mathcal{M}^2 + c)] \tag{12}$$

where $a = 0.158$ and $c = 0.776$ are universal parameters and γ is such that it gives unitary variance to the distribution [32]. Following the Bruce–Wilding mixing parameter method, the ordering operator is a linear combination of the density field ρ and the energy density field u, i.e., $\mathcal{M} = \rho - su$, where s is the so-called mixing parameter. Although in general $s \sim 0$ for isotropic interactions (and hence $\mathcal{M} = \rho$), the role of the mixing field becomes important for particles with anisotropic interactions (*see* also **Note 5**). Indeed it has been shown [33] that for near-critical systems of patchy particles the distribution of density fluctuations is not symmetric; in particular, the lower the valence of the particles, the larger the asymmetry between the two peaks of the distribution, signaling the increasing role of the mixing field. This is shown in Fig. 2 (right panel) where the critical $P(\rho)$ is reported for a patchy-particle model of globular proteins [14, 20], which displays asymmetric peaks in density. However a symmetrical distribution and a perfect superposition of the data with the universal Ising distribution are recovered when $P(\mathcal{M} = \rho - su)$ is considered, as shown in Fig. 3.

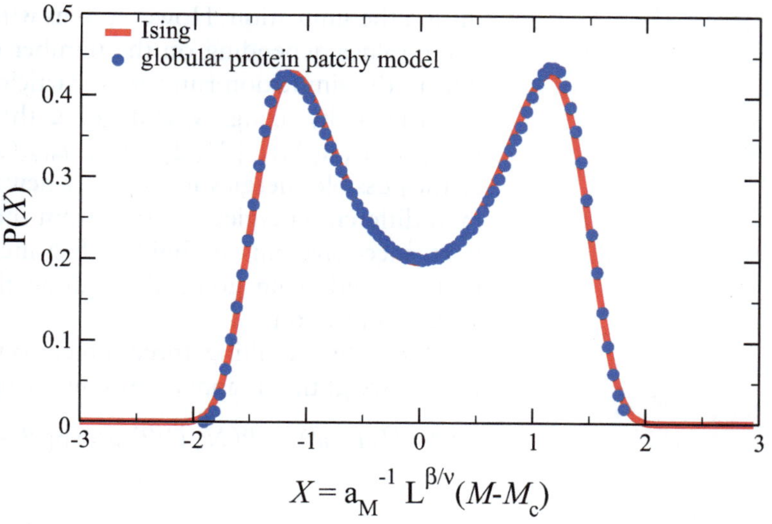

Fig. 3 Distribution of the ordering parameter $\mathcal{M} = \rho - su$ at the critical point for a patchy-particle model of globular proteins (filled symbols) and comparison with the universal distribution of the Ising universality class (solid line). The considered model is the same as in Fig. 2. The mixing parameter is found to be $s = 0.028$

3.4.1 Numerical Estimate of the Critical Point: Least-Square Algorithm and Histogram Reweighting

Having discussed the shape of the density fluctuations and how the ordering parameter distribution must look like at the critical point, it is now important to be able to detect numerically the critical point. To this aim the histogram reweighing technique is employed, which is combined with a fitting procedure corresponding to a least-square based algorithm that finds the best values of the control parameters (T, μ) and of the mixing parameter s such that the distribution at the critical point superimposes to that of the Ising universality class.

We here discuss the implementation of the method. To locate the critical point we first perform several MC-GC simulations at different T and μ in order to bracket the (T, μ) portion of the phase diagram close to the critical point by locating those state points that display two peaks in the numerical joint distribution $P(N, U; T, \mu)$. During the simulation, for each step, the number of particles and the total energy are recorded. The calculation of $P(N)$, rather than $P(N, U; T, \mu)$, is an efficient proxy for the critical point since it also displays a double peak distribution close to criticality (*see* Fig. 2). Once established a close enough state point from which to start the fitting procedure, the joint distribution $P(N, U; T, \mu)$ is calculated. This is done by incrementing the bins of the histogram associated with the combined occurrence of a given number of particles and energy during the simulation. In particular, the bins $[N_i, E_j]$ (where N_i is the particle number and E_j is the energy value) are built by incrementing N_i from the minimum to the maximum value found during the simulation. However, bins with same N_i can have different E values depending on the number of bonds formed/broken during the simulation run for N_i particles. Therefore the bins will be ordered by fixing N_i and going through all the possible E_j, i.e., $[N_i, E_{min}(N_i)]$, $[N_i, E_j]...[N_i, E_{max}(N_i)]$. After accounting for all the possible energies at fixed N_i, a new set of bins with $N = N_i+1$ (and different energies) is built (from the smallest to the highest); next, successive bins are built by first incrementing the number of particles and again going through all the possible energies with such N and so on.

From the resulting three-dimensional distribution the histogram reweighting is applied by employing the relation

$$P(N, U; T', \mu') = P(N, U; T, \mu)\exp[(\beta - \beta')U]\exp[(\beta\mu' - \beta\mu)N] \tag{13}$$

Eq. 13 allows to predict the joint distribution $P(N, U; T', \mu')$ from $P(N, U; T, \mu)$ having the same potential energy U, i.e., to find the values of the control parameters T', μ' whose joint distribution best fit the Ising curve. The fitting procedure is described as follows:

1. Given (T, μ), change their values to (T', μ') and calculate the histogram reweighting factor $\exp[(\beta - \beta')U]\exp[(\beta\mu' - \beta\mu)N$.
2. Multiply $P(N, U; T, \mu)$ by the previous factor in order to obtain $P(N, U; T', \mu')$.
3. Vary the mixing parameter s and calculate the order parameter $\mathcal{M} = N - sU$ for all the values of N and U in $P(N, U; T', \mu')$. $\mathcal{M}$ is then normalized by V to generate an intensive quantity. From these, build the histogram of $P(\mathcal{M})$ by summing all the contributions of $P(N, U; T', \mu')$ coming from the same values of $N - sU$.
4. Scale $P(\mathcal{M})$ onto the Ising curve. To this aim, calculate the norm $\mathcal{N} = \int P(\mathcal{M})d\mathcal{M}$, the average value $\langle\mathcal{M}\rangle = (1/\mathcal{N})\int \mathcal{M}P(\mathcal{M})d\mathcal{M}$, and its variance $var[\mathcal{M}] = (1/\mathcal{N})\int(\mathcal{M} - \langle\mathcal{M}\rangle)^2 P(\mathcal{M})d\mathcal{M}$ and define the new variable $\mathcal{X} = (\mathcal{M} - \langle\mathcal{M}\rangle)/\sqrt{var[\mathcal{M}]}$.
5. Since we want to know $P(\mathcal{X})$, impose $P(\mathcal{X})d\mathcal{X} = P(\mathcal{M})d\mathcal{M}$, where $d\mathcal{M}/d\mathcal{X} = \sqrt{var[\mathcal{M}]}$ since $\mathcal{M} = \mathcal{X}\sqrt{var[\mathcal{M}]} + \langle\mathcal{M}\rangle$. It follows that $P(\mathcal{X}) = P(\mathcal{M})\sqrt{var[\mathcal{M}]}$. $P(\mathcal{X})$ is a distribution with zero mean and unitary variance which has the correct form to be compared with the universal distribution of the Ising class.
6. Normalize $P(\mathcal{X})$ and perform a least-square fit to minimize the sum of residuals. If the result of the fit is unsatisfactory, start from the beginning of the scheme by changing the value of T, μ, and s as explained above.

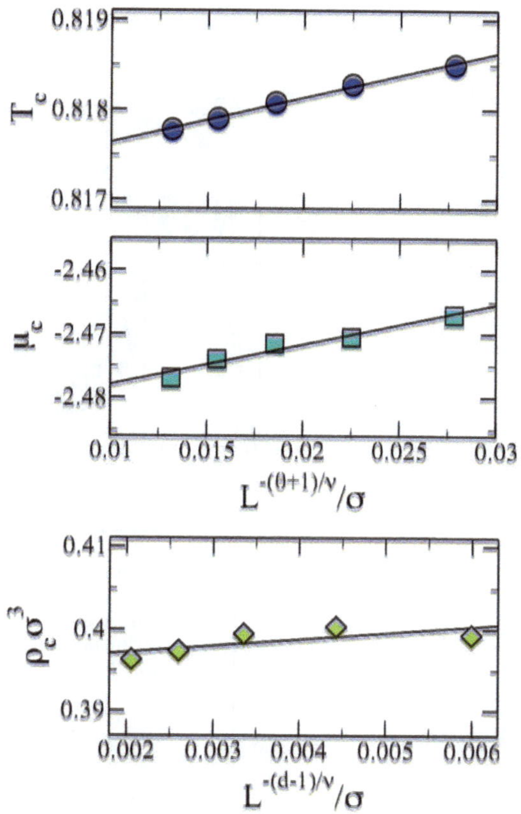

Fig. 4 Size dependence ($L = 5, 5.5, 6, 6.5, 7\sigma$) of the critical temperature T_c (upper panel), the critical chemical potential μ_c (middle panel), and the critical density ρ_c (lower panel) for the same model as in Fig. 2 [14, 20]. Here k_BT and μ are expressed in units of u_0, the depth of the isotropic square-well [20]

3.4.2 Size Effects

Finite size scaling theory predicts that the values of the control parameters at the critical point depend on the size of the system and satisfy scaling laws. In particular, finite size scaling predicts that $T_c \sim L^{-(\theta+1)/\nu}$,$\mu_c \sim L^{-(\theta+1)/\nu}$ and $\rho_c \sim L^{-(d-1)/\nu}$, where $\theta = 0.54$ is the universal correction to the scaling exponent [34] and d is the dimensionality of the system. The size dependence of the critical parameters as a function of the system size is shown in Fig. 4.

3.5 Vapor–Liquid Coexistence Curve Estimation

The coexistence between a low-density and a high-density phase implies the equality of the chemical potential, the temperature, and the pressure of the two phases. While T and μ are fixed in the MC-GC simulation and therefore they are identical for the two phases, the equality of pressure can be found as follows. If phase coexistence is satisfied, then a bimodal distribution of density appears corresponding to the two distinct phases. Equality of pressure for the two phases is obtained when the area below the two peaks is identical. Bimodal distributions with equal area underneath

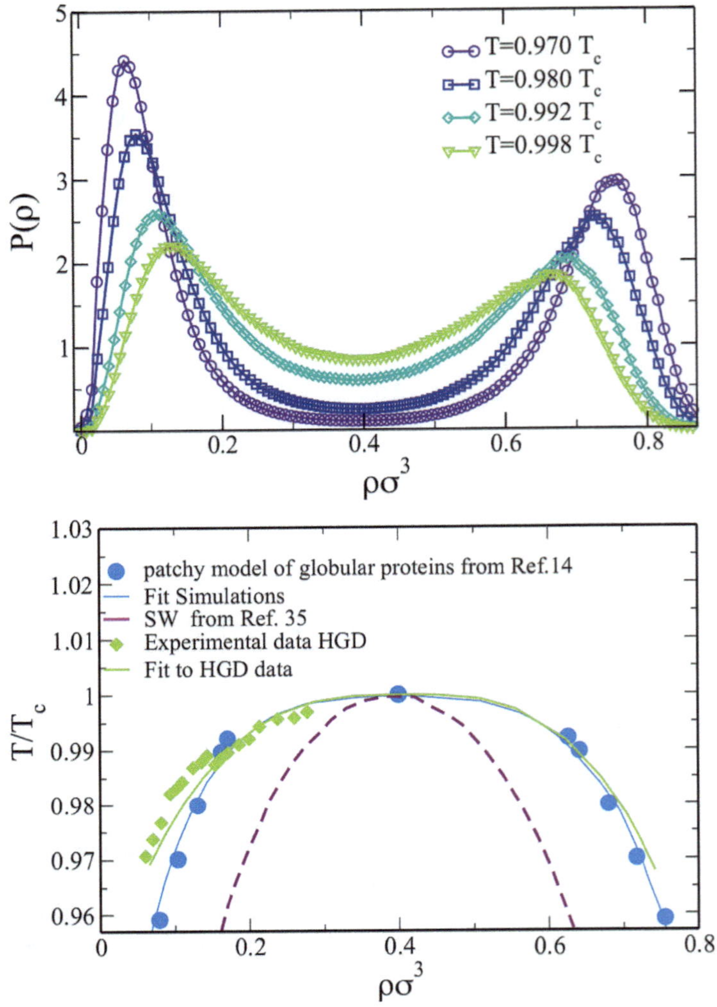

Fig. 5 (Upper panel) Density probability distributions along the coexistence line for the same patchy model as in previous figures and $L = 5\sigma$. (Lower panel) Numerical estimation of the fluid-fluid phase diagram for this model [14, 20] and comparison with experimental results for the low-density coexistence branch of the HGD protein. The dashed line represents the fluid-fluid coexistence curve of an isotropic model of globular proteins [35]

the peaks for temperatures below T_c, corresponding to state points along the coexistence curve, are shown in Fig. 5 (upper panel). An operative way to estimate the coexistence curve is to start from a state point close to the critical point (the same employed for the fitting procedure) and to slightly decrease T to encounter phase separation. Histogram reweighting of $P(\rho)$ (in which T is kept fixed and only μ is changed) allows to find the distribution for which the two areas below the two peaks are equal. The average density, weighed on the left and right part of the central minimum, provides an estimate of the low-density and high-density coexisting phase.

By estimating the critical point and the coexistence line it is possible to draw the fluid-fluid phase diagram of the patchy model. Figure 5 (lower panel) shows the resulting phase diagram for the reference patchy-particle model of globular proteins [14, 20]. Numerical results are compared with experimental data for HDG proteins [14], showing that the patchy model correctly estimates the location of the critical point and fairly captures the amplitude of the coexistence line. On the other hand, isotropic models for proteins [35] give rise to a too narrow coexistence curve (dashed line of Fig. 5), showing that the role of anisotropic interactions is fundamental for the correct estimation of the fluid-fluid phase diagram.

3.5.1 Successive Umbrella Sampling

An efficient sampling close to the critical point and along the coexistence curve of the low- and high-density phases requires the system to overcome the free energy barrier separating the two phases. The free energy profile $F(\rho, T)$ is related to the probability distribution $P(\rho)$ via the expression $F(\rho, T) \sim -k_B T \ln(P(\rho))$. If the barrier separating the two fluid phases is high, it will takes a very long time for phase-crossing to happen spontaneously in the course of the simulation. However, a good sampling of states across the free energy barrier is fundamental for building the correct $P(\mathcal{M})$ and an efficient sampling method is required. We here discuss one of such methods named successive umbrella sampling (SUS) [36] that allows to flatten the free energy barriers and thus to sample all the relevant states by dividing them into small windows that are sampled separately. Instead of performing a single MC-GC simulation at a given (T, μ), the simulation run will be split into several MC-GC windows with the same (T, μ), each of them with the constraint that the number of particles N can oscillate in a range ΔN within the simulation box. For instance, by choosing $\Delta N = 2$, the first window W_0 will explore the fluctuations of particles number within the interval $\in [0, 1]$, the second window $W_1 \in [1, 2]$, and the $k + 1$ window $W_k \in [k, k + 1]$. It is important to note that, in the example above, consecutive windows are built such that they superimpose of $\delta_{\text{ov}}^{k,k+1}$ that when $\Delta N = 2$ coincides with 1. Within each simulation window W_k the histogram H_k is calculated as in standard MC-GC simulations. After a given number of MC steps, the total $P(N)$ is obtained merging the histogram windows. This is done by recursively calculating $r_{k,k+1} = (\sum H_k^R H_{k+1}^L / \sum (H_{k+1}^L)^2)$ where R and L denote, respectively, the right and left region of the histogram window that superimpose with the consecutive windows. If the overlap region is $\delta_{\text{ov}}^{k,k+1} = 1$, then $r_{k,k+1} = H_k^R / H_{k+1}^L$. Hence for each window it is possible to reconstruct piecewise the (not normalized) particle distribution as follows:

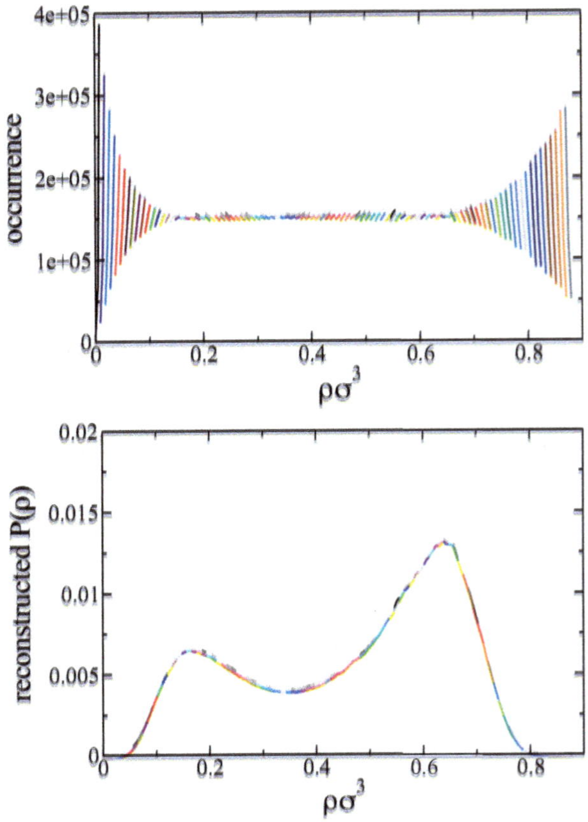

Fig. 6 (Upper panel) Histograms generated from several MC-GC simulations at the same (T, ν) with the constraint that each window has a fixed number of particles and hence the density is fixed in a given interval. (Lower panel) Reconstructed normalized $P(\rho)$ from SUS sampling shown in the upper panel

$$P(N_{W_0}) = H_0$$
$$P(N_{W_1}) = r_{0,1} H_1$$
$$P(N_{W_2}) = r_{0,1} \cdot r_{1,2} H_2$$
$$\vdots$$
$$P(N_{W_j}) = \left[\prod_{k=0}^{j-1} r_{k,k+1} \right] H_j$$

where N_{W_k} indicates the interval of particles sampled in the window k. An example of the SUS histograms calculated in several windows (*see* also **Note 6**) and the resulting piecewise reconstructed probability $P(\rho)$ are shown in Fig. 6.

Analogously, the joint distribution $P(N_k, U_k, T, \mu)$ is obtained by keeping track in each window of the combined occurrence of a given N associated with an energy E as described in Subheading 3.4 for standard MC-GC simulations.

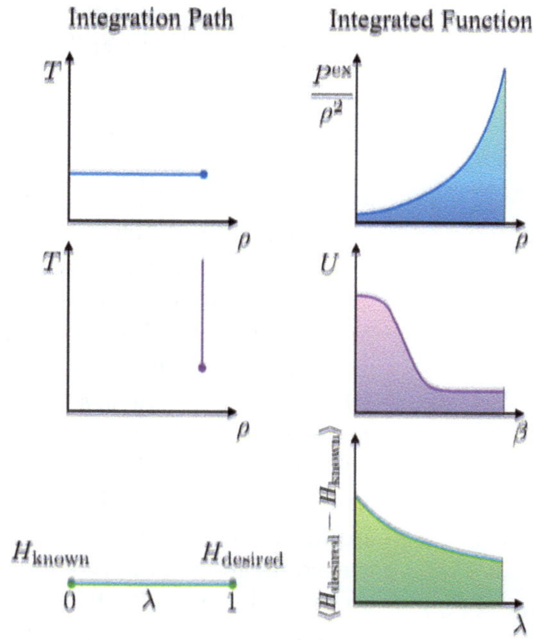

Fig. 7 Sketch of the different methods employed in numerical evaluation of the free energy: (upper panels) thermodynamic integration from infinite V to the desired state point along an isotherm. The area below the P^{ex}/ρ^2 vs. ρ measures the excess free energy (Eq. 14); (middle panels) thermodynamic integration along isochore from infinite T. Here the area below the potential energy vs. $\beta = 1/k_B T$ provides the required free energy contribution (Eq. 15); (lower panels) Hamiltonian integration, where the system is continuously changed from a known system (H_{known}) to the desired system ($H_{desired}$). Here the area below $\langle H_{desired} - H_{known} \rangle_\lambda$ needs to be calculated (Eq. 18)

3.6 Fluid–Crystal Coexistence

Methods to evaluate the coexistence line between the fluid and the crystal phases require a precise estimate of the system chemical potential μ. Indeed, thermodynamic coexistence is defined by the identity in the two phases of T, P, and μ. Since $N\mu = F + PV$, the evaluation of μ boils down to the evaluation of the Helmholtz free energy F. For the fluid phase, thermodynamic integration along a path that starts from the ideal gas (infinite V) or from the hard-sphere fluid (infinite T) are valid options (*see* Fig. 7). Indeed, in both cases the free energy of the starting state point is known (the ideal gas free energy $F_{\text{ideal gas}}$ [37] and the Carnahan–Starling free energy F_{CS} [38], respectively). In the first case, integration along an isotherm gives

$$\begin{aligned} F(V,T) &= F_{\text{ideal gas}}(V,T) - \int_{\infty}^{V} P^{ex} dV \\ &= F_{\text{ideal gas}}(\rho,T) + N \int_0^{\rho} \frac{P^{ex}(\rho,T)}{\rho^2} d\rho \end{aligned} \quad (14)$$

where ρ is the number density and P^{ex} is the excess pressure, which can be estimated via standard MC techniques. In the second case,

$$F(V,T) = F_{CS}(V,T) + k_B T \int_0^{\beta} U(\beta')d\beta' \tag{15}$$

where $\beta = 1/k_B T$ and $U(\beta)$ is the potential energy.

The commonly chosen reference state in the case of the crystal, the one for which the free energy is analytically known, is the so-called Einstein harmonic crystal [39], complemented with an orientational additional hamiltonian for the case of anisotropic interaction potentials. The interested reader should consult ref. 40, where all the details and tricks of this methodology are discussed in depth. In brief, a path is invented connecting the known reference state in which particles are confined by springs to reside around the lattice positions of the selected crystal and to librate around the crystal orientation to the final state in which springs are turned off and the desired interaction potential is turned on. In this process, the particles have always occupied the lattice position, preventing any symmetry change and thus any unwanted first order transition. This method, introduced by Frenkel and Ladd [41], build on the concept of hamiltonian thermodynamic integration. To grasp the essence of this method, let's consider starting from a hamiltonian H_{known} for which the free energy is known. In Frenkel and Ladd case, H_{known} models a system of independent particles located on the crystal lattice sites and coupled by harmonic springs. Each particle feels only its own spring and does not interact with any other particle. In the case of atoms, this non-interacting particles model coincides with the Einstein model for harmonic solids [39] and its free energy is known. We then consider a system with a λ dependent hamiltonian defined by

$$H_{\text{system}}(\lambda) = (1-\lambda)H_{\text{known}} + \lambda H_{\text{desired}}. \tag{16}$$

Thus, for $\lambda = 0$ the system coincides with the known system, while for $\lambda = 1$ the system becomes identical to the final desired system with hamiltonian H_{desired}. Then, trivially,

$$\beta F_{\text{desired}} = \beta F_{\text{known}} + \int_0^1 \frac{\partial \beta F_\lambda}{\partial \lambda} d\lambda. \tag{17}$$

The derivative of the free energy can be calculated by deriving the partition function to give

$$\beta F_{\text{desired}} = \beta F_{\text{known}} + \int_0^1 \langle H_{\text{desired}} - H_{\text{known}} \rangle_\lambda d\lambda \tag{18}$$

where $\langle H_{\text{desired}} - H_{\text{known}} \rangle_\lambda$ indicates the average value of $H_{\text{desired}} - H_{\text{known}}$ evaluated in a standard simulation with the hamiltonian $H_{\text{system}}(\lambda)$. Performing simulations for different λ values, the integral in Eq. 18 can be numerically performed and the crystal free energy thus evaluated (*see* Fig. 7).

3.7 Binary Mixtures

For completeness we now also briefly discuss the case of binary mixtures of patchy model for proteins. The procedure to evaluate the critical point in binary mixtures is similar to that explained for the monodisperse case. The ordering operator is now given by $\mathcal{M} = \rho_1 + c\rho_2$ where $\rho_{1(2)}$ is the number density of the species 1 (2) and c is the associated mixing parameter (a different one with respect to the one-component case). It follows that the numerical joint distribution will depend only on the number of particles of the two species and on the associated chemical potentials as $P(N^{(1)}, N^{(2)};\mu_1, \mu_2)$.[1] Note that since the joint distribution does not depend on energy, T is a parameter that cannot be varied to locate the critical point. In addition, for each T it is possible to find critical values of μ_1 and μ_2 or, in other words, each T is the critical temperature T_c for a given concentration of species 1 and 2. Thus, among the critical temperatures one has to select the T_c for the desired relative concentration. As for the monodisperse case, to locate the critical point SUS simulations are performed in which the number of particles of the more abundant species, that here we suppose to be species 1, are kept fixed within the interval $N_k^{(1)} \in [k, k+1]$. We stress that for all the SUS windows the parameters T, V, μ_1, and μ_2 are kept fixed during the simulation. For each step the number of particles of the two species is recorded. To build the joint probability distribution we adopt the same scheme as for the monodisperse case where the energy E is replaced by $N^{(2)}$: in each SUS window the bins of the joint histogram are built-in sequence; the first bin is given by $[N^{(1)}_{\text{min}}, N^{(2)}_{\text{min}}]$, being $N^{(1)}_{\text{min}}$ and $N^{(2)}_{\text{min}}$ the smallest values of particles of the two species found in that window, the second $[N^{(1)}_{\text{min}}, N^{(2)}_{\text{min}} + 1]$, the third $[N^{(1)}_{\text{min}}, N^{(2)}_{\text{min}} + 2]$, and so on until we build the bin $[N^{(1)}_{\text{min}}, N^{(2)}_{\text{max}}]$, being $N^{(2)}_{\text{max}}$ the largest values of particles of species 2 found in the simulation. Then the next bins will be $[N^{(1)}_{\text{min}} + 1, N^{(2)}_{\text{min}}]$, $[N^{(1)}_{\text{min}} + 1, N^{(2)}_{\text{min}} + 1]$, ..., $[N^{(1)}_{\text{min}} + 1, N^{(2)}_{\text{max}}]$ and so on until the last bin $[N^{(1)}_{\text{max}}, N^{(2)}_{\text{max}}]$ is built. Arranging the bins of the combined histogram in such a way allows to easily merge the histograms from all the windows consecutively as explained in Subheading 3.5.1, to build the joint probability $P(N^{(1)}, N^{(2)};\mu_1, \mu_2)$. Histogram reweighting can then be applied to evaluate $P(N^{(1)}, N^{(2)};\mu'_1, \mu'_2)$ when searching for the critical point and the vapor–liquid coexistence line as for monodisperse case. For binary mixtures the reweighting relation is generalized as:

[1] In principle the ordering operator depends also on the energy density, but to generate a joint distribution of this kind would require to store a large number of data during the MC run that is not possible due to memory limitations.

$$P(N^{(1)},N^{(2)};\mu_1',\mu_2') = e^{N^{(1)}(\beta\mu_1'-\beta\mu_1)}e^{N^{(2)}(\beta\mu_2'-\beta\mu_2)}P(N^{(1)},N^{(2)};\mu_1,\mu_2). \tag{19}$$

Equation 19 allows to locate the critical point by exploiting the same least squared algorithm explained in Subheading 3.4 for the monodisperse system, to superimpose the $P(N^{(1)}, N^{(2)};\mu_1, \mu_2)$ with the Ising universality curve. For the coexistence line, a good strategy is, for a given T, to fix μ_1 (possibly to a value close to its critical value), and reweight the distribution by changing only μ_2 until a bimodal distribution with equal area under the two peaks is encountered. If the latter condition is not found, it means that we are out of the coexistence region for the chosen μ_1. As already stressed above, for a given T there are critical values of μ_1 and μ_2 that signal the presence of a critical point for a specific concentration of the two species. To find the value of such concentrations, the equal area condition under the two peaks of the near-critical $P(N^{(1)}, N^{(2)};\mu_1, \mu_2)$ is needed. Indeed, only in this case the probability can split up as the sum of two contributions $p_1(N^{(1)};N^{(2)})$ and $p_2(N^{(1)};N^{(2)})$ one for each phase [42]. The concentration can be thus obtained by simply finding the average value of the particles of the two species, performing a numerical integration of the $P(N^{(1)}, N^{(2)};\mu_1, \mu_2)$;

$$\langle N^i \rangle = \frac{\sum_{n_i=0}^{N_{\mathrm{MAX}}^{(i)}} n_i p_i(N^{(1)}, N^{(1)})}{p_1(N^{(1)}, N^{(2)}))p_2(N^{(1)}), N^{(2)})} \qquad i = 1,2 \tag{20}$$

where $p_{1,(2)}(N^{(1)}, N^{(2)}) = \sum_{n_{2,(1)}=0}^{N_{\mathrm{MAX}}^{(2),(1)}} P(N^{(1)}, N^{(2)};\mu_1,\mu_2)$ and then calculating $x_i = \langle N^i\rangle/(\langle N^1\rangle + \langle N^2\rangle)$.

4 Notes

1. In addition to the Kern–Frenkel model, sometimes a different interaction potential between patches is used. Among these, we recall the "sticky spot" model in which $V_{\mathrm{KF}}(r_{ij}, \hat{n}_i^{\alpha}, \hat{n}_j^{\beta})$ is replaced by an attractive well where the bond condition is fulfilled whenever the patch–patch distance $d_{\alpha,\beta}$ is smaller than $\delta_{\alpha,\beta}$:

$$V_{\mathrm{STSP}}(r_{ij}) = \begin{cases} -\varepsilon_{\alpha\beta} & \text{if} \quad d_{\alpha\beta} \leq \delta_{\alpha,\beta} \\ 0 & \text{if} \quad d_{\alpha\beta} > \delta_{\alpha,\beta}. \end{cases} \tag{21}$$

To ensure the single-bond per patch condition, the amplitude of the attractive well must be set to $\delta_{\alpha,\beta} \lesssim 0.119$ [33]. The main difference with respect to the Kern–Frenkel model is the coupling between angular and radial extension of the patch–patch interactions, which instead can be tuned independently for Kern-Frenkel particles.

2. Controlling whether a single patch can form single or multiple bonds with patches of other particles is crucial to tune the valence, i.e., the number of maximum bonds that a particle can form. Indeed, this quantity plays a fundamental role in the phase behavior of patchy particles because the fluid-fluid critical point for low-valence particles moves to very low densities and temperature. In parallel, the coexistence region narrows in a very small region of the phase diagram [33, 43]. A good control of the valence is also of particular importance for a comparison with theoretical predictions based on the thermodynamic perturbation theory introduced by Wertheim [44–47], which describes the association of patchy particles under the constraint of a single-bond per patch.
3. To investigate the phase behavior of a system of N particles in a volume V at temperature T, the best strategy would be to find a way to evaluate numerically the configurational part of the partition function Z which encodes the statistical properties of the system in equilibrium. Neglecting orientation for simplicity,

$$Z = \int_V d\mathbf{r}^{\mathbf{N}} \exp[-\beta U(\{\mathbf{r}^N\})], \tag{22}$$

where $\mathbf{r}^{\mathbf{N}}$ in the N-tuple of all particle position, from which it would be possible to derive all the ensemble averages as for the generic observable A

$$\langle A \rangle = \frac{\int_V d\mathbf{r}^N A(\{\mathbf{r}^N\}) \exp[-\beta U(\{\mathbf{r}^N\})]}{Z} \tag{23}$$

where $\beta = 1/k_B T$, k_B is the Boltzmann constant and $U(\{\mathbf{r}^N\})$ is the configurational energy. A numerical procedure to evaluate this kind of integrals is based on the Monte Carlo (MC) integration. The latter is a statistical sampling which allows to approximate the integral of a function $\int_a^b f(x)dx$ with its mean value $I = (b - a)\langle f(x) \rangle$; this can be done by extracting randomly R variables x within the interval $[a, b]$ and evaluating $f(x)$ R-times to calculate its average value. The estimate of $\langle f(x) \rangle$ will depend only on the number of trials employed to sample the function in the interval $[a, b]$. This kind of procedure is called "random sampling" and it is used to solve a large number of statistical problems [29, 48]. However the random sampling technique is not the fastest procedure to estimate the average in Eq. 23: indeed, given the thermodynamic parameters (N, V, T), not all particle configurations have the same probability to occur but instead follow the Boltzmann distribution [29]. Since the random sampling uniformly explores the phase space, it would also sample, with equal probability, those configurations that poorly contribute to the average of Eq. 23. Therefore to be able to sample all the relevant configurations, the random sampling would require a huge amount of trials R.

A step forward is given by the "importance sampling" which chooses random numbers from a density distribution $w(\mathbf{r}^N)$ in order that the average sampling is concentrated in the phase space that gives significant contribution to the average

$$\langle A \rangle = \int_V d\mathbf{r}^N A(\{\mathbf{r}^N\}) w(\{\mathbf{r}^N\}) = \langle A \rangle_R \tag{24}$$

and hence the observable average coincides with the average value of the observable estimated in R trials. This strategy is at the core of MC simulations described in Subheading 3.2.

4. An open source code with MC algorithms described so far can be found on the web (http://dx.doi.org/10.5281/zenodo.1153959) [49].
5. On a lattice representation of a fluid, the two scaling fields ρ and u are orthogonal and $\mathcal{M} = \rho$. This is due to the so-called particle-hole symmetry, for which an inversion move (i.e., a spin flip) necessarily corresponds to a removal or an insertion move and thus there exists a symmetry between the way a hole and a particle are created. This is not true for off-lattice fluids for which the space filled by a particle can be always transformed in an empty space, but not vice versa. For the case of patchy particles, the anisotropy in the interactions enhances such particle-hole asymmetry, since particle insertion and removal also depend on the way particles are oriented with respect to one another.
6. As discussed in the original work on the SUS method [36], the constraint of the number of particles in a given window must be taken with care in the calculation of the histogram when an attempt is made for increasing (decreasing) the number of particles outside the window range. In that case the move would be rejected leaving the number of particles in the box equal to the maximum (minimum) of the interval in that window, but the corresponding bin still has to be incremented by one in order to fulfill the detailed balance condition.

Acknowledgements

NG and EZ acknowledge support from the European Research Council (ERC Consolidator Grant 681597, MIMIC).

References

1. Glotzer SC, Solomon MJ (2007) Anisotropy of building blocks and their assembly into complex structures. Nat Mater 6(8):557
2. Prasad V, Semwogerere D, Weeks ER (2007) Confocal microscopy of colloids. J Phys Condens Matter 19(11):113102
3. Anderson VJ, Lekkerkerker HN (2002) Insights into phase transition kinetics from colloid science. Nature 416(6883):811
4. Muschol M, Rosenberger F (1997) Liquid–liquid phase separation in supersaturated lysozyme solutions and associated precipitate formation/crystallization. J Chem Phys 107 (6):1953–1962
5. Foffi G, McCullagh GD, Lawlor A, Zaccarelli E, Dawson KA, Sciortino F, Tartaglia P, Pini D, Stell G (2002) Phase equilibria and glass transition in colloidal systems with short-ranged attractive interactions: application to protein crystallization. Phys Rev E 65 (3):031407
6. Schurtenberger P, Chamberlin RA, Thurston GM, Thomson JA, Benedek GB (1989) Observation of critical phenomena in a protein-water solution. Phys Rev Lett 63(19):2064
7. McManus JJ, Charbonneau P, Zaccarelli E, Asherie N (2016) The physics of protein self-assembly. Curr Opin Colloid Interface Sci 22:73–79
8. ten Wolde PR, Frenkel D (1997) Enhancement of protein crystal nucleation by critical density fluctuations. Science 277(5334):1975–1978
9. Pagan D, Gunton J (2005) Phase behavior of short-range square-well model. J Chem Phys 122(18):184515
10. Cardinaux F, Stradner A, Schurtenberger P, Sciortino F, Zaccarelli E (2007) Modeling equilibrium clusters in lysozyme solutions. Europhys Lett 77(4):48004
11. Bianchi E, Blaak R, Likos CN (2011) Patchy colloids: state of the art and perspectives. Phys Chem Chem Phys 13(14):6397–6410
12. Gögelein C, Nägele G, Tuinier R, Gibaud T, Stradner A, Schurtenberger P (2008) A simple patchy colloid model for the phase behavior of lysozyme dispersions. J Chem Phys 129 (8):08B615
13. McManus JJ, Lomakin A, Ogun O, Pande A, Basan M, Pande J, Benedek GB (2007) Altered phase diagram due to a single point mutation in human γd-crystallin. Proc Natl Acad Sci 104 (43):16856–16861
14. Quinn M, Gnan N, James S, Ninarello A, Sciortino F, Zaccarelli E, McManus JJ (2015) How fluorescent labelling alters the solution behaviour of proteins. Phys Chem Chem Phys 17(46):31177–31187
15. Cai J, Townsend J, Dodson T, Heiney P, Sweeney A (2017) Eye patches: protein assembly of index-gradient squid lenses. Science 357 (6351):564–569
16. Cai J, Sweeney AM (2018) The proof is in the pidan: generalizing proteins as patchy particles. ACS Cent Sci 4(7):840–853
17. Kastelic M, Dill KA, Kalyuzhnyi YV, Vlachy V (2017) Controlling the viscosities of antibody solutions through control of their binding sites. J Mol Liq. https://doi.org/10.1016/j.molliq.2017.11.106
18. Skar-Gislinge N, Ronti M, Garting T, Rischel C, Schurtenberger P, Zaccarelli E, Stradner A (2018) A colloid approach to self-assembling antibodies. arXiv preprint, arXiv:1810.01160
19. Lomakin A, Asherie N, Benedek GB (1999) Aeolotopic interactions of globular proteins. Proc Natl Acad Sci 96(17):9465–9468
20. Liu H, Kumar SK, Sciortino F (2007) Vapor-liquid coexistence of patchy models: relevance to protein phase behavior. J Chem Phys 127 (8):084902
21. Roosen-Runge F, Zhang F, Schreiber F, Roth R (2014) Ion-activated attractive patches as a mechanism for controlled protein interactions. Sci Rep 4:7016
22. Fusco D, Headd JJ, De Simone A, Wang J, Charbonneau P (2014) Characterizing protein crystal contacts and their role in crystallization: rubredoxin as a case study. Soft Matter 10 (2):290–302
23. Hansen JP, McDonald IR (2013) Theory of simple liquids: with applications to soft matter. Academic, Amsterdam
24. Bol W (1982) Monte Carlo simulations of fluid systems of waterlike molecules. Mol Phys 45 (3):605–616
25. Kern N, Frenkel D (2003) Fluid–fluid coexistence in colloidal systems with short-ranged strongly directional attraction. J Chem Phys 118(21):9882–9889
26. Hastings WK (1970) Monte Carlo sampling methods using Markov chains and their applications. Biometrika 57(1):97–109
27. Chapman W, Quirke N (1985) Metropolis Monte Carlo simulation of fluids with multi-particle moves. Phys B+ C 131(1–3):34–40
28. Mountain RD, Thirumalai D (1994) Quantitative measure of efficiency of Monte Carlo

simulations. Phys A Stat Mech Appl 210 (3–4):453–460
29. Frenkel D, Smit B (2001) Understanding molecular dynamics simulations. Academic, New York
30. Rapaport DC, Rapaport DCR (2004) The art of molecular dynamics simulation. Cambridge University Press, Cambridge
31. Wilding N, Bruce A (1992) Density fluctuations and field mixing in the critical fluid. J Phys Condens Matter 4(12):3087
32. Tsypin M, Blöte H (2000) Probability distribution of the order parameter for the three-dimensional Ising-model universality class: a high-precision Monte Carlo study. Phys Rev E 62(1):73
33. Bianchi E, Largo J, Tartaglia P, Zaccarelli E, Sciortino F (2006) Phase diagram of patchy colloids: towards empty liquids. Phys Rev Lett 97(16):168301
34. Wilding NB (1997) Simulation studies of fluid critical behaviour. J Phys Condens Matter 9 (3):585
35. Lomakin A, Asherie N, Benedek GB (1996) Monte Carlo study of phase separation in aqueous protein solutions. J Chem Phys 104 (4):1646–1656
36. Virnau P, Müller M (2004) Calculation of free energy through successive umbrella sampling. J Chem Phys 120(23):10925–10930
37. Hill TL (1986) An introduction to statistical thermodynamics. Courier Corporation, Chelmsford
38. Carnahan NF, Starling KE (1969) Equation of state for nonattracting rigid spheres. J Chem Phys 51(2):635–636
39. Kittel C *et al* (1976) Introduction to solid state physics, vol 8. Wiley, New York
40. Vega C, Sanz E, Abascal J, Noya E (2008) Determination of phase diagrams via computer simulation: methodology and applications to water, electrolytes and proteins. J Phys Condens Matter 20(15):153101
41. Frenkel D, Ladd AJ (1984) New Monte Carlo method to compute the free energy of arbitrary solids. Application to the FCC and HCP phases of hard spheres. J Chem Phys 81 (7):3188–3193
42. Rovigatti L, de las Heras D, Tavares JM, Telo da Gama MM, Sciortino F (2013) Computing the phase diagram of binary mixtures: a patchy particle case study. J Chem Phys 138 (16):164904
43. Sciortino F, Zaccarelli E (2017) Equilibrium gels of limited valence colloids. Curr Opin Colloid Interface Sci 30:90–96
44. Wertheim M (1984) Fluids with highly directional attractive forces. I. Statistical thermodynamics. J Stat Phys 35(1–2):19–34
45. Wertheim M (1984) Fluids with highly directional attractive forces. II. Thermodynamic perturbation theory and integral equations. J Stat Phys 35(1–2):35–47
46. Wertheim M (1986) Fluids with highly directional attractive forces. III. Multiple attraction sites. J Stat Phys 42(3–4):459–476
47. Wertheim M (1986) Fluids with highly directional attractive forces. IV. Equilibrium polymerization. J Stat Phys 42(3–4):477–492
48. Allen MP, Tildesley DJ (2017) Computer simulation of liquids. Oxford University Press, Oxford
49. Rovigatti L, Russo J, Romano F (2018) How to simulate patchy particles. Eur Phys J E 41 (5):59

Chapter 15

Obtaining Soft Matter Models of Proteins and their Phase Behavior

Irem Altan and Patrick Charbonneau

Abstract

Globular proteins are roughly spherical biomolecules with attractive and highly directional interactions. This microscopic observation motivates describing these proteins as patchy particles: hard spheres with attractive surface patches. Mapping a biomolecule to a patchy model requires simplifying effective protein–protein interactions, which in turn provides a microscopic understanding of the protein solution behavior. The patchy model can indeed be fully analyzed, including its phase diagram. In this chapter, we detail the methodology of mapping a given protein to a patchy model and of determining the phase diagram of the latter. We also briefly describe the theory upon which the methodology is based, provide practical information, and discuss potential pitfalls. Data and scripts relevant to this work have been archived and can be accessed at https://doi.org/10.7924/r4ww7bs1p.

Key words Soft matter, Phase behavior, Protein crystallization, Coarse-grained simulation

1 Introduction

While all-atom simulations of a single solvated protein are now fairly run-of-the-mill, simulating protein crystallization in a similar way is far beyond computational reach. The operation would require simulating hundreds of copies of the macromolecule, over very long timescales. In addition, such simulations would contain so much information that teasing out the relevant physico-chemical features that drive crystal assembly would itself be challenging. To circumvent these obstacles, coarse-grained models are used to capture protein–protein interactions in an effective manner, and thus to hide from view (and from computations) most of the obfuscating details. The key operations for such coarse-graining are: (1) determining and characterizing the relevant features of protein–protein interactions, and (2) solving the properties of the resulting effective

The original version of this chapter was revised. The correction to this chapter is available at https://doi.org/10.1007/978-1-4939-9678-0_18

Jennifer J. McManus (ed.), *Protein Self-Assembly: Methods and Protocols*, Methods in Molecular Biology, vol. 2039, https://doi.org/10.1007/978-1-4939-9678-0_15, © Springer Science+Business Media, LLC, part of Springer Nature 2019

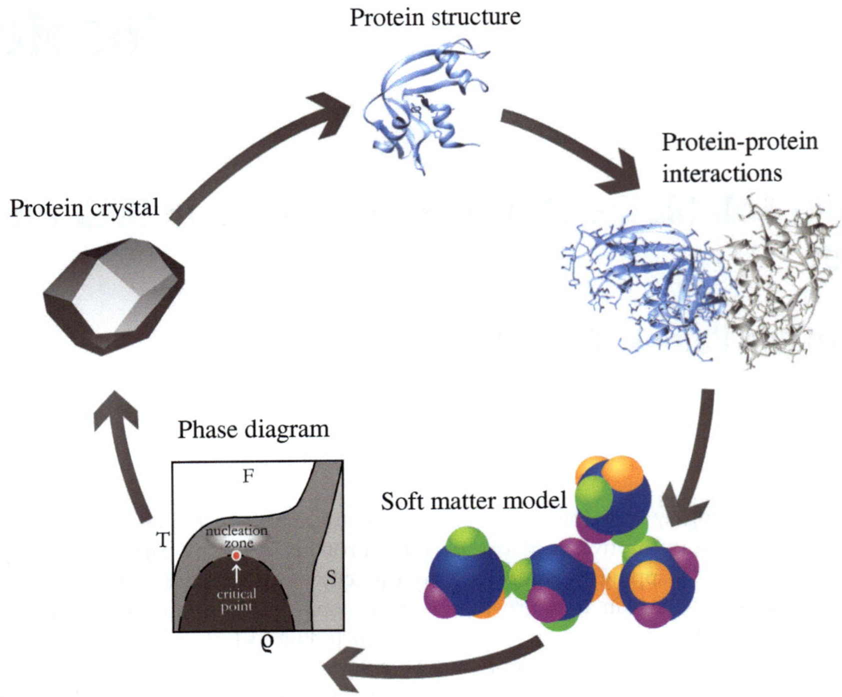

Fig. 1 The vicious cycle of protein crystallization (adapted from ref. 15). The phase diagram of a given patchy model can be straightforwardly determined. The resulting phase information can in turn be used to optimize crystallization screens to reliably obtain protein crystals. However, constructing patchy models requires knowing protein–protein interactions, for which the protein structure itself is needed. Going through this process nevertheless results in a better understanding of how the microscopic properties of the protein control its phase behavior

model. The first is done using all-atom simulations, and the second with the coarse-grained model alone. From a conceptual viewpoint, the key difficulties consist of identifying the relevant features and choosing an appropriate coarse-graining scheme. Here, we describe a procedure developed over the last few years that focuses on crystal contacts between proteins for the former, and uses patchy models for the latter.

An attentive reader will notice that this approach gives information about the crystal assembly of a protein of known structure, and thus that has presumably already been crystallized. This is the vicious cycle of protein crystallization (*see* Fig. 1). Despite this obvious drawback for the study of a specific protein, such a scheme can be employed to understand the generic features that control protein crystallization, and thus the different classes of macromolecular assembly.

In this chapter, we detail the methodology for obtaining effective representations of protein–protein interactions, patchy models, and the steps involved in determining their phase diagram. For each of these tasks, we also briefly summarize the underlying theory. We conclude by discussing common complications and workarounds.

2 Materials

It may sound paradoxical but the most essential information needed to study protein crystal assembly is the protein crystal structure itself (Fig. 1). While high quality crystal structures are preferable, if the positions of some of the side chains cannot be resolved, it is still possible to add them on using tools such as KiNG [1], and then minimize the energy of the resulting configuration in order to avoid steric clashes. Once that has been resolved, the expensive computational work can begin.

In order to run all-atom simulations, molecular dynamics (MD) packages, such as Gromacs [2] or Amber [3], are essential. These packages include a variety of protein force fields and water models and are designed for sharing the computational load with graphical processing units (GPUs). This is especially useful for simulating systems that contain a large number of non-bonded interactions, e.g., interactions that involve solvent molecules.

Once model parameters have been determined from all-atom MD simulations, the remainder of the work uses Monte Carlo (MC) simulations. Note that no generic MC code distribution is widely available, but the relevant methods for patchy models can be straightforwardly implemented based on ref. 4. Rovigatti et al. have also recently published a detailed review of the specific MC methods used for simulating patchy particles, along with an educational package for performing various such simulations [5, 6].

3 Methods

In this section we first describe how effective protein–protein interactions are obtained (Subheading 3.1), and how the nature of these interactions leads to a minimal patchy model. We then detail the process of obtaining the phase diagram of this model (Subheading 3.2). For the sake of concreteness, we use Gromacs for the first step and illustrate the overall process with a specific rubredoxin mutant [7] (Protein data bank [8] ID: 1YK4 [9]).

3.1 Effective Protein–Protein Interactions Through Umbrella Sampling

The change in free energy upon forming or destroying a protein–protein contact is determined from simulations that mimic experimental conditions and thus include solvent molecules and ions. In general, the free energy difference between two states along a reaction coordinate, ξ, is called the potential of mean force (PMF). For protein–protein interactions in particular, one needs to determine the PMF as a function of distance between two proteins, given a specific crystal contact. The natural choice for the reaction coordinate is then the protein–protein distance.

At equilibrium, the probability that the system is found at a given ξ is [10]

$$Q(\xi)\mathrm{d}\xi = \left[\frac{\int \delta(\tilde{\xi}(\mathbf{r}^N) - \xi)e^{-\beta U(\mathbf{r}^N)}\mathrm{d}\mathbf{r}^N}{\int e^{-\beta U(\mathbf{r}^N)}\mathrm{d}\mathbf{r}^N}\right]\mathrm{d}\xi, \tag{1}$$

where δ is the Dirac delta function, $\tilde{\xi}(\mathbf{r}^N)$ is a function that relates the reaction coordinate to particle positions, $\mathbf{r}^N$ denotes the coordinates of the N particles of the system, and $U(\mathbf{r}^N)$ is the potential energy of a given configuration. This configuration is observed with a probability proportional to its Boltzmann weight, i.e., $e^{-\beta U(\mathbf{r}^N)}$ at inverse temperature $\beta \equiv 1/k_{\mathrm{B}}T$ where k_{B} is the Boltzmann constant. The constrained (Helmholtz) free energy, $A(\xi)$, as a function of the reaction coordinate, $-\beta A(\xi) = \ln Q(\xi)$, corresponds to the PMF. While it is theoretically possible to sample $Q(\xi)$ in a single molecular dynamics (MD) simulation, in practice the small Boltzmann weight of the dissociated configurations makes sampling exceedingly difficult. It is thus advantageous to introduce a series of biasing potential, $w_i(\xi)$, and to simulate the system with modified energy functions,

$$U_i'(\mathbf{r}^N) = U(\mathbf{r}^N) + w_i(\xi). \tag{2}$$

Sampling all ξ is then possible because the original energy barriers are lowered, or equivalently, the Boltzmann weight of the associated configurations is increased. A convenient choice of bias is a harmonic potential, i.e., a spring,

$$w_i(\xi) = k(\xi_i - \xi_0)^2, \tag{3}$$

where ξ_0 is the imposed protein–protein distance and k is the spring constant. Using this potential, we separate the protein–protein center of mass distance into M umbrella sampling windows, with different ξ_i (Fig. 2).

In what follows, we detail the computational steps involved in this procedure for a given protein. Note that a detailed tutorial for the process is available for Gromacs (*see* **Note 1**). We here provide

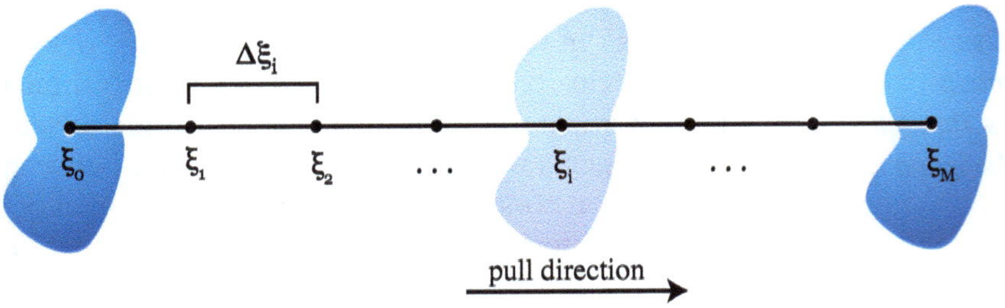

Fig. 2 Two proteins are pulled away from each other to generate umbrella sampling windows centered at ξ_i. Each window is sampled by MD simulations using a biased interaction potential $U_i'(\mathbf{r}^N)$. The results for the different windows are then joined to reconstruct the overall PMF as a function of ξ

the details that are not mentioned in that tutorial or that differ for crystal contacts.

1. **Determine contacts.** Crystal contacts can be determined using PISA [11], an online tool that identifies protein–protein interfaces for a given `.pdb` (protein data bank) file. PISA takes into account protein symmetry and crystal periodicity, in addition to using a distance cutoff for determining contacts. For each contact, it also lists the residues involved in hydrogen bonding and salt bridges, and provides an estimate of the contact free energy. This estimate, however, is fairly rough, because many contributions, such as interactions with the crystallization cocktail or side chain and backbone conformational changes, are not explicitly considered.
2. **Add any missing or incomplete residues.** When starting from a crystal structure, it is possible that entire residues or some of their side chains might be missing because they could not be crystallographically resolved. These should be added manually to the protein structure. Note that the accuracy of the orientation and the conformation of these residues is not critical at this stage (within chemical reasonableness), because the protein structure is minimized in subsequent steps, eliminating any steric clash that might arise. The interaction strength between patches also depends on the protonation states of the contained residues. A rough estimate for the protonation states can be obtained with propKa [12], keeping in mind that in most cases the solution conditions for crystallization experiments are such that the protein carries no net charge.
3. **Generate input files for each contact.** For each contact, separate `.pdb` files with two copies of the protein forming the contact should be generated. These `.pdb` files should then be converted to `.gro` files (the default structure file format for Gromacs) using the Gromacs command `pdb2gmx`. This command also prompts the user to choose a force field and a water model. Once the `.gro` file is created, it is convenient to rotate the protein pair so that the *z*-axis corresponds to the pull direction and to center the pair in the simulation box. The assembly should be at least 1 nm away from the box sides, in order to prevent one protein from interacting with its copy across the (periodic) box boundary, given that the cutoff for neighbor lists, electrostatic, and van der Waals interactions is less than 2 nm. This centering and resizing can be achieved by the following command:

```
gmx editconf -f box.gro -o newbox.gro -c -d 1.0
```

The next step is to elongate the box along the pull direction, making sure that the box is at least twice the pull direction plus the original box size. Suppose that the box size in `newbox.gro` is 10 × 10 × 10 nm. In order to pull one of the protein copies 5 nm away one should run

```
gmx editconf -f newbox.gro -o newbox2.gro -center 5 5 5
    -box 10 10 20
```

for the box to be resized to 10 × 10 × 20 nm.

4. **Add solvent and ions.** Once the box dimensions are selected and the proteins are positioned, solvent and ions are inserted with the commands `gmx solvate` and `gmx genion`, respectively. Ideally, one should use the same ion type and concentration as in the crystallization cocktail. If the force field parameters for these specific ions are not readily available, however, one might consider replacing them with simple generic ions such as sodium and chloride. This replacement at least matches the ionic strength of the solvent and thus the extent of charge screening in the experimental setup.
5. **Minimize energy and equilibrate.** The energy of the resulting system should be minimized before running any simulation, because the positions of the water molecules and ions placed in the box, as well as the conformation and orientation of the inserted residues and side chains, need to be relaxed. To further relax the system, a short additional simulation should be performed in which the number of particles, pressure, and temperature are kept constant (constant *NPT*), before generating input configurations for umbrella sampling. Note that keeping the center of mass of the protein–protein complex fixed for this step facilitates the remainder of the procedure.
6. **Generate configurations for umbrella sampling.** In order to generate initial configurations for umbrella sampling, a simulation is run in which one protein is pulled away from the other at a constant rate, using a harmonic potential with force constant k, while the other protein is restrained. (A convenient approach is to restrain three or four backbone atoms.) Typically, k ranging from 1000 to 10,000 kJ mol^{-1} nm^{-2} is appropriate for pulling the proteins apart. Since the resulting configurations are not necessarily equilibrated they need to be further relaxed after the pull, before being used as inputs to umbrella sampling simulations.
7. **Choose umbrella sampling windows and run simulations.** The M chosen configurations should cover the whole distance interval of interest, and their pair separation $\Delta\xi_i$ (Fig. 2) should be such that the resulting umbrella sampling windows overlap

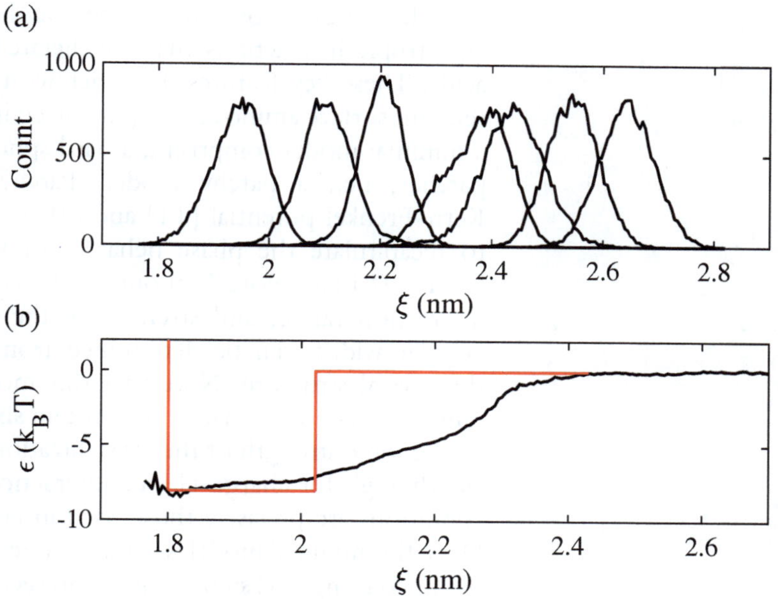

Fig. 3 (**a**) Histograms generated by the weighted histogram analysis method (`gmx wham`) [13]. The distribution for each window overlaps well with those of their neighbors. (**b**) The PMF for a contact of rubredoxin as a function of pull distance, ξ, at $T_{\text{ref}} = 300$ K [7]. Infinite separation sets the zero of the energy. The square-well potential generated from this PMF by fitting the second virial coefficient is shown in red (*see* Subheading 3.2.1)

sufficiently. In particular, a significant portion of the tails of distributions of ξ for each neighboring pair of windows should overlap (*see* Fig. 3a). The constant k for the umbrella simulations should be strong enough to keep the proteins at roughly the desired separation, but not so strong that the resulting distributions are overly narrow. Additional windows, as permitted by the box size, can be added after the PMF is generated, if any pair of windows does not sufficiently overlap. Note that if the overlap is poor, one often observes discontinuities in the resulting PMF. This serves as a diagnostic.

8. **Generate PMF.** The PMF is constructed from the output of each umbrella sampling simulation using the command `gmx wham` (Fig. 3). The input files necessary for this step are the `.tpr` (the Gromacs executables for individual umbrella sampling simulations) and `pullf*.xvg`, which contain the force information for each window.

3.2 Phase Diagram

Our ultimate goal is to capture the solution and the assembly behavior of a protein from the simplest possible physical model. This is not only useful in making the simulations computationally tractable, but also serves as a consistency check for the microscopic features we previously identified as relevant. In that context, we consider globular proteins to be roughly spherical objects with anisotropic interactions that are dictated by their surface amino acids. These key features, together with the assumption that the relevant surface amino acids are involved in crystal contacts, suggest a minimal model comprising a hard sphere with attractive surface patches, i.e., a patchy model. Patchy models based on the Kern–Frenkel potential [14] and others have indeed been shown to recapitulate the phase behavior of various globular proteins [7, 15–18] (*see* **Note 7** about model complexity). The location, interaction range, and strength of the patches, as well as their angular width, can be determined from all-atom simulations of the crystal structure. Note that this model is suitable for short-range interactions. That is, the protein should either be uncharged or the ionic strength of the crystallization cocktail should be sufficiently high for charge–charge interactions to be screened. These conditions are precisely those used in crystallization experiments. Once this minimal model is parameterized, its phase diagram can be obtained using MC simulations. This section first describes a model that captures the properties of the effective protein–protein interactions computed in Subheading 3.1 (Subheading 3.2.1). The notion of thermodynamic integration, which is used for calculating the free energy of a system from a reference state, is then introduced (Subheading 3.2.2). The calculation of fluid (Subheading 3.2.3) and crystal (Subheading 3.2.4) free energies are subsequently described, as well as the procedure for obtaining a coexistence point between these two phases, and the Gibbs–Duhem integration scheme (Subheading 3.2.5) for obtaining the complete crystal solubility curve. Finally, we discuss how Gibbs Ensemble simulations can be used to obtain the (metastable) gas–liquid binodal (Subheading 3.2.6).

3.2.1 Patchy Models

In a patchy model, the interaction potential between two patchy particles i and j is

$$U(r_{ij}, \Omega_i, \Omega_j) = U_{\mathrm{HS}} + \sum_{\alpha,\beta=1}^{n} U_{\alpha\beta}(r_{ij}, \Omega_i, \Omega_j), \tag{4}$$

where U_{HS} is the hard sphere repulsion, which prohibits overlaps between particles of diameter σ, α and β label one of the n surface patches, r_{ij} is the distance between particles, and Ω_1 and Ω_2 describe the particle orientations (either in terms of Euler angles or quaternions). The attractive part of the potential, $U_{\alpha\beta}$, is then

$$U_{\alpha\beta} = v_{\alpha\beta}(r_{ij}) f_{\alpha\beta}(\Omega_i, \Omega_j), \tag{5}$$

with

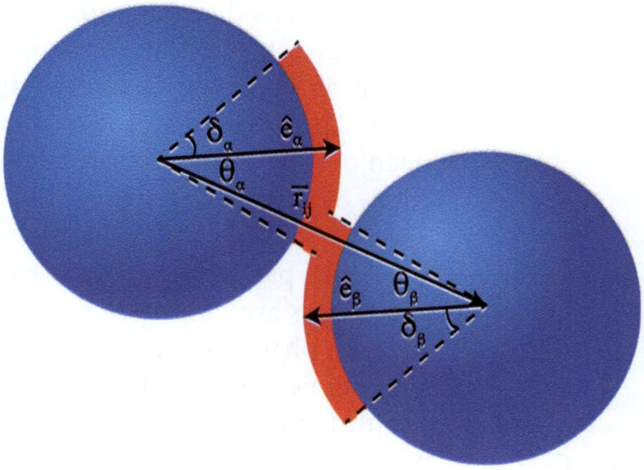

Fig. 4 Two particles do not interact unless their patches overlap in distance and orientation. Here, $\hat{e}_\alpha$ and $\hat{e}_\beta$ point to the centers of patches α and β, respectively, hence they interact if both $\hat{e}_\alpha \cdot \hat{r}_{ij} = \cos\theta_\alpha \geq \cos\delta_\alpha$ and $\hat{e}_\beta \cdot \hat{r}_{ij} = -\cos\theta_\beta \leq \cos\delta_\beta$, with $r_{ij} \leq \lambda_{\alpha\beta}\sigma$

$$v_{\alpha\beta}(r_{ij}) = \begin{cases} -\varepsilon_{\alpha\beta}, & \sigma < r_{ij} \leq \lambda_{\alpha\beta}\sigma \\ 0, & \text{otherwise} \end{cases};$$
$$f_{\alpha\beta}(\Omega_i,\Omega_j) = \begin{cases} 1, & \cos\theta_\alpha \geq \cos\delta_\alpha \text{ and } \cos\theta_\beta \geq \cos\delta_\beta \\ 0, & \text{otherwise} \end{cases}. \quad (6)$$

In other words, a square-well potential of range $\lambda_{\alpha\beta}\sigma$ controls the radial part of the attraction, and the widths of patches α and β, δ_α and δ_β, respectively set their angular range (Fig. 4). That is, patches only attract when the vector joining their centers of mass passes through the patch of both particles.

The parameters for the radial part of the attraction are obtained from the PMF computed in Subheading 3.1 (Fig. 3a). The depth of the square-well potential, $\varepsilon_{\alpha\beta}$, is that of the PMF for contact between patches α and β. The range of the square-well attraction (*see* **Note 6**) is obtained by matching its contribution to B_2, the second virial coefficient, to that of the PMF, where

$$B_2 = -\frac{1}{2}\int (e^{-\beta U(\mathbf{r})} - 1)\mathrm{d}\mathbf{r}. \quad (7)$$

This integral is evaluated numerically for the PMF (its value denoted I) and analytically for the α-β contact. For a given contact, the interaction range, $\lambda_{\alpha\beta}$, is found by equating the two results,

$$\lambda_{\alpha\beta} = \left(\frac{3I}{e^{\beta\varepsilon_{\alpha\beta}} - 1} + 1\right)^{1/3}. \quad (8)$$

Finally, the angular breadth of the interaction is set by running simulations that fix the distance between the two proteins, but not their relative orientation. This is achieved by constraining the center of mass distance with a harmonic spring, at the equilibrium bonding distance. The deviation of the patch vectors from the center of mass axis is tracked in terms of the angle δ between them. The angular breadth, $\cos\delta_\alpha$, for patch α is taken to be the mean of the computed distribution for that angle, and the same for $\cos\delta_\beta$ of patch β.

3.2.2 Thermodynamic Integration

Once the patchy model is parameterized, various types of MC simulations are employed to trace out its phase diagram. For two or more phases to be in coexistence, their temperature, pressure, and chemical potential, μ, must all be equal. While P and T can be straightforwardly enforced, μ is more challenging. Simulations, like experiments, can only determine the *change* in free energy along a transformation, not its absolute value. One thus needs a reference state of known free energy and a transformation from that reference to the system of interest to calculate its free energy [4, 19]. Reference states that are of particular interest for us are the ideal gas and the Einstein crystal. Integrating from any of these states along an isotherm yields for the Helmholtz free energy

$$A(\rho, T) = A(\rho_0, T) + N\int_{\rho_0}^{\rho} \frac{P(\rho')}{\rho'^2}\mathrm{d}\rho', \tag{9}$$

where ρ_0 is the density of the reference system and $P(\rho)$ is the equilibrium pressure of the system at a density, ρ. Here, we consider the number density, such that $\rho \equiv N/V$, where V is the volume of the system. For numerical convenience, if the reference system is an ideal gas, this expression is rewritten as [20]

$$A^{\text{fluid}}(\rho, T) = A^{\text{ideal}}(\rho) + N\int_0^{\rho}\left[\frac{P}{\rho'^2} - \frac{1}{\beta\rho'}\right]\mathrm{d}\rho', \tag{10}$$

where A^{ideal} is the free energy of the ideal gas (*see* **Note 2**)

$$\frac{A^{\text{ideal}}(\rho)}{N} = \frac{1}{\beta}\left[\log(\rho\Lambda^3) - 1 + \frac{1}{N}\log(2\pi N)\right]. \tag{11}$$

where the de Broglie wavelength, Λ^3, is set to unity without loss of generality. Another option is to integrate at constant pressure, i.e., along an isobar, varying the temperature of the system

$$\beta\mu(T, P) = \beta\mu(T_0, P) - \int_{T_0}^{T}\frac{H(T')}{Nk_{\text{B}}T'^2}\mathrm{d}T', \tag{12}$$

where H is the enthalpy of the systems, or along an isochore,

$$\beta\frac{A(T, V)}{N} = \beta\frac{A(T_0, V)}{N} - \int_{T_0}^{T}\frac{U(T')}{Nk_{\text{B}}T'^2}\mathrm{d}T', \tag{13}$$

where U is the internal energy of the system.

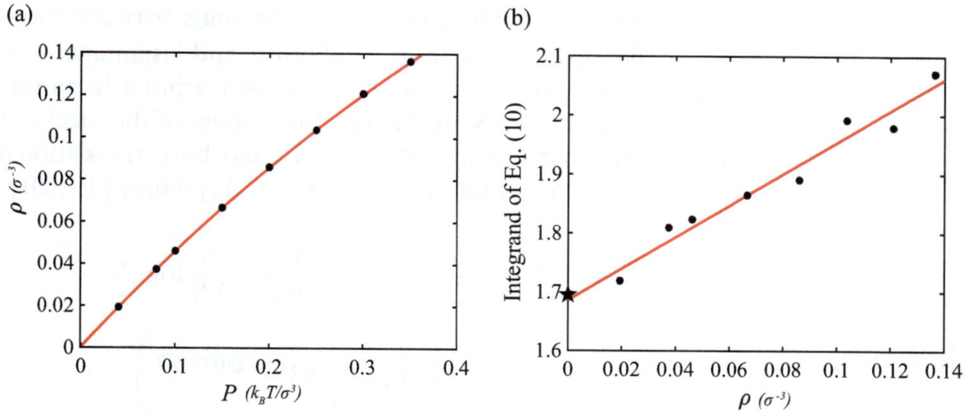

Fig. 5 (**a**) Equation of state (density as a function of pressure), and (**b**) the integrand of Eq. 10 for the fluid phase of the patchy model of rubredoxin at $\beta = 0.2$ (in units of $1/k_B T_{ref}$) [7]. The star denotes B_2, the $\rho \to 0$ limit of the integrand of Eq. 10. Curves are polynomial fits to the data

3.2.3 Free Energy of the Fluid Phase

Using the principles of thermodynamic integration introduced above, the following steps summarize how we obtain the free energy of the fluid phase. To integrate along the isotherm from the ideal gas, we run *NPT* simulations at a set of pressures $\{P_1, \ldots, P_m\}$, where P_1 is a very low pressure and $\rho(P_m)$ is the density of interest. Figure 5a shows the numerical equation of state of the fluid phase for the patchy model of rubredoxin [7]. The integrand of Eq. 10 is calculated from these data points (Fig. 5b). The integrand gets noisy as pressure decreases, because both $1/\rho^2$ and $1/\rho$ diverge as $\rho \to 0$, hence the numerical error is then amplified. In that regime, one can use the fact that the integrand converges to B_2, Eq. 7, as $\rho \to 0$ to increase numerical accuracy. There are three options for the rest of the thermodynamic integration: (1) continue integrating along the same isotherm to obtain the free energy as a function of pressure, (2) integrate along an isobar using Eq. 12, or (3) integrate along an isochore using Eq. 13.

3.2.4 Free Energy of the Crystal Phase

The Frenkel–Ladd method [21] is a thermodynamic integration scheme to obtain the free energy of a crystal using an Einstein crystal with a fixed center of mass as a reference. Particles are then restrained to their equilibrium lattice positions and orientations, and do not otherwise interact. The interaction energy of this system is

$$U_E = \sum_{i=1}^{N} \kappa(\mathbf{r}_i - \mathbf{r}_{i,0})^2 + \sum_{i=1}^{N} \kappa\eta g(\Omega_i). \tag{14}$$

The first term restrains the positions $\mathbf{r}_i$ of the N particles to their positions $\mathbf{r}_{i,0}$ by a harmonic potential with spring constant $\kappa \in [0, \infty)$. The second term penalizes particles that deviate from their equilibrium orientations, by the potential $g(\Omega) = 1 - \cos(\psi_{ia}) +$

$1 - \cos(\psi_{i\beta})$ [7], where $\psi_{i\alpha}$ ($\psi_{i\beta}$) is the angle between the vector that defines patch α (β) in its equilibrium and instantaneous orientations, and patches α and β are chosen arbitrarily among the surface patches (*see* **Note 3**). The free energy of the ideal Einstein crystal with fixed center of mass, A_E, has both translational and orientational contributions $A_E = A_{E,t} + A_{E,o}$ where [19, 20]

$$\beta\frac{A_{E,t}}{N} = -\frac{3}{2}\frac{N-1}{N}\ln\left(\frac{\pi}{\beta\kappa}\right) - \frac{3}{2N}\log(N) \tag{15}$$

$$\beta\frac{A_{E,o}}{N} = -\ln\left(\frac{1}{8\pi^2}\int d\Omega\, e^{-\kappa\eta g(\Omega)/k_B T}\right). \tag{16}$$

This reference system is converted to the interacting protein crystal in three steps.

1. **Switch on interactions.** The free energy change in this step is

$$\Delta A_1 = -\ln\left\langle e^{-\beta(\tilde{U}-U_0)}\right\rangle_E + U_0, \tag{17}$$

where $\langle\cdot\rangle_E$ denotes an averaging over ideal Einstein crystal configurations, $\tilde{U}$ is the energy of the interacting Einstein crystal without the harmonic spring contributions, and U_0 is the interacting crystal ground state energy. For large enough κ (denoted κ_{max}) the contribution from the thermal average vanishes, i.e., $\Delta A_1 \approx U_0$. This condition sets κ_{max}.

2. **Turn off position and orientation restraints.** In this step, the springs are turned off, i.e., $\kappa \to 0$. The change in free energy of this process is

$$\begin{aligned}\Delta A_2 &= -\int_0^{\kappa_{max}} d\kappa' \left\langle\frac{\partial U_E}{\partial\kappa'}\right\rangle_{NVT\kappa'} \\ &= -\int_0^{\kappa_{max}} d\kappa' \left\langle\sum_{i=0}^{N}(\mathbf{r}_i - \mathbf{r}_{i,0})^2 + \eta\sum_{i=0}^{N} g(\Omega_i)\right\rangle_{NVT\kappa'}.\end{aligned} \tag{18}$$

Note that because κ_{max} can be very large, it is often more convenient to use $\ln\kappa$ as the integration variable

$$\Delta A_2 = -\int_{-\infty}^{\ln\kappa_{max}} d(\ln\kappa')\, \kappa' \left\langle\sum_{i=0}^{N}(\mathbf{r}_i - \mathbf{r}_{i,0})^2 + \eta\sum_{i=0}^{N} g(\Omega_i)\right\rangle_{NVT\kappa'}, \tag{19}$$

where the integrand is evaluated by running NVT simulations at different κ (Fig. 6a). The first term in the integrand is the mean square displacement and the second term is a measure of the orientational displacement, which are both easily measured in simulations. The integral can then be evaluated using

Gaussian quadrature [22] with 20–40 logarithmically spaced points. Because the integrand vanishes when $\ln \kappa \to -\infty$ the integration can start from a small (non-zero) κ. Evaluating the translational contribution to the Einstein crystal free energy in Eq. 15 is straightforward, and although the orientational part Eq. 16 cannot be calculated analytically, for large $\kappa_{\max}\eta$, one can use the saddle point approximation

$$\int d\Omega e^{-\beta\kappa_{\max}\eta g(\Omega)} \approx e^{-\beta\kappa_{\max}\eta g(\Omega_0)} \sqrt{\frac{2\pi}{\beta\kappa_{\max}\eta g''(\Omega_0)}}. \tag{20}$$

Here we have approximated $g(\Omega)$ with its second order Taylor expansion, and Ω_0 is the orientation that minimizes $g(\Omega)$. The orientational contribution to the Einstein crystal free energy then becomes

$$\beta\frac{A_{E,o}}{N} \approx \frac{3}{2}\ln(\beta\kappa_{\max}\eta) + \frac{1}{2}\{8\pi \det(H[g(\Omega_0)])\}, \tag{21}$$

where $\det(H[g(\Omega_0)])$ is the determinant of the Hessian computed at the minimum of $g(\Omega)$ [7]. Note that one should check whether $\kappa_{\max}$ is indeed large enough by verifying that higher order terms in the Taylor expansion are negligible.

An estimate for the plateau value for the integrand of Eq. 19 can be analytically estimated for sufficiently large κ, and thus serves as a consistency check. The orientational contribution to this quantity is calculated using the saddle point approximation to $A_{E,o}$, Eq. 21,

$$\kappa\eta\left\langle\sum_{i=0}^{N} g(\Omega_i)\right\rangle_\kappa = \kappa\eta\frac{\partial A_{E,o}}{\partial(\kappa\eta)} = \kappa\eta\frac{3}{2\beta}\frac{\beta}{\beta\kappa\eta} = \frac{3N}{2\beta}, \tag{22}$$

and the translational contribution can be estimated using the expression derived for the hard sphere mean squared displacement [21]. In the limit of very large κ, this result should be exact because translational and orientational displacements are then too small to break any bond.

3. **Release the crystal center of mass.** Removing the constraint over the center of mass finally gives

$$\Delta A_3 = \frac{1}{\beta}\ln(\rho). \tag{23}$$

Cumulating these results gives the absolute free energy of the crystal of patchy particles

$$A = A_E + \Delta A_1 + \Delta A_2 + \Delta A_3 \tag{24}$$

at a given density and temperature. Integration along an isobar, isotherm, or an isochore within the crystal phase can then be

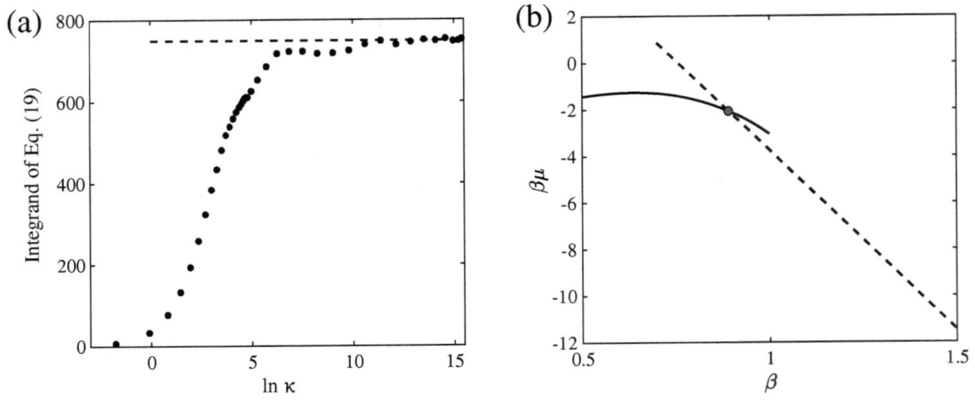

Fig. 6 (**a**) The integrand of Eq. 19. The dashed line is the plateau value estimated from the translational and orientational displacements of particles at large κ (*see* Eq. 22 and ref. 21). (**b**) The chemical potential, $\mu = \frac{A}{N} - \frac{P}{\rho}$ of the fluid phase obtained by integrating Eq. 10 with the data of Fig. 5 (solid line) and the crystal phase (dashed line) for a fixed $P = 0.35$ ($k_BT_{\text{ref}}/\sigma^3$). Their intersection gives a coexistence point between the two phases (red point)

performed to explore a broad range of conditions within that phase.

3.2.5 Gibbs–Duhem Integration

Given a coexistence point (Fig. 6b) between the crystal and the liquid, the Gibbs–Duhem relation,

$$t(P(\beta),\beta) \equiv \left(\frac{\mathrm{d}P}{\mathrm{d}\beta}\right)_{\mathrm{coex}} = -\frac{H_{\mathrm{crys}}/N - H_{\mathrm{liq}}/N}{\beta(1/\rho_{\mathrm{crys}} - 1/\rho_{\mathrm{liq}})} = -\frac{\Delta H/N}{\beta\Delta(1/\rho)}, \tag{25}$$

can be integrated to obtain coexistence points at different temperatures [23]. This can be done using a numerical method, such as predictor–corrector algorithms, and evaluating the thermodynamic quantities via MC simulations. The general idea is as follows.

1. Start from a known coexistence point (P_0, β_0), consider a system at $\beta_1 = \beta_0 + \Delta\beta$, where $\Delta\beta$ is small (see below).
2. Guess the coexistence pressure at this temperature according to the appropriate predictor formula.
3. Run *NPT* simulations of the two phases simultaneously to equilibrate $\Delta H/N$ and $\Delta(1/\rho)$. Correct the pressure prediction using these quantities according to the appropriate corrector formula.
4. Repeat 2 and 3 until convergence.

The chosen integration scheme depends on how many coexistence points are known (*see* Table 1). At the start of the process, only one such point, (P_0, β_0), is known. A short simulation is run

Table 1
Predictor–corrector algorithms, where $t_i \equiv t(P_i, \beta_i)$ for $\beta_i = \beta_0 + i\,\Delta\beta$, is defined in Eq. 25

Method	Predictor	Corrector
Trapezoid	$P_{i+1} = P_i + \Delta\beta t_i$	$P_{i+1} = P_i + \frac{\Delta\beta}{2}(t_i + t_{i+1})$
Midpoint	$P_{i+2} = P_i + 2\,\Delta\beta t_{i+1}$	$P_{i+2} = P_i + \frac{\Delta\beta}{3}(t_{i+2} + 4t_{i+1} + t_i)$
Adams	$P_{i+4} = P_{i+3} + \frac{\Delta\beta}{24}(55t_{i+3} - 59t_{i+2} + 37t_{i+1} - 9t_i)$	$P_{i+4} = P_{i+3} + \frac{\Delta\beta}{24}(9t_{i+4} + 19t_{i+3} - 5t_{i+2} + t_{i+1})$

for both phases to obtain $t(P_0, \beta_0)$. The guess for the pressure, P_1, for the next coexistence temperature, $\beta_1 = \beta_0 + \Delta\beta$, and its correction are then given by the trapezoid rule, after calculating $t(P_1, \beta_1)$ with the initial guess. The third and fourth coexistence points can be calculated using the midpoint rule. Once four points are obtained, additional ones can be found iteratively using Adams rule (*see* **Note 4**). While $\Delta\beta$ should be large enough to allow for an efficient tracing, using a too large a value causes numerical instability [24]. One way to validate the resulting coexistence line is to repeat this procedure starting from different, well-separated coexistence points.

3.2.6 Obtaining the Gas–Liquid Binodal

Coexistence points on the gas–liquid binodal can certainly be obtained by slightly modifying the above procedure, but a more straightforward approach is to use Gibbs Ensemble simulations [4, 25], which are specifically designed for identifying coexistence between homogeneous phases of intermediate density. In this method, two boxes of fluid are simulated simultaneously. Their total volume and number of particles are kept constant but boxes can exchange volume as well as particles between each other. The density of each box then converges to the gas or the liquid density, ρ_g and ρ_l, respectively (*see* **Note 5**) (Fig. 7). The binodal is then obtained as follows.

1. **Obtain a few coexistence points from Gibbs Ensemble simulations.** Gibbs Ensemble simulations are run for a set of temperatures below the estimated critical temperature, T_c, from generalized law of corresponding states [26], starting from an intermediate fluid density, $\rho \approx 0.3$.
2. **Fit coexistence data to obtain the full binodal.** The physical universality of the gas–liquid transition allows for tracing the binodal using only a few coexistence points. The full binodal, including the critical point, can then be calculated by fitting two universal equations: a scaling law and the law of rectilinear diameters [4]. The former gives an estimate of T_c,

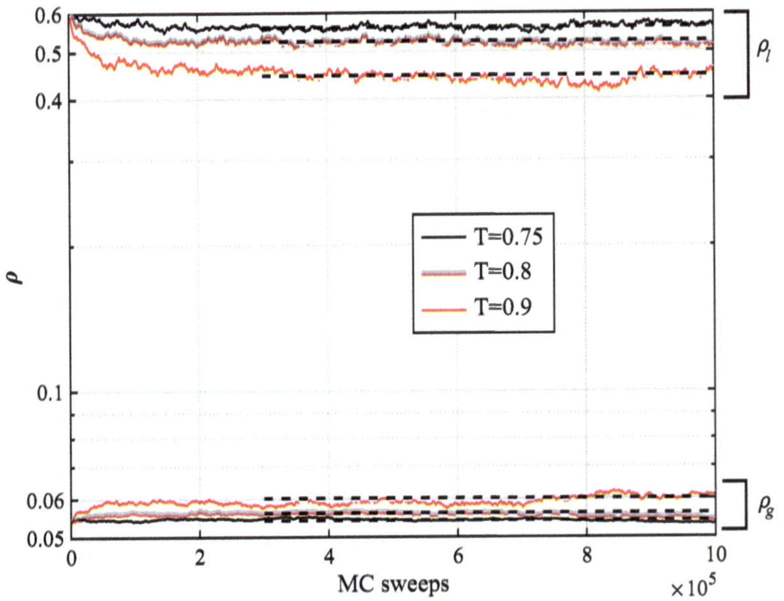

Fig. 7 Evolution of ρ_l and ρ_g throughout the Gibbs Ensemble simulations for various temperatures. Note that the densities converge to their coexistence values after roughly 2×10^5 MC sweeps. Average densities are calculated (dashed lines) after the equilibration period. These three pairs of data points are those that appear in the final phase diagram (Fig. 8)

$$\log(\rho_l - \rho_g) = \log B + \beta \log(T - T_c), \tag{26}$$

where $\beta = 0.32$ (not to be confused with the inverse temperature) is the magnetization exponent for the Ising universality class, and B is a proportionality constant [27]. Once T_c is found, the critical density, ρ_c, can be obtained from the law of rectilinear diameters, which describes the asymmetry of the gas–liquid binodal away from T_c,

$$\frac{\rho_l + \rho_g}{2} = \rho_c + A(T - T_c), \tag{27}$$

where A and ρ_c are determined by fitting. The coexistence binodal is then given by

$$\rho = \rho_c + A(T - T_c) \pm \frac{B(T - T_c)^\beta}{2}. \tag{28}$$

Putting together all of these results, we obtain the final phase diagram shown in Fig. 8.

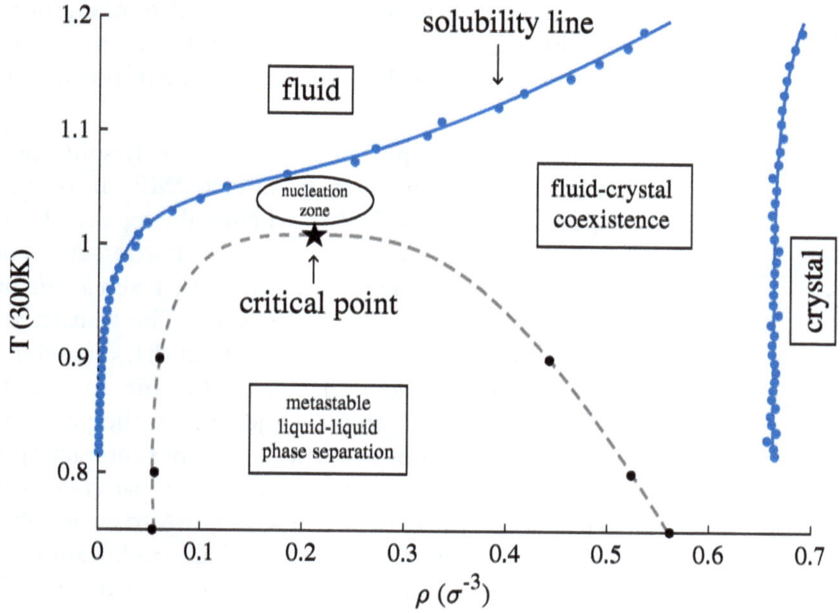

Fig. 8 The $T - \rho$ phase diagram of a patchy model of rubredoxin [7]. Blue points denote the solubility line and are obtained by Gibbs–Duhem integration starting from the previously obtained coexistence point (*see* Fig. 6b). Black points are gas–liquid coexistence points obtained from the Gibbs Ensemble simulations. The fit to the gas–liquid binodal (gray dashed line) terminates at the resulting critical point (black star). Below this line, the system exhibits a metastable liquid–liquid phase coexistence regime in which protein solutions often gel in experiments. This long-lived state often precludes crystallization. The region between the solubility line and T_c is called the nucleation zone because supersaturated solutions in this region are more likely to produce crystals by avoiding gelation

4 Notes

The above procedure results in the phase diagram for a simple globular protein as shown in Fig. 8. In what follows we discuss a number of geometric issues and how they can be avoided, as well as briefly mention possible ways of increasing the model complexity.

1. The tutorial prepared by Justin Lemkul can be accessed at http://www.bevanlab.biochem.vt.edu/Pages/Personal/justin/gmx-tutorials/umbrella/index.html
2. While the simplest method for obtaining the free energy of a fluid is Widom insertion [4], at high densities it is more accurate to use Eq. 10.
3. The parameter η is a proportionality constant chosen such that the strengths of both restraints can be tuned by κ alone.
4. It is often not advantageous to use higher order predictor–corrector formulas. These not only require a higher number of coexistence points but also exhibit stability issues.

5. If temperature is close to the critical temperature, the small density difference can cause the boxes to switch identity (liquid↔gas), which limits the efficacy of the approach in this regime.
6. Because globular proteins are not actually spherical, the onset of harsh repulsion for each contact PMF can occur at slightly different distances. In the scheme above, the chosen particle diameter should be the same for all contacts. The smallest center of mass distance should then be taken as the hard sphere diameter and the other PMFs should be translated such that the onset of attraction coincides with that diameter.

   Another problem that can arise due to deviations from sphericity is that a simple projection of the patch position on the sphere may not result in all patches interacting within the relevant crystal symmetry. In this case, patch vectors and unit cell parameters can be perturbed slightly to ensure that all bonds are satisfied in the crystal phase. This modification is known to have but a very limited effect on the phase diagram [28].
7. The simple patchy model described here does not capture the phase behavior of certain proteins. Numerous enhanced patchy models have been proposed to capture these effects. The impact of shape anisotropy [29–31], patch mobility [32], and the interaction potential form [33, 34] have been investigated in the context of general self-assembly. Such features can be considered if the microscopic properties of the protein of interest suggest that more complex models are required. The application of these features to specific protein systems is still an open area of research.

Acknowledgments

We thank Diana Fusco for guiding discussions. The authors acknowledge support from National Science Foundation Grant no. NSF DMR-1749374. This work used the Extreme Science and Engineering Discovery Environment (XSEDE), which is supported by National Science Foundation grant number ACI-1548562, as well as the Duke Compute Cluster.

References

1. Chen V, Davis I, Richardson D (2009) KiNG (Kinemage, next generation): a versatile interactive molecular and scientific visualization program. Protein Sci 18(11):2403–2409
2. Berendsen H, van der Spoel D, van Drunen R (1995) Gromacs: a message-passing parallel molecular dynamics implementation. Comput Phys Commun 91(1):43–56
3. Case D, Cerutti D, Cheatham T III, Darden T, Duke R, Giese T, Gohlke H, Goetz A, Greene D, Homeyer N, Izadi S, Kovalenko A, Lee T, LeGrand S, Li P, Lin C, Liu J, Luchko T,

Luo R, Mermelstein D, Merz K, Monard G, Nguyen H, Omelyan I, Onufriev A, Pan F, Qi R, Roe D, Roitberg A, Sagui C, Simmerling C, Botello-Smith W, Swails J, Walker R, Wang J, Wolf R, Wu X, Xiao L, York D, PA K (2017) Amber 2017. Tech. rep., University of California, San Francisco

4. Frenkel D, Smit B (2001) Understanding molecular simulation: from algorithms to applications. Academic, Orlando
5. Rovigatti L, Russo J, Romano F (2018) How to simulate patchy particles. Eur Phys J E 41 (5):59
6. Rovigatti L, Romano F, Russo J (2018) lorenzo-rovigatti/patchyparticles v1.0.1. https://doi.org/10.5281/zenodo.1171695
7. Fusco D, Headd J, De Simone A, Wang J, Charbonneau P (2014) Characterizing protein crystal contacts and their role in crystallization: rubredoxin as a case study. Soft Matter 10 (2):290–302
8. Berman HM, Westbrook J, Feng Z, Gilliland G, Bhat TN, Weissig H, Shindyalov IN, Bourne PE (2000) The protein data bank. Nucleic Acids Res 28(1):235–242
9. Bönisch H, Schmidt C, Bianco P, Ladenstein R (2005) Ultrahigh-resolution study on Pyrococcus abyssi rubredoxin. I. 0.69 Å X-ray structure of mutant W4L/R5S. Acta Crystallogr D 61(7):990–1004
10. Kästner J (2011) Umbrella sampling. Wiley Interdiscip Rev Comput Mol Sci 1(6):932–942
11. Krissinel E, Henrick K (2007) Inference of macromolecular assemblies from crystalline state. J Mol Biol 372(3):774–797
12. Rostkowski M, Olsson M, Søndergaard C, Jensen J (2011) Graphical analysis of pH-dependent properties of proteins predicted using PROPKA. BMC Struct Biol 11(1):1
13. Hub J, De Groot B, Van Der Spoel D (2010) g_wham – a free weighted histogram analysis implementation including robust error and autocorrelation estimates. J Chem Theory Comput 6(12):3713–3720
14. Kern N, Frenkel D (2003) Fluid–fluid coexistence in colloidal systems with short-ranged strongly directional attraction. J Chem Phys 118(21):9882–9889
15. Fusco D, Charbonneau P (2016) Soft matter perspective on protein crystal assembly. Colloids Surf B 137:22–31
16. Sear R (1999) Phase behavior of a simple model of globular proteins. J Chem Phys 111 (10):4800–4806
17. Wentzel N, Gunton J (2008) Effect of solvent on the phase diagram of a simple anisotropic model of globular proteins. J Phys Chem B 112 (26):7803–7809
18. Dixit N, Zukoski C (2002) Crystal nucleation rates for particles experiencing anisotropic interactions. J Chem Phys 117 (18):8540–8550
19. Vega C, Sanz E, Abascal J, Noya E (2008) Determination of phase diagrams via computer simulation: methodology and applications to water, electrolytes and proteins. J Phys Condens Matter 20(15):153101
20. Romano F, Sanz E, Sciortino F (2010) Phase diagram of a tetrahedral patchy particle model for different interaction ranges. J Chem Phys 132(18):184501
21. Frenkel D, Ladd A (1984) New Monte Carlo method to compute the free energy of arbitrary solids. Application to the FCC and HCP phases of hard spheres. J Chem Phys 81 (7):3188–3193
22. Riley KF, Hobson MP, Bence SJ (2006) Mathematical methods for physics and engineering: a comprehensive guide. Cambridge University Press, Cambridge
23. Kofke D (1993) Gibbs-Duhem integration: a new method for direct evaluation of phase coexistence by molecular simulation. Mol Phys 78(6):1331–1336
24. Kofke D (1993) Direct evaluation of phase coexistence by molecular simulation via integration along the saturation line. J Chem Phys 98(5):4149–4162
25. Panagiotopoulos A (1987) Direct determination of phase coexistence properties of fluids by Monte Carlo simulation in a new ensemble. Mol Phys 61(4):813–826
26. Noro MG, Frenkel D (2000) Extended corresponding-states behavior for particles with variable range attractions. J Chem Phys 113(8):2941–2944
27. Blote H, Luijten E, Heringa J (1995) Ising universality in three dimensions: a Monte Carlo study. J Phys A 28(22):6289
28. Fusco D, Charbonneau P (2013) Crystallization of asymmetric patchy models for globular proteins in solution. Phys Rev E 88(1):012721
29. Tang Z, Zhang Z, Wang Y, Glotzer S, Kotov N (2006) Self-assembly of CdTe nanocrystals into free-floating sheets. Science 314 (5797):274–278
30. Ye X, Chen J, Engel M, Millan J, Li W, Qi L, Xing G, Collins J, Kagan C, Li J et al. (2013)

Competition of shape and interaction patchiness for self-assembling nanoplates. Nat Chem 5(6):466

31. Glotzer S, Solomon M (2007) Anisotropy of building blocks and their assembly into complex structures. Nat Mater 6(8):557

32. Bianchi E, Capone B, Kahl G, Likos C (2015) Soft-patchy nanoparticles: modeling and self-organization. Faraday Discuss 181:123–138

33. de las Heras D, da Gama M (2016) Temperature (de)activated patchy colloidal particles. J Phys Condens Matter 28(24):244008

34. Wilber AW, Doye JP, Louis AA, Noya EG, Miller MA, Wong P (2007) Reversible self-assembly of patchy particles into monodisperse icosahedral clusters. J Chem Phys 127 (8):08B618

Chapter 16

Binding Free Energies of Conformationally Disordered Peptides Through Extensive Sampling and End-Point Methods

Matthew G. Nixon and Elisa Fadda

Abstract

The ability to obtain binding free energies from molecular simulation techniques provides a valuable support to the interpretation and design of experiments. Among all methods available, the most widely used equilibrium free energy methods range from highly accurate and computationally expensive perturbation theory-based methods, such as free energy perturbation (FEP), or thermodynamic integration (TI), through end-point methods, such as molecular mechanics with generalized Born and surface area solvation (MM/GBSA) or MM/PBSA, when the Poisson–Boltzmann method is used instead of GB, and linear interaction energy (LIE) methods, to scoring functions, which are relatively simple empirical functions widely used as part of molecular docking protocols. Because the use of FEP and TI approaches is restricted to cases where the perturbation leading from an initial to final state is negligible or minimal, their application to cases where large conformational changes are involved between bound and unbound states is rather complex, if not prohibitive in terms of convergence. Here we describe a protocol that involves the use of extensive conformational sampling through molecular dynamics (MD) in combination with end-point methods (MM/GB(PB)SA) with an additional quasi-harmonic entropy component, for the calculation of the relative binding free energies of highly flexible, or intrinsically disordered, peptides to a structured receptor.

Key words Binding free energy, MM/GBSA, Protein–protein interactions, Intrinsically disordered proteins, Prestructuring, Molecular recognition, Conformational sampling, Molecular dynamics

1 Introduction

Conformational disorder is now widely recognized as an essential functional trait of many proteins [1, 2]. In apparent contrast with the structure–function relationship, high conformational flexibility, or intrinsic disorder, can facilitate different biomolecular roles [3, 4]. For example, the degree of conformational disorder can contribute to modulate the binding affinity in protein–protein interactions (PPI) as a direct consequence of the nonnegligible entropic penalty associated to the restriction of the conformational

Jennifer J. McManus (ed.), *Protein Self-Assembly: Methods and Protocols*, Methods in Molecular Biology, vol. 2039, https://doi.org/10.1007/978-1-4939-9678-0_16,

degrees of freedom upon binding. This trademark feature makes intrinsically disordered protein (IDP) regions ideal counterparts in transient and reversible PPIs, thus essential players in signaling and regulatory pathways [1, 5, 6] and as scaffolding proteins, coordinating the reversible formation of macromolecular assemblies [4, 7].

The high flexibility characterizing IDP regions does not directly or necessarily translate into a complete lack of structure. Indeed, from a structural biology perspective, any system that undergoes conformational interconversions at a timescale below the experimental observation timeframe can be defined as conformationally disordered. Recent work from our lab [8] and from other groups [9–12] suggests that in some cases the time-averaged overlay of quickly interconverting different prestructured motifs embedded within random coils can appear as conformational disorder at higher timescales. These unstable, or short-lived, conformers can carry 3D structural features complementary to the binding site architecture of specific receptors [8, 13, 14], potentially functioning as nucleation sites in the molecular recognition process [5, 15]. The degree of prestructuring of the unbound IDP region can also modulate the entropic penalty upon binding [11], ranging from a maximum penalty for a completely disordered region, where binding is largely driven by induced fit, to a minimum penalty for an ID region with well-defined conformational propensity, where recognition is likely based on conformational selection [16]. Recognition and binding of highly flexible motifs that also have a nonnegligible conformational propensity will be likely based on a combination of induced fit and conformational selection events [15, 16].

The structural discretization of IDP conformational ensembles can fall easily beyond the capabilities of experimental techniques, especially in case of high structural interconversion rates and high degrees of flexibility. Nevertheless, this is one area where computing with the extensive degree of sampling achievable nowadays can provide a great deal of information [5]. Understanding how conformational disorder modulates PPI binding affinity is not a trivial task from a simulation point of view either [17]. In principle, monitoring the binding free energy from the molecular recognition stage through a partial or complete folding-in-place pathway is possible by atomistic simulations [18], but, because the reaction coordinate is not usually known or obvious, it is extremely demanding in terms of computational resources. Also, because of the structural instability of IDP targets, highly accurate binding free energy calculations based on perturbation theory are very difficult, if not impossible, to converge. The free-energy calculation approach we describe here involves the combination of (1) *extensive sampling via MD simulations*, aimed at identifying the IDP regions' residual structural propensity/degree of disorder, with (2) *end-*

point free energy calculation methods, more specifically the molecular mechanics (MM) generalized Born (GB), or Poisson–Boltzmann (PB), solvent accessible surface area (SASA), widely known as MM/GBSA (MM/PBSA) for short [19, 20], with an additional (3) *entropic contribution calculated based on the quasi-harmonic approximation* [21, 22]. This approach represents a valuable alternative for the calculation of relative binding affinities of PPIs involving IDPs, as it is both feasible in terms of computational resources and informative in terms of accuracy.

1.1 Part 1: Extensive Sampling

The extensive sampling via MD simulations can be either achieved through multiple parallel and independent conventional trajectories or through enhanced sampling schemes, such as replica exchange [23], the choice of which method depending on the specific system at hand. This step is aimed at identifying stable conformers or lack thereof within the conformational ensemble of the ID peptide in its free and bound state. This information helps us to (a) characterize the peptide level of disorder and (b) identify recurrent and distinct short 3D motifs that can potentially act as nucleation sites in the molecular recognition process. Furthermore, converged MD simulations of the free peptide in solution are essential to calculate correctly the entropic contribution to the binding free energy. MD simulations of the bound state can be run from starting structured based on structural data, that is, NMR or X-ray when available, or from a set of different (potential) "recognition complexes" generated either by molecular docking or by structural alignment of the short 3D motifs, identified by clustering analysis of the MD trajectories of the free peptide.

1.2 Parts 2 and 3: Free Energy Calculations

Within the MM/GBSA (MM/PBSA) approximation the binding free energy is determined as

$$\Delta G_{\text{bind}}^{\text{MM/GB(PB)SA}} = \Delta H_{\text{gas}}^{\text{MM}} + \Delta G_{\text{solv}}^{\text{GB(PB)SA}} \tag{1}$$

where each contribution is time averaged over the MD trajectory obtained for the complex. The enthalpic term is obtained directly from a force field-based evaluation of the potential energies, while the second term represents the solvation free energy as the sum of a polar $\Delta G_{\text{electr}}^{\text{GB(PB)}}$ and of a nonpolar $\Delta G_{\text{nonp}}^{\text{SA}}$ contribution. $\Delta G_{\text{electr}}^{\text{GB(PB)}}$ is obtained by solving the generalized Born (GB), or the Poisson--Boltzmann (PB), equations, and $\Delta G_{\text{nonp}}^{\text{SA}}$ from a linear equation that depends on the solvent accessible surface area [19, 20, 24]. Because end-point free energy calculations are based on a thermodynamic cycle linking an "initial state" where both receptor (R) and ligand (L) are unbound (R + L), to a "final state" where R and L are bound in a complex (RL), ideally the $\Delta G_{\text{bind}}^{\text{MM/GB(PB)SA}}$ term should represent an average over three separate MD trajectories, two for the unbound isolated species (R and L), and one for the complex

(RL). For practical reasons this is rarely (if ever) done, and the calculation is an average based on a single MD trajectory of the complex [20], which is a fair approximation when both R and L are conformationally restrained structures, although it is not a safe bet in general [25]. Extensive sampling were large structural changes are explored and taken into account and that allows for conformational convergence, minimizes the limitations of the method when flexibility is, or can be, an issue. Because of the conformational disorder of the system at hand, we include a conformational entropy term ($\Delta S_{\text{conf}}^{\text{PCA}}$) in the free energy estimate, *see* Eq. 2, determined by a quasi-harmonic approach [21, 22]. The total binding free energy that we calculate is then given then by,

$$\Delta G_{\text{bind}}^{\text{MM/GB(PB)SA}} = \Delta H_{\text{gas}}^{\text{MM}} + \Delta G_{\text{solv}}^{\text{GB(PB)SA}} - T\Delta S_{\text{conf}}^{\text{PCA}} \quad (2)$$

where T is the absolute temperature.

1.3 Test Case

Based on the approximation to the binding free energy expression shown in Eq. 2, we show here to the case of the binding ID peptides derived from different sequences of the nucleotide excision repair (NER) scaffolding protein *Xeroderma pigmentosum* group A (XPA) [7, 26] to the excision repair cross-complementation group 1 (ERCC1) central domain. The 14 residue peptides tested correspond to the ERCC1-binding sequence ($XPA_{67–80}$) found in *H. sapiens*, *R. norvegicus*, *C. lanigera*, and *X. laevis*. Structural data are available only for the complex between the ERCC1 central domain and *H. sapiens* $XPA_{67–80}$ peptide (PDBid 2JNW) [26]. Extensive sampling of the *H. sapiens* $XPA_{67–80}$ peptide unbound in solution shows distinct propensity to form a stable hairpin that closely resembles the bound conformation [27, 28], see Fig. 1. The technical details on the methods used to perform these calculations are described in the following sections.

2 Materials

All simulations were carried out at 300 K with AMBER99SB-ILDN [29] to represent the protein atoms and counterions, and the TIP4P-Ew model [30] to represent water molecules. Counterions were added to neutralize the total electrostatic charge. For a brief discussion on the force field's choice *see* **Note 1**. In the specific study case presented here, conformational sampling has been achieved by means uncorrelated conventional MD simulations run in parallel. All calculations were run with GROMACS v. 4.6.3 [31] (GMX) and with Amber v.12 [32]. Computational resources were provided by the Irish Centre for High-End Computing (ICHEC). All calculations were run on the ICHEC supercomputer

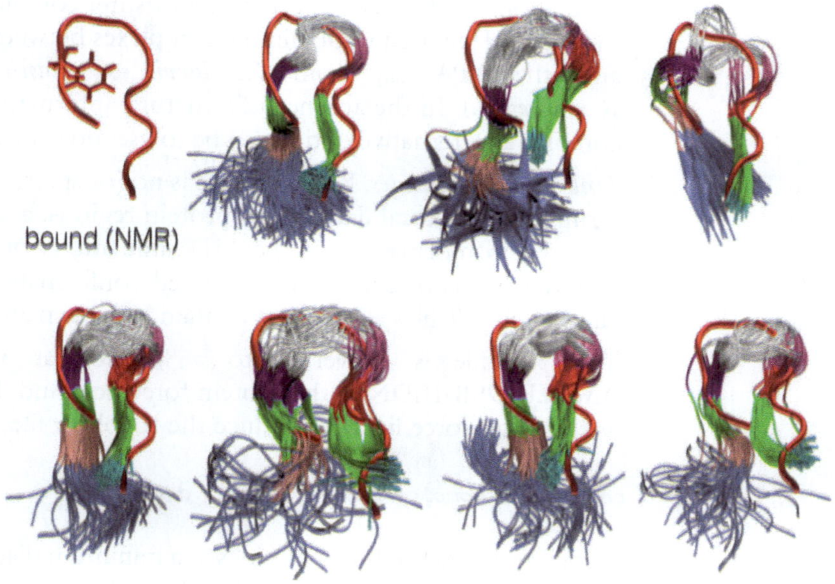

Fig. 1 Structural alignment of the XPA_{67-77} peptide (*H. sapiens*) in the bound conformation, shown in red (PDBid 2JNW) [26] and the highest populated conformers obtained from ten parallel MD runs of the peptide unbound in solution for a cumulative simulation time of 10 μs [27, 28]. The coloring scheme corresponds to the XPA_{67-77} peptide sequence

fionn "Thin" partition, an SGI ICE-X system of 320 nodes, where each node has 24 core 2.4 GHz Intel Ivy Bridge processors, 64 GiB of RAM and an FDR InfiniBand network adaptor. Minimization and initial 500 ps equilibration steps in the NVT and NPT ensemble were carried out using two nodes (i.e., 48 cores). All further equilibration steps and production simulations were run on four nodes (i.e., 96 cores). We estimated that the final cost for all MD simulations in this project reached approximately 1,000,000 cpu hours. For visualizations and analysis we have used the Visual Molecular Dynamics (VMD) software [33].

3 Methods

The following protocol was used to set up and run the MD simulations of both the ligand unbound in solution (i.e., the free ID peptide) and the complex.

3.1 MD Simulation Protocol

1. *Receptor/ID-peptide complexes.* A *pdb* file of complex can be obtained from either from structural data, when available, in our case the 2JNW PDBid corresponds to the NMR structure of the XPA_{67-80} (*H. sapiens*) in complex with ERCC1. Alternatively, a structure file can be obtained from the structural alignment of the short 3D motifs identified from the MD

simulations of the free peptide to an existing complex. This is the method we used to obtain the complexes between ERCC1 and the XPA_{67-80} from *X. laevis*, *C. lanigera*, and *R. norvegicus*. In the absence of structural information on the complex an alternative option can be to use molecular docking.

2. *Unbound ID peptides.* Because there is no (or sparse and undetermined) structural data for ID protein regions, as a starting, unbiased conformation of the MD simulations of unbound peptides, we considered fully extended conformations. Note: all starting *pdb* files should not contain hydrogen atoms.

3. The *pdb* file is converted to a *gro* format and select AMBER99SB-ILDN as the protein force field and TIP4P-Ew as the water force field to produce the topology file,

```
> pdb2gmx -f complex.pdb -o complex.gro
```

4. Generate a periodic box with sides at a minimum distance from the complex of 1.2 nm. This value depends on the size of the system at hand and the chosen cut-off values of long range interactions. A rhombohedric dodecahedral simulation box minimizes the number of water molecules included in the calculation.

```
> editconf -f complex.gro -o complex_box.gro -c -d 1.2 -bt
dodecahedron
```

To add the explicit solvent,

```
> genbox -cs tip4p.gro -cp complex_box.gro -o complex_solv.gro
-p topol.top
```

5. The following two commands generate a GMX-type topology file (*.tpr*) that is the input for the *genion* tool to add enough counterions to reach neutrality, in the example below three Na^+ and no Cl^- have been added. Ionic strength can be adjusted at this stage to reach the desired counterion concentration.

```
> grompp -f ions.mdp -c complex_solv.gro -o ions.tpr -p
topol.top
> genion -s ions.tpr -o complex_ions.gro -p topol.top -pname
NA -nname CL -np 3 -nn 0
```

6. At this point the system is solvated and neutralized and ready for the setup stage of the MD simulation. The first step involves an energy minimization of the positions of the solvent, ions, and hydrogen atoms, with all protein heavy atoms restrained. The application of position restraints is turned on by the

statement *-define -DPOSRES* flag at the very top of the *.mdp* input file. This flag calls the *posre.itp* restraints file, which was generated earlier at **step 2**. Our energy minimizations were carried out through 500,000 steps of steepest descent, with a force-based convergence threshold of 100 kJ/mol/nm. Long range electrostatics were represented through periodic boundary conditions within the particle mesh Ewald (PME) framework, with order 6 and a cutoff of 1.2 nm. van der Waals interactions were calculated using a cutoff method, with a cutoff value of 1.2 nm. All hydrogen bonds were constrained using the LINCS [34] approach with order 12. The energy minimization is initiated with the following set of commands,

```
> grompp -f min_steep.mdp -c complex_ions.gro -o min_steep.
tpr -p topol.top
> mdrun -s min_steep.tpr -deffnm min_steep.
```

7. After minimization we carried out an equilibration of 500 ps in the NVT ensemble restraining the position of all solute heavy atoms. This step is aimed at filling out potential volumes with low solvent density. To integrate the equation of motion we used a leap-frog stochastic dynamics (*sd*) integrator, with a friction coefficient corresponding to the inverse of tau-*t* equal to 0.1 ps, where tau-*t* is the time constant for coupling. The *sd* integrator was set to maintain a target temperature of 300 K.

```
> grompp -f nvt.mdp -c min_steep.gro -o min_steep.tpr -p
topol.top
> mdrun -s nvt.tpr -deffnm nvt
```

8. A second restrained equilibration of 500 ps follows, this time in the NPT ensemble, with a Berendsen barostat set to a target pressure of 1 bar.

```
> grompp -f npt.mdp -c nvt.gro -o npt.tpr -p topol.top
> mdrun -s npt.tpr -deffnm npt
```

9. Now that the system has reached the target temperature of 300 K and the target pressure of 1 bar with the correct density, the conformational equilibration stage can be started. The specific steps depend on the system. For the *receptor/ID-peptide complex* this stage involves three consecutive equilibration steps of 5 ns each, first with the heavy atoms of the receptor and the ligand backbone atoms restrained, then restraining only the backbone atoms of ligand and receptor restrained, and finally with the ligand atoms free and the receptor backbone atoms restrained.

```
> grompp -f eq_all_rec_lig_backbone.mdp -c npt.gro -o
eq_all_rec_lig_backbone.tpr -p topol.top
> mdrun -s eq_all_rec_lig_backbone.tpr -deffnm eq_all_re-
c_lig_backbone
```

For the *free ID-peptide* only one 5 ns equilibration step is required, with all heavy atoms free.

10. The length of the production phase is also system dependent and it is dictated by the structural and/or thermodynamic parameters that are targeted. In our case, all *receptor/ID-peptide complexes (XPA$_{67-80}$/ERCC1)* were analyzed for 2 μs of conventional MD simulations.

```
> grompp -f md1.mdp -c eq_rec_backbone.gro -t eq_rec_back-
bone.trr -e eq_rec_backbone.edr -o md1.tpr -p topol.top
> mdrun -s md1.tpr -deffnm md1
```

In case of the *free ID-peptide* the initial production step was extended to 100 ns. This simulation was used uniquely to generate ten uncorrelated snapshots that were used as starting point of ten independent MD simulations, from which data were collected. Each of these ten simulations was run for 1 μs.

A clustering analysis provides a strategy to analyze the conformational propensity of the peptide during an MD trajectory. Clustering organizes the whole trajectory based on their structural similarity into separate bins, with each structure only being placed into a single cluster. The trajectory is first compared to a reference structure, here chosen as first NMR structure in 2JNW by means of an index (*.ndx*) file that indicates the XPA$_{67-77}$ backbone atoms selected to run the structural comparison.

```
> g_rms -f X_ns.trr -n index.ndx -m rmsd.xpm -s 2JNW_nmr.pdb
```

We then performed clustering of the structures using the *gromos* method [35] and a RMSD cutoff of 0.15 nm, chosen as optimal, in terms of providing structurally distinct clusters, after testing cutoff values in a range between 0.05 and 0.20 nm.

```
> g_cluster -s 2JNW_nmr.pdb -f X_ns.trr -dm rmsd.xpm -dist
rmsd_distribution.xvg -o clusters.xpm -sz cluster_sizes.xvg
-tr cluster_transitions.xvg -clid cluster_id_overtime.xvg
-cl clusters.pdb -cutoff 0.15 -method gromos -n index.ndx
```

3.2 MM/GB(PB)SA Calculations

1. The complex/peptide dynamics from GMX was separated from the solvent and counterion dynamics with the with the *trjconv* command trajectory files obtained from the MD production were stripped from the solvent use to remove solvent, ions, and periodic boundary conditions from starting *gro* and *trr* files.

```
> trjconv -f eq_rec_backbone.gro -s md1.trr -o md1_nopbc.trr
-c -pbc mol
```

2. Use trjcat to combine multiple trajectory files into one continuous file. The -settime flag puts input trr files in order when prompted.

```
> trjcat -f md1_nopbc.trr md2_nopbc.trr ... mdx_nopbc.trr -o
X_ns.trr -settime
```

3. Load *.gro* file into VMD and save starting conformation for ligand receptor and complex as *.pdb* files.
4. Load *.trr* file into VMD and save trajectory in AMBER *.crd* format.
5. Prepare AMBER parameters (*.prm7*) and coordinate (*.rst7*) files of the ligand, receptor and complex.

```
> tleap
> source leaprc.ff99SBildn
> ligand = loadpdb ligand.pdb
> receptor = loadpdb receptor.pdb
> complex = loadpdb complex.pdb
> saveamberparm ligand ligand.prm7 ligand.rst7
> saveamberparm receptor receptor.prm7 receptor.rst7
> saveamberparm complex complex.prm7 complex.rst7
```

6. Use the MMPBSA.py script to perform MMGBSA. This produces an output file that provides you with an average of the free energy over the whole simulation trajectory.

```
>$AMBERHOME/exe/MMPBSA.py -O -i mmgbsa.in -o mmgbsa.dat -cp
complex.prm7 -rp receptor.prm7 -lp ligand.prm7 -y X_ns.crd
```

7. Extract the values for the van der Waals contribution (VDW), the electrostatic energy (EEL), the electrostatic contribution to the solvation free energy calculated by GB (or PB) (EGB) and the nonpolar contribution to the solvation free energy calculated by an empirical model (ESURF) from output files using the 'grep' bash command for ligand, receptor and complex. For further information on ESURF, please *see* **Note 3**.

Table 1
Average binding free energy after convergence of the MD trajectories, *see* Notes 2 and 3 for details

	C. lanigera	*H. sapiens*	*X. laevis*	*R. norvegicus*
$\Delta G_{\mathrm{MM/GBSA}}$	−49.1 ± 3.5	−40.7 ± 4.3	−41.9 ± 4.5	−42.7 ± 4.3
$\Delta G_{\mathrm{MM/PBSA}}$	−47.9 ± 4.4	−40.2 ± 5.5	−40.8 ± 4.9	−40.5 ± 4.7

All values are in kcal/mol, errors are calculated as standard deviations

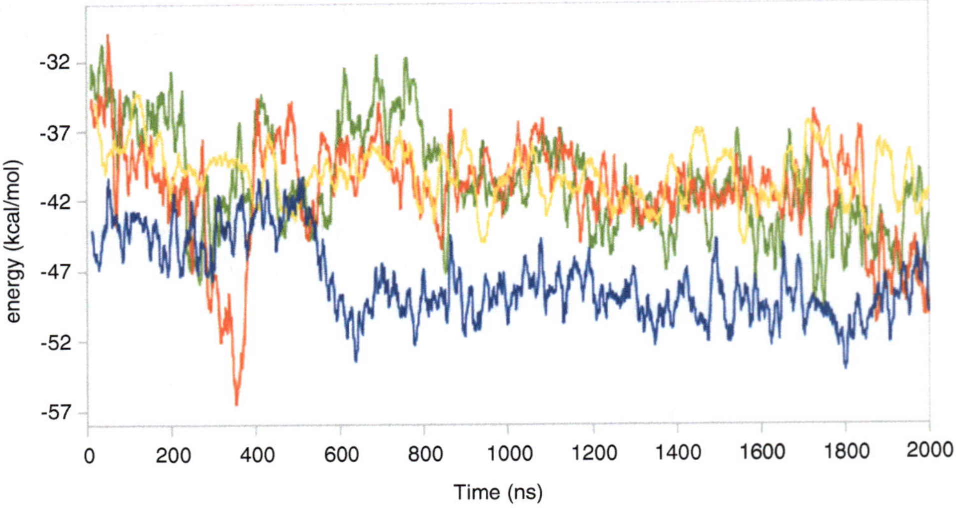

Fig. 2 Total MM/GBSA binding free energy values corresponding to the three different XPA_{67-80} sequences, namely, *C. lanigera* (in blue), *H. sapiens* (in yellow), *X. laevis* (in orange), and *R. norvegicus* (in green)

8. The free energy values, shown in Table 1 and plotted in Fig. 2 in function of the simulation time, are obtained from the sum of these contributions calculated for ligand and the receptor separately, minus the sum of the contributions for the complex.

3.3 Conformational Entropy

Principal component analysis (PCA) can be carried out using GMX. In the PCA calculation the motion of the ligand is divided into a set of independent eigenvectors using *g_covar*. This calculation removes the overall rotational and translational modes, leaving only the vibrational modes, then the total entropy of the system is calculated, *see* Table 2, and added to the component from the MM/GBSA calculation shown in Table 1.

3.3.1 Determination of the Entropy Term of Ligand Bound to Receptor

1. Create an index file, using *make_ndx*, corresponding to the backbone of the ligand molecule.

```
> make_ndx -f ligand.pdb -o index.ndx
```

Table 2
Conformational entropy contributions calculated through the quasi-harmonic approximation implemented in GMX v. 4.6.5

	C. lanigera	*H. sapiens*	*X. laevis*	*R. norvegicus*
TΔS	−30.3 ± 3.6	−14.3 ± 5.5	−17.9 ± 4.7	−20.1 ± 5.4

All values are in kcal/mol, errors are calculated as standard deviations

2. Use *g_covar* to diagonalize the covariance matrix. When prompted to select a group for least square fits and covariance analysis choose the group created above. This produces an output file of the eigenvectors as a trajectory file (*eigenval.xvg*).

```
> g_covar -s eq_rec_backbone.gro -f X_ns.trr -n index.ndx
```

3. Use *g_anaeig* to determine the conformational entropy. *g_anaeig* will use *eigenval.xvg* as an input implicitly but if this saved under another name the -v flag can be used to specify the name of the eigenvector file. *g_anaeig* along with the *-entropy* and *-temp* flags will print out the total conformational entropy at a specified temperature using quasi-harmonic formula and Schlitter's method [22].

```
> g_anaeig -s eq_rec_backbone.gro -f X_ns.trr -entropy -temp
300 > entropy.txt
```

3.3.2 Determination of Total Entropy of Ligand Free in Solution

1. As described in the MM/GBSA method section, generate a single continuous trajectory for each snapshot and repeat **steps 2** and **3**.

4 Notes

1. The optimal choice of force field for molecular simulations is an ongoing debate, and even more so in the case of simulations of IDPs [36, 37]. In the specific case of the XPA_{68-80} peptide the choice of AMBER99SB-ILDN/TIP4P-Ew has been proven as suitable. However, as for all other force fields, we are aware of its limitations and invite the users to make their own choice based on the literature and *in-house* testing.
2. The simulation needs to be long enough in order for the system to reach convergence. We found that for the system examined here a 2 μs trajectory for the receptor–ligand complex is sufficient. Convergence was measured using RMSD average correlation (RAC), *see* Fig. 3. The free energy calculation is carried out after the system has reached convergence; all

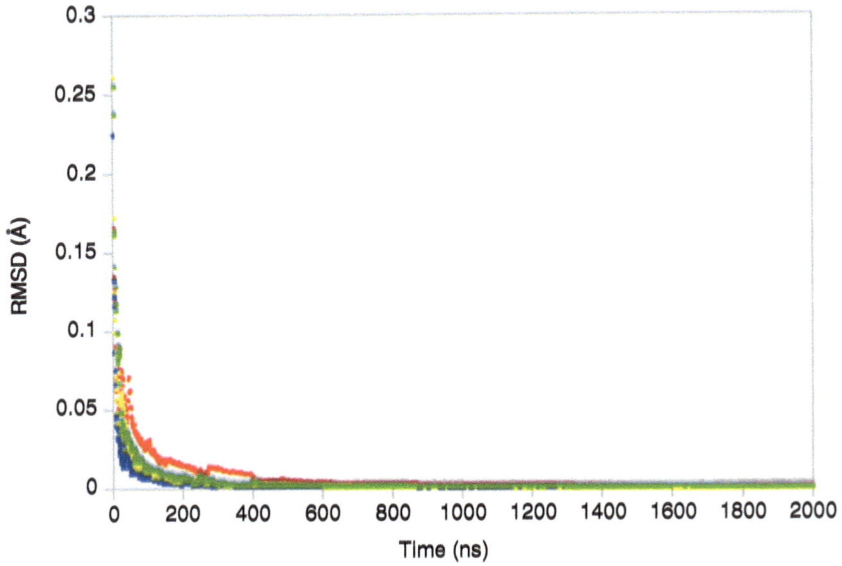

Fig. 3 RMSD average correlation (RAC) data obtained for the bound systems, namely, *C. lanigera*, shown in blue, *H. sapiens*, shown in yellow, *X. laevis*, shown in orange, and *R. norvegicus*, shown in green. All systems were considered converged after a 1 μs threshold

systems appeared to have reached convergence after 600 ns. However, based on the binding free energies it appears that the systems are not converged at this time. To ensure the systems are fully convergence the free energy was calculated beginning at 1 μs.

3. We also performed MM/PBSA and found that the polar solvation component using the PB equation agreed within ±3 kcal/mol of the GB values; however, the nonpolar contribution is calculated differently for MM/PBSA compared to MM/GBSA [38]. This difference is approximately 30 kcal/mol. This discrepancy in the nonpolar solvation term reduced the binding free energy to almost zero in the MM/PBSA methodology.

References

1. Wright PE, Dyson HJ (2015) Intrinsically disordered proteins in cellular signaling and regulation. Nat Rev Mol Cell Biol 16(1):18–29
2. Dunker AK, Obradovic Z, Romero P, Garner EC (2000) Intrinsic protein disorder in complete genomes. Genome Inform 11:161–171
3. Tompa P (2005) The interplay between structure and function in intrinsically unstructured proteins. FEBS Lett 579:3346–3354
4. Babu M, van der Lee R, de Groot NS, Gsponer J (2011) Intrinsically disordered proteins: regulation and disease. Curr Opin Struct Biol 21:432–440
5. Mollica L, Bessa LM, Hanoulle X, Jensen MR, Blackledge M, Schneider R (2016) Binding mechanisms of intrinsically disordered proteins: theory, simulation, and experiment. Front Mol Biosci 3:1–18
6. Iakoucheva LM, Radivojac P, Brown CJ, O'Connor TR, Sikes JG, Obradovic Z, Dunker AK (2004) The importance of intrinsic disorder for protein phosphorylation. Nucleic Acids Res 32:1037–1049
7. Fadda E (2016) Role of the XPA protein in the NER pathway: a perspective on the function of

structural disorder in macromolecular assembly. Comput Struct Biotechnol J 14:78–85
8. Fadda E, Nixon MG (2017) The transient manifold structure of the p53 extreme C-terminal domain: insight into disorder, recognition, and binding promiscuity by molecular dynamics simulations. Phys Chem Chem Phys 19:21287–21296
9. Wayment-Steele HK, Hernandez CX, Pande VS (2018) Modelling intrinsically disordered protein dynamics as networks of transient secondary structure. bioRxiv. https://doi.org/10.1101/377564
10. Choi UB, McCann JJ, Weninger KR, Bowen ME (2011) Beyond the random coil: stochastic conformational switching in intrinsically disordered proteins. Structure 19:566–576
11. Fuxreiter M, Simon I, Friedrich P, Tompa P (2004) Preformed structural elements feature in partner recognition by intrinsically unstructured proteins. J Mol Biol 338:1015–1026
12. Salmon L, Nodet G, Ozenne V, Yin G, Jensen MR, Zweckestetter M, Blackledge M (2010) NMR characterization of long-range order in intrinsically disordered proteins. J Am Chem Soc 132:8407–8418
13. Tompa P, Szász C, Buday L (2005) Structural disorder throws new light on moonlighting. Trends Biochem Sci 30:484–489
14. Oldfield CJ, Meng J, Yang JY, Yang MQ, Uversky VN, Dunker AK (2008) Flexible nets: disorder and induced fit in the associations of p53 and 14-3-3 with their partners. BMC Genomics 9:S1
15. Arai M, Sugase K, Dyson HJ, Wright PE (2015) Conformational propensities of intrinsically disordered proteins influence the mechanism of binding and folding. Proc Natl Acad Sci 112:9614–9619
16. Csermely P, Palotai R, Nussinov R (2010) Induced fit, conformational selection and independent dynamic segments: an extended view of binding events. Trends Biochem Sci 35:539–546
17. Meirovitch H, Cheluvaraja S, White DP (2009) Methods for calculating the entropy and free energy to problems involving protein flexibility and ligand binding. Curr Protein Pept Sci 10:229–243
18. Perez A, Morrone JA, Simmerling C, Dill KA (2016) Advances in free-energy-based simulations of protein folding and ligand binding. Curr Opin Struct Biol 36:25–31
19. Kollman PA, Massova I, Reyes C, Kuhn B, Huo S, Chong L, Lee M, Lee T, Wang W, Donini O, Cieplak P, Srinivasan J, Case DA, Cheatham TE III (2000) Calculating structures and free energies of complex molecules: combining molecular mechanics and continuum models. Acc Chem Res 33:889–897
20. Genheden S, Ryde U (2015) The MM/PBSA and MM/GBSA methods to estimate ligand-binding affinities. Expert Opin Drug Discovery 10:449–461
21. Brooks BB, Janežič D, Karplus M (1995) Harmonic analysis of large systems. I. Methodology. J Comput Chem 16:1522–1542
22. Schlitter J (1993) Estimation of absolute and relative entropies of macromolecules using the covariance matrix. Chem Phys Lett 215:617–621
23. Sugita Y, Okamoto Y (1999) Replica-exchange molecular dynamics method for protein folding. Chem Phys Lett 314:141–151
24. Hilser VJ, Garcia-Moreno EB, Oas TG, Kapp G, Whitten ST (2006) A statistical thermodynamic model of the protein ensemble. Chem Rev 106:1545–1558
25. Grünberg R, Nilges M, Leckner J (2006) Flexibility and conformational entropy in protein-protein binding. Structure 14:683–693
26. Tsodikov OV, Ivanov D, Orelli B, Staresincic L, Shoshani I, Oberman R, Scharer OD, Wagner G, Ellenberger T (2007) Structural basis for the recruitment of ERCC1-XPF to nucleotide excision repair complexes by XPA. EMBO J 26:4768–4776
27. Fadda E (2013) Conformational determinants for the recruitment of ERCC1 by XPA in the nucleotide excision repair (NER) pathway: structure and dynamics of the XPA binding motif. Biophys J 104:2503–2511
28. Fadda E (2015) The role of conformational selection in the molecular recognition of the wild type and mutants XPA_{67-80} peptides by ERCC1. Proteins 83:1341–1351
29. Lindorff-Larsen K, Piana S, Palmo K, Maragakis P, Klepeis JL, Dror RO, Shaw DE (2010) Improved side-chain torsion potentials for the Amber99SB protein force field. Proteins 78:1950–1958
30. Horn HW, Swope WC, Pitera JW, Madura JD, Dick TJ, Hura GL, Head-Gordon T (2004) Development of an improved four-site water model for biomolecular simulations: TIP4P-Ew. J Chem Phys 120:9665–9678
31. Hess B, Kutzner C, van der Spoel D, Lindahl E (2008) GROMACS 4: algorithms for highly efficient, load-balanced, and scalable molecular simulation. J Chem Theory Comput 4:435–447
32. Case DA, Darden TA, Cheatham IIITE, Simmerling CL, Wang J, Duke RE, Luo R, Walker RC, Zhang W, Merz KM, Roberts B, Hayik S, Roitberg A, Seabra G, Swails J, Götz AW,

Kolossváry I, Wong KF, Paesani F, Vanicek J, Wolf RM, Liu J, Wu X, Brozell SR, Steinbrecher T, Gohlke H, Cai Q, Ye X, Wang J, Hsieh M-J, Cui G, Roe DR, Mathews DH, Seetin MG, Salomon-Ferrer R, Sagui C, Babin V, Luchko T, Gusarov S, Kovalenko A, Kollman PA (2012) Amber 12 reference manual. University of California, San Francisco, CA
33. Humphrey W, Dalke A, Schulten K (1996) VMD – visual molecular dynamics. J Mol Graph 14:33–38
34. Hess B, Bekker H, Berendsen HJ, Fraaije JG (1997) LINCS: a linear constraint solver for molecular simulations. J Comput Chem 18:1463–1472
35. Daura X, Gademann K, Jaun B, Seebach D, van Gunsteren WF, Mark AE (1999) Peptide folding: when simulation meets experiment. Angew Chem Int Ed 38:236–240
36. Rauscher S, Gapsys V, Gajda MJ, Zweckstetter M, de Groot BL, Grubmüller H (2015) Structural ensembles of intrinsically disordered proteins depend strongly on force field: a comparison to experiment. J Chem Theory Comput 11:5513–5524
37. Robustelli P, Piana S, Shaw DE (2018) Developing a molecular dynamics force field for both folded and disordered protein states. Proc Natl Acad Sci. https://doi.org/10.1073/pnas.1800690115
38. Miller IIIBR, McGee TD Jr, Swails JM, Homeyer N, Gohlke H, Roitberg AE (2012) MMPBSA.py: an efficient program for end-state free energy calculations. J Chem Theory Comput 8:3314–3321

Chapter 17

Atomistic Simulation Tools to Study Protein Self-Aggregation

Deniz Meneksedag-Erol and Sarah Rauscher

Abstract

Aberrant aggregation of proteins into poorly soluble, toxic structures that accumulate intracellularly or extracellularly leads to a range of disease states including Alzheimer's, Parkinson's, Huntington's, prion diseases, and type II diabetes. Many of the disease-associated amyloidogenic proteins are intrinsically disordered, which makes their experimental investigation challenging due to a limited number of experimental observables to effectively characterize their ensemble of conformations. Molecular dynamics simulations provide dynamic information with atomistic detail, and are increasingly employed to study aggregation processes, offering valuable structural and mechanistic insights. In this chapter, we demonstrate the use of all-atom molecular dynamics simulations to model the self-aggregation of a six-residue amyloidogenic peptide derived from amyloid β, a 39–43 residue-long peptide associated with the pathogenesis of Alzheimer's disease. We provide detailed instructions on how to obtain the initial monomer conformations and build the multichain systems, how to carry out the simulations, and how to analyze the simulation trajectories to investigate the peptide self-aggregation process.

Key words Protein self-aggregation, Alzheimer's disease, Amyloid β, Intrinsically disordered proteins, Molecular dynamics simulations

1 Introduction

Protein aggregation is the assembly of misfolded proteins, unfolded proteins or folding intermediates into energetically stable insoluble structures. Aberrant aggregation behavior, that is the accumulation of "toxic" aggregates intracellularly and/or extracellularly [1, 2], is associated with a number of disease states including systemic amyloidoses [3]; neurodegenerative disorders such as Alzheimer's, Parkinson's, Huntington's, and spongiform encephalopathies [3, 4]; and other localized diseases such as type II diabetes [3, 5]. In addition, aggregates of the tumor suppressor protein p53 have been observed in vitro and have been proposed to inactivate the wild-type p53 function [6–8]. In vivo aggregation of p53 has recently been linked to chemoresistance in high-grade serous

Jennifer J. McManus (ed.), *Protein Self-Assembly: Methods and Protocols*, Methods in Molecular Biology, vol. 2039,
https://doi.org/10.1007/978-1-4939-9678-0_17,

ovarian carcinoma [9]. From a pharmaceutical standpoint, the fact that aberrant aggregation has a wide range of pathogenic consequences makes it particularly important to study the mechanistic and structural details of aggregate formation.

The pathology of the protein aggregation diseases typically presents amyloid fibrils, which are enriched in β-sheet structures. Although the mature fibrils could be toxic contingent on the type of the amyloid disease [10], the early oligomeric states have been identified as the primary source of toxicity [11–14]. Toxicity has been proposed to originate from the undesired interactions of the aggregates with cells, mainly with cellular membranes, followed by disruption of membrane integrity [15, 16]. Interestingly, the formation of toxic amyloid structures is not specific to disease-related amyloidogenic proteins. Proteins without any known disease association, such as the SH3 domain of bovine phosphatidyl-inositol-3′-kinase (PI3-SH3) and acylphosphatase, were demonstrated to form fibrils [17, 18], which led to the proposal that the formation of amyloid fibrils is a common feature of polypeptide chains, dependent on the environmental conditions [18].

The propensity to aggregate, on the other hand, is multifactorial in origin and dependent on amino acid sequence as well as a combination of environmental factors such as temperature [19], pH, and solution ionic strength [20, 21]. An alternating sequence of polar and nonpolar amino acids has been shown to favor the formation of β-strands, as well as fibrils [22]; the latter attributed to the need for stabilizing the amphiphilic β-strands via interstrand hydrogen bonds and burial of hydrophobic segments away from the aqueous phase by assembling into a more ordered structure [23]. β-Aggregation does not appear to have a strong preference for particular amino acids, provided a certain level of hydrophobicity and/or neutral charge is maintained within the structure [24, 25]; the formation of amyloid fibrils is more sequence-selective, albeit more permissive to the inclusion of polar and charged amino acids [24]. In this context, an extensive sequence comparison between a wide range of elastomeric and amyloidogenic peptides has shown that high combined content of proline and glycine, as seen in elastomeric proteins, precludes the formation of amyloid-like structures [26].

Amyloidogenic proteins associated with several diseases are intrinsically disordered. Examples to this include α-synuclein (α-syn, Parkinson's and Alzheimer's), amyloid β (Aβ, Alzheimer's), prion (spongiform encephalopathies), amylin (type II diabetes), and p53 (cancer) [3, 27]. Intrinsically disordered proteins (IDPs) inherently lack fixed, folded structures; instead, they exchange between a multitude of conformational states. The fact that IDPs have an ensemble of structures makes it particularly challenging to obtain structural descriptions of their aggregates during different stages in the aggregation process. Experimental techniques such as nuclear magnetic resonance (NMR), small-angle X-ray scattering (SAXS), circular dichroism (CD), and fluorescence spectroscopy provide structural information about the aggregates [28]. However, these techniques have limitations in investigating the dynamics of aggregation due to a limited number of observables measured over large conformational ensembles. Molecular dynamics (MD) simulations can provide dynamic information with atomistic detail, and bridge the gap in interpreting the experimental observables by means of a molecular model [29]. In an MD simulation, the motions of atoms are modeled with classical molecular mechanics; bonded and nonbonded interactions are defined with a set of parameters along with a functional form of the potential energy; the energy function and the parameters are together termed the force field. The accuracy of empirical force fields in reproducing the experimentally determined structural properties of IDPs has been the subject of intense investigation (*see* [30–35] for selected studies). In a recent systematic force field comparison by Rauscher et al., conformational ensembles of IDPs were reported as highly sensitive to the force field of choice; CHARMM22* performed the best in reproducing the experimentally determined structural ensembles of IDPs, while CHARMM36 exhibited a bias toward a left-handed α-helix structure, inconsistent with NMR and SAXS data [34]. These findings led to a refined CHARMM36 force field, CHARMM36m, which shows improved conformational sampling of IDP backbones [36].

The time scale needed to study early aggregation events with MD simulations is typically within the range of a few to tens of microseconds, while more complex processes such as formation of higher order aggregates or interaction of aggregate structures with membranes require time scales at least one or two orders of magnitude higher. Thus, a variety of different techniques have been exploited to overcome the challenges in simulating complex aggregation processes. These include enhanced sampling techniques such as replica exchange MD (REMD) and metadynamics; and/or varying levels of representation, such as coarse-graining (CG) or implicit solvent. A list of selected modeling studies is given in Table 1, demonstrating the diversity in the simulation methods, level of representation, and force field combinations used to model protein aggregation.

In efforts to find therapeutic strategies for the treatment and prevention of amyloid diseases, a thorough understanding of the structural and mechanistic details of protein aggregation is crucial. By means of different levels of representation and sampling techniques, simulations can provide insights into several aspects of the protein aggregation process ranging from the structural basis of oligomerization (e.g., formation and arrangement of β-strands in the aggregates) to mechanistic details of aggregation, (e.g., mechanism of β-sheet extension, fibril nucleation, fibril elongation). For

Table 1
Selected simulation studies of protein aggregation published within the last 5 years

Reference	Simulation method	Level of representation	Force field/water model	Protein or peptide	Number of monomers	Phenomena studied
[37]	REMD	All-atom/explicit solvent	OPLS-AA/TIP3P	$A\beta_{25-35}$	4	Oligomerization
[38]	MD	All-atom/explicit solvent	AMBER99SB*-ILDN/TIP3P AMBER03/TIP3P CHARMM22*/TIP3P CHARMM36/TIP3P GROMOS96 43A1/SPC GROMOS96 54A7/SPC OPLS-AA/L/TIP4P	KLVFFAE ($A\beta_{16-22}$) AIIGLM ($A\beta_{30-35}$) NNQQNY($Sup35p_{8-13,wt}$) VIQVVY($Sup35p_{8-13,mut}$) VQIVYK($hTau40_{306-311}$) GSRSRT($hTau40_{207-212}$)	12	Oligomerization
[39]	STDR[a], MD	All-atom/explicit solvent	OPLS-AA/L/TIP3P	KLVFFAE ($A\beta_{16-22}$)	1–16	Inositol binding to $A\beta_{16-22}$ monomers and aggregates
[40]	REMD	All-atom/explicit solvent	CHARMM27/TIP3P	GGVVIA ($A\beta_{37-42}$)	16	Oligomerization
[41]	Bias-exchange metadynamics	All-atom/explicit solvent	AMBER99/TIP3P	VVVVVVVV	18	Fibril nucleation
[42]	MD	All-atom/implicit solvent	OPLS-AA/GB/SA implicit	$A\beta_{1-42}$	20	Oligomerization

[43]	REMD, MD	All-atom/explicit solvent	CHARMM22*/ CHARMM modified TIP3P CHARMM22*/TIP4P-D	$(GVPGV)_7$	1, 27	Aggregation of an elastin-like peptide
[44]	MD	All-atom/explicit solvent CG	AMBER99SB*-ILDN/ TIP4P-Ew Off-lattice bead model potentials	$A\beta_{1–40}$	3–15 6–30	Fibril stability Fibril nucleation and elongation
[45]	DMD[b]	CG/implicit solvent	PRIME20	SNQNNF($PrP_{170–175}$) SSTSAA($RNaseA_{15–20}$) MVGGVV($A\beta_{35–40}$) GGVVIA($A\beta_{37–42}$) MVGGVVIA($A\beta_{35–42}$)	48	Effect of sequence on fibril formation
[46]	DMC[c]	CG/implicit solvent	–	A generic amyloidogenic peptide with a (1) random coil, and (2) β-sheet conformation	600	Amyloid formation

[a]Simulated tempering distributed replica sampling algorithm
[b]Discontinuous MD
[c]Dynamic Monte Carlo

example, based on the results of all-atom REMD simulations of 16 $A\beta_{37-42}$ peptide chains in explicit water, Nguyen and Derreumaux [40] suggested that hydrophobic collapse of the random coils initiates the aggregation process, followed by the formation of highly dynamic aggregates enriched in both parallel and antiparallel β-strands. The aggregated system at equilibrium was observed to be dominated by larger aggregates ranging from a 14-mer to a 16-mer. Free monomers and smaller aggregates comprising 2- to 3-mers were also observed.

In this chapter, we describe the use of all-atom MD simulations to study the oligomerization of a six-residue fragment of the intrinsically disordered Aβ peptide (KLVFFA, $A\beta_{16-21}$). $A\beta_{16-21}$ contains the central hydrophobic region of the full-length Aβ peptide, LVFFA—a region of critical importance in the formation of fibrillar structures [47]. Previous studies have shown that microcrystals of $A\beta_{16-21}$ possess an antiparallel β-strand arrangement; at 200 μM concentration in phosphate buffered saline (PBS), fibrils formed from $A\beta_{16-21}$ were detected by electron microscopy after 5 days of incubation [48]. Here, we provide detailed instructions on how to obtain initial monomer conformations of $A\beta_{16-21}$ and build/simulate the multichain aggregated system, along with representative analyses to characterize the structure and dynamics of peptide self-aggregation.

2 Methods

2.1 Obtaining the Initial Conformations of Single Peptide Chains

1. Experimentally resolved structures of folded proteins/peptides can be found in the Protein Data Bank (PDB, http://www.rcsb.org). Intrinsically disordered peptides, on the other hand, require structure generation from scratch, which could be done with software such as the UCSF Chimera program [49]. We generate an initial structure in this case with φ and ψ angles corresponding to an α-helix structure (*see* **Note 1**), which was chosen to eliminate any conformational bias toward the formation of a β-strand. After saving the coordinates in PDB format, the generated PDB file can be loaded into any molecular visualization program such as VMD [50]. The starting structure of the KLVFFA peptide is shown in Fig. 1.
2. All simulations reported in this chapter are carried out using GROMACS version 2016.3 [51]; we outline the steps involved in the simulation setup and analysis using the built-in programs of the GROMACS package. To prepare the initial structure of the peptide and generate the initial topology, the "gmx pdb2gmx" tool of GROMACS can be invoked from the command line as follows:

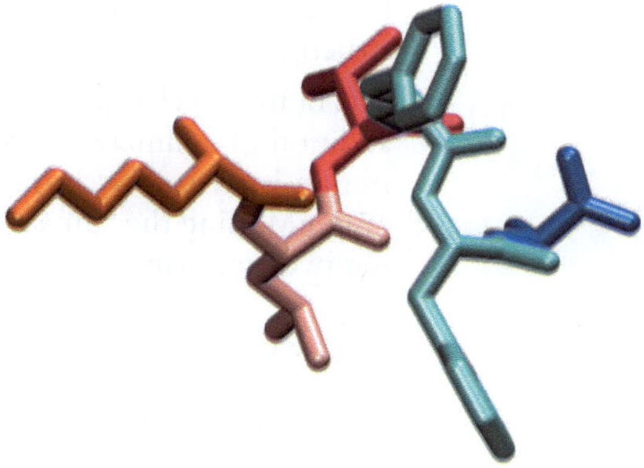

Fig. 1 The initial structure of the KLVFFA peptide rendered using VMD [50]. The color coding indicates the residue type: K, orange; L, pink; V, red; F, cyan; A, blue

```
gmx pdb2gmx -f KLVFFA.pdb -o init.gro -p init.top -ignh -ter -v
```

Executing the command above will prompt interactive selections for the force field, water model, and termini caps (*see* **Note 2**). Here, the CHARMM36m force field [36], and CHARMM-modified TIP3P model [52] are used as the force field and water model, respectively, and the peptide is capped by neutral groups on both ends ($-NH_2$ at the N-terminus and $-COOH$ at the C-terminus).

3. **Step 2** generates a structure file, init.gro, and a system topology file, init.top. To set up the simulation box for the initial structure, the following command can be executed from the command line:

```
gmx editconf -f init.gro -o editconf.gro -c -d 1.5 -bt
dodecahedron
```

Here "gmx editconf" takes the input structure init.gro, and centers it in a rhombic dodecahedron box with a 15 Å distance to the box edge on all sides (*see* **Note 3**).

4. With the simulation box set up, the peptide is now ready for solvation. To do so, the "gmx solvate" tool can be invoked by the following command:

```
gmx solvate -cp editconf.gro -cs spc216 -o solvate.gro -p
init.top
```

gmx solvate reads the coordinates of the peptide and box information from the structure file editconf.gro and takes the coordinates of the solvent from the spc216.gro in GROMACS' library. Upon solvation, the topology is updated (with the -p flag) to include the number of water molecules in the structure file.

5. The non-zero charge of the system can be neutralized with the addition of counterions, Na^+ and Cl^-. Here, the ion concentration is set to match the physiological salt concentration of 150 mM using the "gmx genion" tool by executing the following two commands (*see* **Note 4**):

```
gmx grompp -f em.mdp -c solvate.gro -p init.top -o solvate.
tpr
gmx genion -s solvate.tpr -o solvate_genion.pdb -p init.top
-neutral -conc 0.15
```

6. Energy minimization is carried out using the steepest descent algorithm; *see* **Note 5** for the details of the run parameters used for minimizing the system. The run input file for the minimization is generated with "gmx grompp" as follows:

```
gmx grompp -f em.mdp -c solvate_genion.pdb -p init.top -o
em.tpr
```

   The minimization is run with the following command:

```
gmx mdrun -s em.tpr
```

   The resulting structure of the minimization process is a geometry optimized structure at the nearest local minimum.

7. To generate a pool of alternate conformations of the peptide chain in solution, one approach is to carry out high temperature simulations. Thermal denaturation is an effective way to sample a wide range of conformational states without the need for extensive computational power; it has been mostly adopted to study the protein unfolding processes (for example works *see* [53–55]). Here, to sample different conformational states of the single peptide chain, we carry out three independent, 50 ns-long simulations of the peptide in explicit water at 450 K in the NVT ensemble, started directly from the energy-minimized conformation. *See* **Note 6** for the run parameters (md_NVT.mdp) used for these simulations; the following are brief explanations on selected parameters with the corresponding mdp options given in parentheses: A time step (dt) of 2 fs is used. Periodic boundary conditions (pbc) are applied. The short-range electrostatic interactions and van der Waals interactions are calculated with a cutoff of 0.95 nm (rcoulomb and rvdw, respectively). Long-range electrostatic interactions (coulombtype) are evaluated using particle-mesh Ewald summation [56] with 0.12 nm grid spacing

(fourierspacing) and a fourth order interpolation (pme_order). The Verlet cutoff scheme (cutoff-scheme) is used for neighbor searching. The bonds involving hydrogen atoms (constraints) are constrained using the LINCS algorithm (constraint_algorithm) [57]. Water molecules are constrained using the SETTLE algorithm [58] (*see* **Note 7**). The temperature is maintained at 450 K (ref_t) using the velocity rescaling thermostat (tcoupl) [59].

8. The simulations reported here are carried out with MPI parallelization by invoking "gmx_mpi mdrun" (*see* **Note 8**). The "gmx tune_pme" tool was used to determine the number of PME nodes for optimal performance.
9. Upon completion of the three independent runs, the trajectories are corrected for periodic boundary conditions (*see* **Note 9**) and concatenated (*see* **Note 10**). Using the "gmx trjconv" tool, conformations of the peptide at each frame can then be saved. Here, from a trajectory of 150 ns in total, 15,000 conformations are extracted.
10. From this pool of 15,000 conformations, a subset should then be selected at random to build the aggregate simulations. Four-hundred peptide single chain conformations (shown in the left panel of Fig. 2) are randomly chosen out of 15,000 structures.

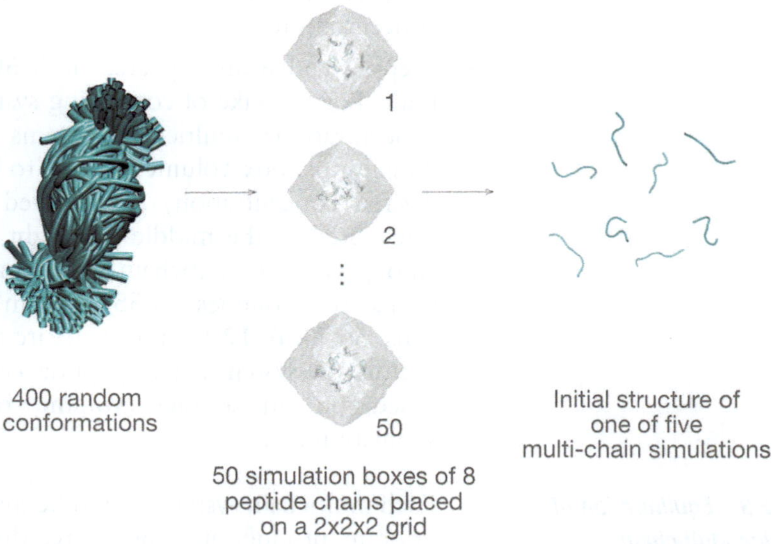

Fig. 2 General protocol to build the multichain systems. (Left) 400 conformations of the single KLVFFA peptide chain selected randomly from three independent high temperature simulations of 50 ns. (Middle) 50 multichain systems are built by placing eight random single chains on a $2 \times 2 \times 2$ grid, and the systems are placed in a rhombic dodecahedron simulation box with a 15 Å distance to the box edge on all sides. (Right) One of the five conformations with a simulation box volume closest to 555 nm^3, selected out of 50 multichain systems

2.2 Building the Multichain System

1. The multichain systems studied in this chapter are composed of eight single peptide chains. It should be noted that for a multichain system comprising a different number of single peptide chains, these selections will differ; thus, the following steps should be taken as a general guide.

   The first step in building the multichain systems is to place and center each randomly selected peptide chain in a rhombic dodecahedron box with 1 Å distance to the box edge on all sides with "gmx editconf".
2. Eight randomly chosen simulation boxes (each with one peptide) are then concatenated into one structure file. Using the "gmx genconf" tool, these simulation boxes are placed on a $2 \times 2 \times 2$ grid (specified with the -nbox flag) by executing the following command:

   ```
   gmx genconf -f frame${framenumber}_editconf.gro -trj
   conf_8.gro -o conf_8_rd.gro -nbox 2 2 2
   ```

   See **Note 11** for brief descriptions of the structure files mentioned in the command above.
3. The multichain system is then centered in a rhombic dodecahedron box with a 15 Å distance to all box edges by invoking "gmx editconf"; the unit cell dimensions and box volume information are recorded to determine the peptide concentration.
4. **Steps 2** and **3** are repeated until 50 multichain systems are built. For the sake of comparing systems with similar peptide concentrations, multichain systems possessing the unit cell dimensions/box volumes closest to one another (and to the desired concentration) are selected for further simulations (illustrated in the middle and right panels of Fig. 2). Here, among the 50 multichain configurations, five systems with similar box volumes (~555 ± 5 nm^3) are chosen for further runs. *See* **Note 12** for the structure file of one of the selected multichain systems. The peptide concentration is ~24 mM based on an average volume of 555 nm^3 for each simulation box.

2.3 Equilibration of the Multichain Systems and Production Runs

1. Each multichain system should be minimized and equilibrated prior to production runs. Here, the multichain systems are subjected to a two-step equilibration at 1 bar and 298 K in the NPT ensemble; first for 10 ns using Berendsen pressure coupling [60], followed by 10 ns using the Parrinello–Rahman barostat [61]. The steps to follow for the preparation of the input files for the equilibration runs are the same as **steps 4–6** in Subheading 2.1 above. One important difference is the preparation of the system topology. To prepare the correct

topology of the multichain system, the topology of the single peptide chain is manually modified by deleting the lines specifying the number of solvent molecules (SOL) and counterions (NA, CL) and changing the number of molecules to "8" in the "Protein_chain_A" line. *See* **Note 13** for the details of the topology of the multichain systems. *See* **Note 14** for the run parameters used in the NPT equilibration steps.

2. The resulting structure from the second NPT equilibration step is the input structure for the production runs in the NPT ensemble using the Parrinello–Rahman barostat (1 bar and 298 K). *See* **Note 15** for the run parameters used in the long NPT simulations. Here, each multichain system is simulated for 900 ns, for a total sampling time of 4.5 μs for the five independent systems.

2.4 Trajectory Analysis

Below are some examples of the types of analysis that can be carried out using the GROMACS built-in tools to investigate the self-aggregation of the KLVFFA peptide. The following analysis represents some of the metrics to assess the sampling and convergence of the aggregation simulations.

1. To monitor the hydrogen bonds formed between the peptide chains throughout the course of the simulations, "gmx hbond" can be invoked by executing the following command and, when prompted, selecting "protein" for the two groups to carry out the analysis:

```
gmx hbond -s md.tpr -f traj_comp.xtc
```

   The output of this analysis is an xvg file containing the total number of peptide–peptide hydrogen bonds at each time frame. Figure 3 shows the number of hydrogen bonds per residue, X_{HB}, for the five multichain systems. The curves are plotted separately to monitor the convergence of the five independent simulations. As is evident from the lack of a plateau, the simulations have not converged within the simulation time (900 ns). Given that oligomerization events require time scales of a few to tens of microseconds, these simulations should be run for much longer in order to obtain adequate sampling.

2. The peptide self-aggregation dynamics can be investigated by monitoring the oligomer size throughout the simulations with the GROMACS tool "gmx clustsize" by executing the following command (*see* **Note 16**):

```
gmx clustsize -s md.tpr -f MC1-traj_comp.xtc -n MC1-prot_
only.ndx
```

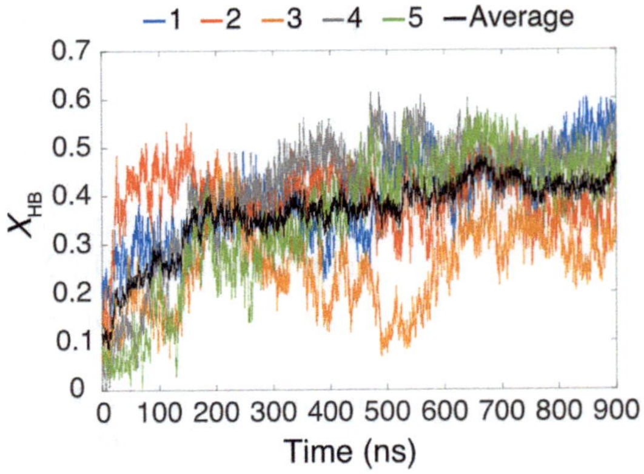

Fig. 3 The number of hydrogen bonds per residue, X_{HB}, for the five multichain systems. The average of five systems is shown in black. X_{HB} is calculated every 100 ps, and the running average over every 25 data points is plotted. Each system is colored individually

The number of clusters formed and the maximum number of peptide chains the clusters contain are plotted in Fig. 4a, b, respectively. The number of clusters display a decreasing trend in the early stages of the simulations and fluctuate around 1–3 clusters within the last 400 ns. The corresponding number of peptide chains forming the clusters, on the other hand, tends to increase and fluctuates around 6–8 peptides within the last 400 ns, indicating the formation of a larger oligomer. However, the largest oligomer (comprising eight peptide chains) is not stable and has a transient lifetime, i.e., it either spontaneously disassembles into smaller size oligomers or decreases in size by losing a peptide chain or two, seen as sudden changes in both the number and the size of the aggregates observed along the trajectory. To obtain converged populations of the different cluster sizes, much more extensive simulations would be necessary.

3. As the formation of β-sheet structures are of interest in the oligomerization of amyloid forming proteins/peptides, one can visually examine the simulation trajectory with a molecular visualization software such as VMD [50] to monitor the changes in the secondary structure content (*see* **Note 17**). Figure 5 illustrates the formation of β-sheet structures in one of the multichain systems.

4. To assess the secondary structure content of the aggregates quantitatively, we use the DSSP algorithm [62]. The "gmx do_dssp" tool can be invoked as follows:

```
gmx do_dssp -f MC1-traj_comp.xtc -s md.tpr -o ss.xpm -sc
ss.xvg
```

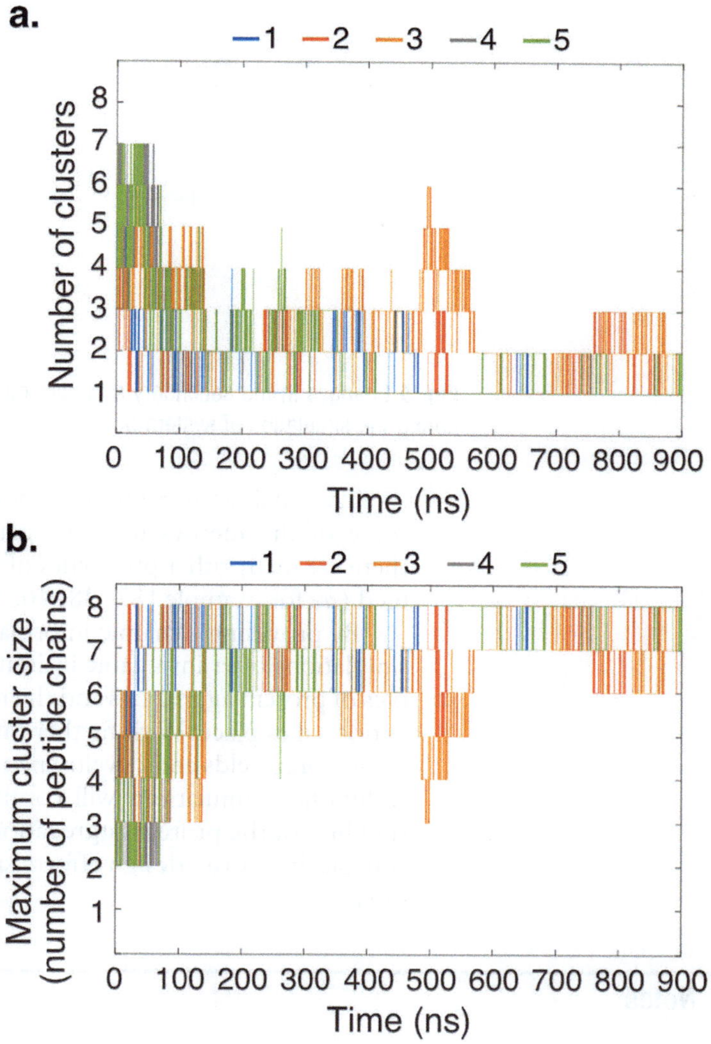

Fig. 4 Time evolution of the (**a**) number of clusters, (**b**) maximum number of peptide chains within the clusters. The data is obtained at every 100 ps and each system is colored individually

An interactive selection will be prompted for the selection of a part of the system (i.e., protein) to run the DSSP algorithm [62]. Figure 6 depicts the running average of the fraction of residues assigned to β-sheet and β-bridges or turns. The aggregate's β-sheet and β-bridge content generally increases as the simulations progress (with the exception of system 3 where the fraction of residues possessing β-sheet and β-bridge structures decreases significantly between 350–500 ns). Turns are significantly less populated and the probability of their formation tends to decrease after 750 ns.

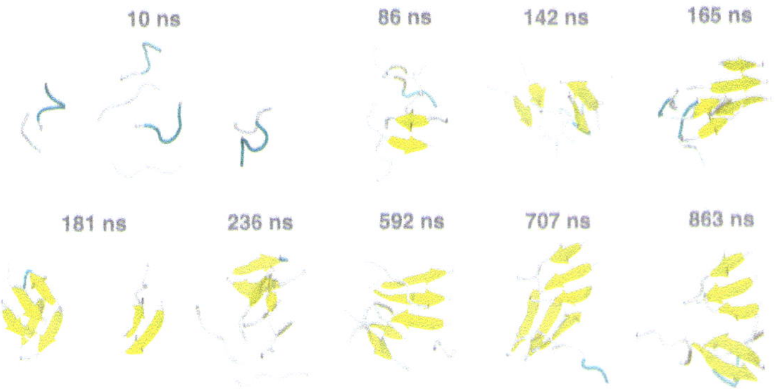

Fig. 5 Changes in the secondary structure content of the aggregates observed during the simulation of system 5

The analysis presented is meant to be representative of some of the metrics to investigate peptide self-aggregation; there are many other properties of interest that could be examined (*see* for example [37, 38, 40, 43]).

By providing dynamic information in all-atom detail, MD simulations offer invaluable insights into the structural ensembles of protein aggregates and the mechanisms by which aggregation takes place. With further improvements in the accuracy of the force fields and development of more effective sampling techniques, simulations will continue to increase our understanding of the protein aggregation processes and, in the long run, facilitate drug design efforts targeting protein aggregation diseases.

3 Notes

1. The "Build Structure" tool of UCSF Chimera can be found under the tab Tools > Structure Editing. Selecting the "Start Structure" from the drop-down menu and "peptide" from the "add" section opens the "Peptide Parameters" window to enter the peptide sequence. Upon entering the peptide sequence, a pop-up window will open with options to specify the backbone φ and ψ angles for different secondary structures.
2. The "gmx pdb2gmx" tool of GROMACS prepares the initial structure further (e.g., capping of the termini, adding missing hydrogens, selecting the protonation states of amino acids) and generates the initial topology of the structure. GROMACS has an assortment of force fields in its library; if one wishes to use a different force field or modify an existing one, new/modified <forcefield>.ff directories can be sourced from the working directory.

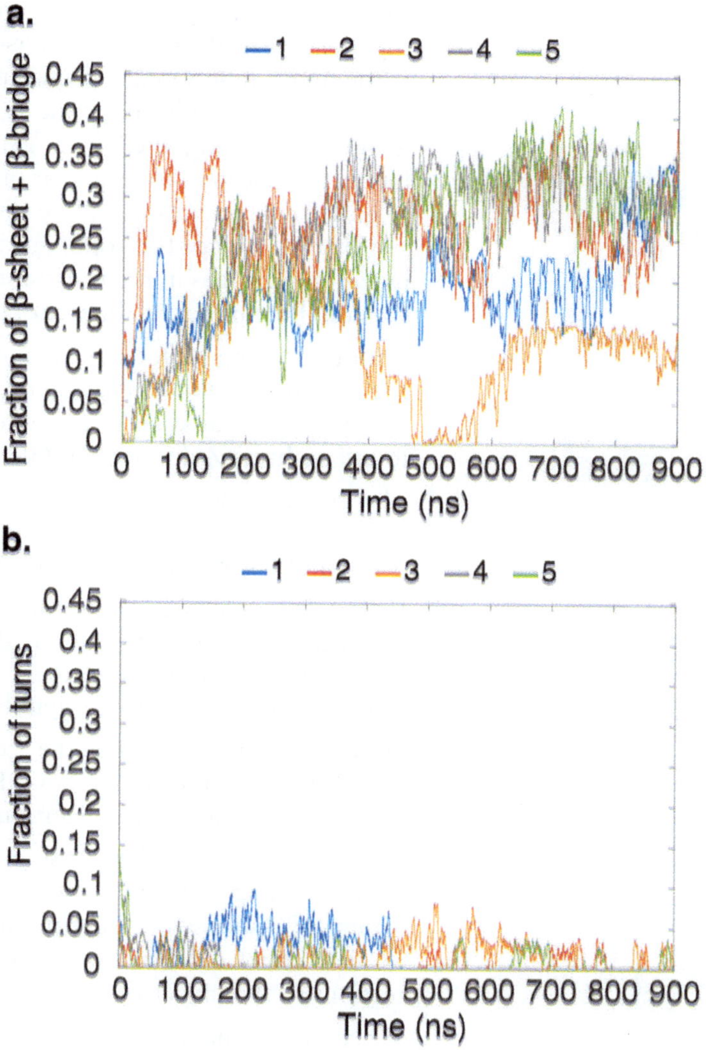

Fig. 6 The fraction of residues assigned to (**a**) β-sheet and β-bridge, (**b**) turn structures using the DSSP algorithm [62]. The data is obtained every 1 ns, and the running average over every 25 data points is plotted. Each system is colored individually

3. The simulation box should be sufficiently large to prevent interactions between the periodic images. A distance of 10–15 Å to the box edge on all sides is typical. However, complex many-body systems may require larger boxes. One should also consider that the number of solvent molecules increase proportionally to the size of the simulation box and thus have an impact on the computing time.
4. The “gmx genion” tool accepts input files in binary run input file (tpr file) format. To generate the solvate.tpr file, the GROMACS preprocessor program “gmx grompp” combines the

run parameter file (mdp file), the coordinates, and the topology together. Here, the run parameters in the mdp file are irrelevant as the resulting solvate.tpr file will not be used as an input to an actual run. The solvate.tpr file is used as an input to "gmx genion". Invoking the "gmx genion" tool will prompt an interactive selection to choose a continuous group of molecules (e.g., solvent) to tag and replace with the counterions. With the use of the -p flag, the topology is automatically updated to reflect the changes in the number of solvent molecules, and to include the number of counterions added into the system.

5. Prior to running MD, energy minimization is necessary to relax the system and eliminate the steric clashes between atoms. The run parameter file used in the minimization (em.mdp) can be found at: https://doi.org/10.6084/m9.figshare.5966872.
6. The run parameter file used in the high temperature NVT simulations (md_NVT.mdp) can be found at: https://doi.org/10.6084/m9.figshare.5966884.
7. Unlike the constraint algorithms for the solute (LINCS or SHAKE) which are specified in the run parameter file, constraints on water molecules are set in the water topology. Water molecules are defined to be rigid by default.
8. For detailed information on the acceleration and parallelization in GROMACS, the reader can refer to http://www.gromacs.org/Documentation/Acceleration_and_parallelization.
9. The visualization artifacts arising from the periodicity can be adjusted using the "gmx trjconv" tool. The -pbc flag has a variety of options for the type of the correction; depending on the desired outcome, one can apply a combination of corrections to the trajectory.
10. When concatenating trajectories obtained from independent runs, the "-cat" flag can be passed onto the "gmx trjcat" to prevent discarding the identical time frames.
11. The structure file frame${framenumber}_editconf.gro contains a single peptide chain, conf_8.gro has eight peptide chains in no specific arrangement, and conf_8_rd.gro has eight peptide chains placed on a 2x2x2 grid.
12. The structure file of one of the multichain systems can be found at https://doi.org/10.6084/m9.figshare.5975386.
13. The end of the topology file for the initial conformations of the multichain systems should look as follows:

```
----init.top (multi-chain system)----
[molecules]
; Compound #mols
Protein_chain_A 8
```

This topology is further updated upon solvation and addition of counterions (150 mM NaCl) with the -p flag to include the number of solvent molecules and counterions. The topology of the multichain system can be found at https://doi.org/10.6084/m9.figshare.5975377.

14. The run parameter files used in the equilibration steps in the NPT ensemble with Berendsen (md_equil_Berendsen.mdp) and Parrinello–Rahman (md_equil_Parrinello-Rahman.mdp) barostats can be found respectively at https://doi.org/10.6084/m9.figshare.5966893 and https://doi.org/10.6084/m9.figshare.5966905.
15. The run parameter file used in the production runs in the NPT ensemble (md_Parrinello-Rahman.mdp) can be found at https://doi.org/10.6084/m9.figshare.5966911.
16. The index file of the protein can be created by executing the following command and selecting the protein when prompted:

```
gmx make_ndx -f md.tpr -o MC1-prot_only.ndx
```

17. To investigate the time evolution of the peptide secondary structure content, peptides should be colored according to their secondary structure, and the secondary structure assignment information should be updated at each frame, which could be done with the script sscache.tcl. This script can be downloaded from VMD's script library at http://www.ks.uiuc.edu/Research/vmd/script_library/scripts/sscache/.

References

1. Eisenberg D, Nelson R, Sawaya MR, Balbirnie M, Sambashivan S, Ivanova MI, Madsen AØ, Riekel C (2006) The structural biology of protein aggregation diseases: Fundamental questions and some answers. Acc Chem Res 39:568–575
2. Aguzzi A, O'Connor T (2010) Protein aggregation diseases: Pathogenicity and therapeutic perspectives. Nat Rev Drug Discov 9:1–12
3. Chiti F, Dobson CM (2006) Protein misfolding, functional amyloid, and human disease. Annu Rev Biochem 75:333–366
4. Ross CA, Poirier MA (2004) Protein aggregation and neurodegenerative disease. Nat Rev 10:S10–S17
5. DeToma AS, Salamekh S, Ramamoorthy A, Lim MH (2012) Misfolded proteins in Alzheimer's disease and type II diabetes. Chem Soc Rev 41:608–621
6. Ishimaru D, Andrade LR, Teixeira L, Quesado PA, Maiolino LM, Lopez PM, Cordeiro Y, Costa LT, Heckl WM, Weissmuller G, Foguel D, Silva JL (2003) Fibrillar aggregates of the tumor suppressor p53 core domain. Biochemist 42:9022–9027
7. Rigacci S, Bucciantini M, Relini A, Pesce A, Gliozzi A, Berti A, Stefani M (2008) The (1–63) region of the p53 transactivation domain aggregates in vitro into cytotoxic amyloid assemblies. Biophys J 94:3635–3646
8. Silva JL, Rangel LP, Costa DCF, Cordeiro Y, De Moura Gallo CV (2013) Expanding the prion concept to cancer biology: dominant-negative effect of aggregates of mutant p53 tumour suppressor. Biosci Rep 33:593–603
9. Yang-Hartwich Y, Soteras MG, Lin ZP, Holmberg J, Sumi N, Craveiro V, Liang M, Romanoff E, Bingham J, Garofalo F, Alvero A, Mor G (2014) p53 protein aggregation promotes platinum resistance in ovarian cancer. Oncogene 34:3605–3616

10. Novitskaya V, Bocharova OV, Bronstein I, Baskakov IV (2006) Amyloid fibrils of mammalian prion protein are highly toxic to cultured cells and primary neurons. J Biol Chem 281:13828–13836
11. Lambert MP, Barlow AK, Chromy BA, Edwards C, Freed R, Liosatos M, Morgan TE, Rozovsky I, Trommer B, Viola KL, Wals P, Zhang C, Finch CE, Krafft GA, Klein WL (1998) Diffusible, nonfibrillar ligands derived from Aβ1–42 are potent central nervous system neurotoxins. Proc Natl Acad Sci U S A 95:6448–6453
12. Conway KA, Lee S-J, Rochet J-C, Ding TT, Williamson RE, Lansbury PT Jr (2000) Acceleration of oligomerization, not fibrillization, is a shared property of both α-synuclein mutations linked to early-onset Parkinson's disease: Implications for pathogenesis and therapy. Proc Natl Acad Sci U S A 97:571–576
13. Sousa MM, Cardoso I, Fernandes R, Guimarães A, Saraiva MJ (2001) Deposition of transthyretin in early stages of familial amyloidotic polyneuropathy. Am J Pathol 159:1993–2000
14. Bucciantini M, Giannoni E, Chiti F, Baroni F, Formigli L, Zurdo J, Taddei N, Ramponi G, Dobson CM, Stefani M (2002) Inherent toxicity of aggregates implies a common mechanism for protein misfolding diseases. Nature 416:507–511
15. Anderluh G, Gutirrez-Aguirre I, Rabzelj S, Čeru S, Kopitar-Jerala N, Maček P, Turk V, Žerovnik E (2005) Interaction of human stefin B in the prefibrillar oligomeric form with membranes. FEBS J 272:3042–3051
16. Rabzelj S, Viero G, Gutiérrez-Aguirre I, Turk V, Dalla Serra M, Anderluh G, Žerovnik E (2008) Interaction with model membranes and pore formation by human stefin B - studying the native and prefibrillar states. FEBS J 275:2455–2466
17. Guijarro JI, Sunde M, Jones JA, Campbell ID, Dobson CM (1998) Amyloid fibril formation by an SH3 domain. Proc Natl Acad Sci U S A 95:4224–4228
18. Chiti F, Webster P, Taddei N, Clark A, Stefani M, Ramponi G, Dobson CM (1999) Designing conditions for in vitro formation of amyloid protofilaments and fibrils. Proc Natl Acad Sci U S A 96:3590–3594
19. Kusumoto Y, Lomakin A, Teplow DB, Benedek GB (1998) Temperature dependence of amyloid β-protein fibrillization. Proc Natl Acad Sci U S A 95:12277–12282
20. Zurdo J, Guijarro JI, Jiménez JL, Saibil HR, Dobson CM (2001) Dependence on solution conditions of aggregation and amyloid formation by an SH3 domain. J Mol Biol 311:325–340
21. Marek PJ, Patsalo V, Green DF, Raleigh DP (2012) Ionic strength effects on amyloid formation by amylin are a complicated interplay among Debye screening, ion selectivity, and Hofmeister effects. Biochemist 51:8478–8490
22. West MW, Wang W, Patterson J, Mancias JD, Beasley JR, Hecht MH (1999) De novo amyloid proteins from designed combinatorial libraries. Proc Natl Acad Sci U S A 96:11211–11216
23. Hecht MH (1994) De novo design of β-sheet proteins. Proc Natl Acad Sci U S A 91:8729–8730
24. Rousseau F, Schymkowitz J, Serrano L (2006) Protein aggregation and amyloidosis: confusion of the kinds? Curr Opin Struct Biol 16:118–126
25. Fernandez-Escamilla A-M, Rousseau F, Schymkowitz J, Serrano L (2004) Prediction of sequence-dependent and mutational effects on the aggregation of peptides and proteins. Nat Biotechnol 22:1302–1306
26. Rauscher S, Baud S, Miao M, Keeley FW, Pomes R (2006) Proline and glycine control protein self-organization into elastomeric or amyloid fibrils. Structure 14:1667–1676
27. Uversky VN, Oldfield CJ, Dunker AK (2008) Intrinsically disordered proteins in human diseases: introducing the D^2 concept. Annu Rev Biophys 37:215–246
28. Dobson CM (2004) Experimental investigation of protein folding and misfolding. Methods 34:4–14
29. Best RB (2017) Computational and theoretical advances in studies of intrinsically disordered proteins. Curr Opin Struct Biol 42:147–154
30. Gerben SR, Lemkul JA, Brown AM, Bevan DR (2014) Comparing atomistic molecular mechanics force fields for a difficult target: a case study on the Alzheimer's amyloid β-peptide. J Biomol Struct Dyn 32:1817–1832
31. Piana S, Donchev AG, Robustelli P, Shaw DE (2015) Water dispersion interactions strongly influence simulated structural properties of disordered protein states. J Phys Chem B 119:5113–5123
32. Hoffmann KQ, McGovern M, Chiu C-C, de Pablo JJ (2015) Secondary structure of rat and human amylin across force fields. PLoS One 10:e0134091
33. Henriques J, Cragnell C, Skepö M (2015) Molecular dynamics simulations of intrinsically disordered proteins: force field evaluation and comparison with experiment. J Chem Theory Comput 11:3420–3431

34. Rauscher S, Gapsys V, Gajda MJ, Zweckstetter M, de Groot BL, Grubmüller H (2015) Structural ensembles of intrinsically disordered proteins depend strongly on force field: a comparison to experiment. J Chem Theory Comput 11:5513–5524
35. Carballo-Pacheco M, Strodel B (2016) Comparison of force fields for Alzheimer's Aβ42: a case study for intrinsically disordered proteins. Protein Sci 26:174–185
36. Huang J, Rauscher S, Nawrocki G, Ran T, Feig M, de Groot BL, Grubmüller H, MacKerell AD (2017) CHARMM36m: an improved force field for folded and intrinsically disordered proteins. Nat Methods 14:71–73
37. Larini L, Shea J-E (2012) Role of β-hairpin formation in aggregation: the self-assembly of the amyloid-β(25–35) peptide. Biophys J 103:576–586
38. Matthes D, Gapsys V, Brennecke JT, de Groot BL (2016) An atomistic view of amyloidogenic self-assembly: structure and dynamics of heterogeneous conformational states in the pre-nucleation phase. Sci Rep 6:33156
39. Li G, Pomes R (2013) Binding mechanism of inositol stereoisomers to monomers and aggregates of Aβ(16–22). J Phys Chem B 117:6603–6613
40. Nguyen PH, Derreumaux P (2013) Conformational ensemble and polymorphism of the all-atom Alzheimer's Aβ37–42 amyloid peptide oligomers. J Phys Chem B 117:5831–5840
41. Baftizadeh F, Biarnes X, Pietrucci F, Affinito F, Laio A (2012) Multidimensional view of amyloid fibril nucleation in atomistic detail. J Am Chem Soc 134:3886–3894
42. Barz B, Olubiyi OO, Strodel B (2014) Early amyloid β-protein aggregation precedes conformational change. Chem Commun 50:5373–5375
43. Rauscher S, Pomes R (2017) The liquid structure of elastin. elife 6:e26526–e26521
44. Sasmal S, Schwierz N, Head-Gordon T (2016) Mechanism of nucleation and growth of Aβ40 fibrils from all-atom and coarse-grained simulations. J Phys Chem B 120:12088–12097
45. Wagoner VA, Cheon M, Chang I, Hall CK (2014) Impact of sequence on the molecular assembly of short amyloid peptides. Proteins 82:1469–1483
46. Bieler NS, Knowles TPJ, Frenkel D, Vácha R (2012) Connecting macroscopic observables and microscopic assembly events in amyloid formation using coarse grained simulations. PLoS Comput Biol 8:e1002692
47. Wood SJ, Wetzel R, Martin JD, Hurle MR (1995) Prolines and amyloidogenicity in fragments of the Alzheimer's peptide β/A4. Biochemist 34:724–730
48. Colletier J-P, Laganowsky A, Landau M, Zhao M, Soriaga AB, Goldschmidt L, Flot D, Cascio D, Sawaya MR, Eisenberg D (2011) Molecular basis for amyloid-β polymorphism. Proc Natl Acad Sci U S A 108:16938–16943
49. Pettersen EF, Goddard TD, Huang CC, Couch GS, Greenblatt DM, Meng EC, Ferrin TE (2004) UCSF Chimera - a visualization system for exploratory research and analysis. J Comput Chem 25:1605–1612
50. Humphrey W, Dalke A, Schulten K (1996) VMD: Visual Molecular Dynamics. J Mol Graph 14:33–38
51. Abraham MJ, Murtola T, Schulz R, Páll S, Smith JC, Hess B, Lindahl E (2015) GROMACS: High performance molecular simulations through multi-level parallelism from laptops to supercomputers. SoftwareX 1–2:19–25
52. MacKerell AD Jr, Bashford D, Bellott M, Dunbrack RL Jr, Evanseck JD, Field MJ, Fischer S, Gao J, Guo H, Ha S, Joseph-McCarthy D, Kuchnir L, Kuczera K, Lau FTK, Mattos C, Michnick S, Ngo T, Nguyen DT, Prodhom B, Reiher WE, Roux B, Schlenkrich M, Smith JC, Stote R, Straub J, Watanabe M, Wiórkiewicz-Kuczera J, Yin D, Karplus M (1998) All-atom empirical potential for molecular modeling and dynamics studies of proteins. J Phys Chem B 102:3586–3616
53. Paci E, Karplus M (2000) Unfolding proteins by external forces and temperature: the importance of topology and energetics. Proc Natl Acad Sci U S A 97:6521–6526
54. Mayor U, Johnson CM, Daggett V, Fersht AR (2000) Protein folding and unfolding in microseconds to nanoseconds by experiment and simulation. Proc Natl Acad Sci U S A 97:13518–13522
55. Day R, Bennion BJ, Ham S, Daggett V (2002) Increasing temperature accelerates protein unfolding without changing the pathway of unfolding. J Mol Biol 322:189–203
56. Essmann U, Perera L, Berkowitz ML, Darden T, Lee H, Pedersen LG (1995) A smooth particle mesh Ewald method. J Chem Phys 103:8577–8593
57. Hess B, Bekker H, Berendsen HJC, Fraaije JGEM (1997) LINCS: A linear constraint solver for molecular simulations. J Comput Chem 18:1463–1472
58. Miyamoto S, Kollman PA (1992) SETTLE: An analytical version of the SHAKE and RATTLE algorithm for rigid water molecules. J Comput Chem 13:952–962

59. Bussi G, Donadio D, Parrinello M (2007) Canonical sampling through velocity rescaling. J Chem Phys 126:014101

60. Berendsen HJC, Postma JPM, van Gunsteren WF, DiNola A, Haak JR (1984) Molecular dynamics with coupling to an external bath. J Chem Phys 81:3684–3690

61. Parrinello M, Rahman A (1981) Polymorphic transitions in single crystals: a new molecular dynamics method. J Appl Phys 52:7182–7190

62. Kabsch W, Sander C (1983) Dictionary of protein secondary structure: pattern recognition of hydrogen-bonded and geometrical features. Biopolymers 22:2577–2637

Correction to: Obtaining Soft Matter Models of Proteins and their Phase Behavior

Irem Altan and Patrick Charbonneau

Correction to:
Chapter 15 in: Jennifer J. McManus (ed.), *Protein Self-Assembly: Methods and Protocols*, Methods in Molecular Biology, vol. 2039, https://doi.org/10.1007/978-1-4939-9678-0_15

The acknowledgement section text has been updated in the chapter.

"We thank Diana Fusco for guiding discussions. The authors acknowledge support from National Science Foundation Grant no. NSF DMR-1749374. This work used the Extreme Science and Engineering Discovery Environment (XSEDE), which is supported by National Science Foundation grant number ACI-1548562, as well as the Duke Compute Cluster."

The updated online version of this chapter can be found at
https://doi.org/10.1007/978-1-4939-9678-0_15

Jennifer J. McManus (ed.), *Protein Self-Assembly: Methods and Protocols*, Methods in Molecular Biology, vol. 2039,
https://doi.org/10.1007/978-1-4939-9678-0_18,

Index

Jennifer J. McManus (ed.), *Protein Self-Assembly: Methods and Protocols*, Methods in Molecular Biology, vol. 2039, https://doi.org/10.1007/978-1-4939-9678-0, © Springer Science+Business Media, LLC, part of Springer Nature 2019

MIX
Papier aus verantwortungsvollen Quellen
Paper from responsible sources
FSC® C105338

If you have any concerns about our products,
you can contact us on
ProductSafety@springernature.com

In case Publisher is established outside the EU,
the EU authorized representative is:
Springer Nature Customer Service Center GmbH
Europaplatz 3, 69115 Heidelberg, Germany

Printed by Libri Plureos GmbH
in Hamburg, Germany